MIMO Wireless Communications over Generalized Fading Channels

MIMO Wireless Communications over Generalized Fading Channels

Brijesh Kumbhani
Rakhesh Singh Kshetrimayum

CRC Press
Taylor & Francis Group
Boca Raton London New York

CRC Press is an imprint of the
Taylor & Francis Group, an **Informa** business

CRC Press
Taylor & Francis Group
6000 Broken Sound Parkway NW, Suite 300
Boca Raton, FL 33487-2742

Printed on acid-free paper
Version Date: 20170508

International Standard Book Number-13: 978-1-1380-3300-9 (Hardback)

Visit the Taylor & Francis Web site at
http://www.taylorandfrancis.com

and the CRC Press Web site at
http://www.crcpress.com

Brijesh Kumbhani would like to dedicate this book
to his elder brother
Late Chaturbhai Kumbhani.

Rakhesh Singh Kshetrimayum would like to dedicate
this book
to his mom for her indomitable courage.

Contents

List of Figures

List of Tables

Preface

Generalized fading channels are rather recent channel models. The beauty of the generalized fading distributions, viz. $\eta - \mu$, $\kappa - \mu$ and $\alpha - \mu$, is that all the previous fading distributions like Rayleigh, Rician, Weibull, etc. are particular cases of them. Muliple-input multiple-output (MIMO) wireless communications is one of the biggest advancements in the wireless communications. MIMO systems can achieve rate and diversity gains over single-input single-output (SISO) systems without penalty in bandwidth and signal power. There has been a sizable number of research papers in MIMO wireless communications over generalized fading channels. This is the first book which discusses in depth about the MIMO wireless communications over generalized fading channels.

Chapter 1 gives a brief introduction to MIMO wireless communications. In this, we will discuss about the evolution of MIMO systems from receiver diversity combining schemes. After a short introduction to MIMO system and channel models, a detailed discussion on open and closed loop MIMO systems will be presented.

Chapter 2 describes about the generalized fading channels. After a brief discussion on fading channels and classical fading channels, thorough analyses on the probability density function (PDF), cumulative distribution function (CDF) and moment generating function (MGF) of generalized fading distributions like Nakagami-m, $\eta - \mu$, $\kappa - \mu$ and $\alpha - \mu$ fading channels are presented.

MIMO gives spatial multiplexing gain over SISO systems. It is the subject of focus in the third chapter. We will introduce the reader to diversity-multiplexing trade-off for MIMO system design and analyze the spatially multiplexed MIMO systems over Nakagami-m, $\eta - \mu$, $\kappa - \mu$ and $\alpha - \mu$ fading channels.

Spatial modulation, as the name suggests, does modulation over space. It is a relatively new MIMO technique with multiple advantages. Chapter 4 provides performance analysis of spatial modulated MIMO systems over Nakagami-m, $\eta - \mu$, $\kappa - \mu$ and $\alpha - \mu$ fading channels. A brief introduction to generalized spatial modulation will also be given at the end of this chapter.

Antenna selection is widely explored in the MIMO literature. In Chapter 5, after a brief discussion on antenna selection criteria, a thorough performance analysis of transmit antenna selection (TAS) over Nakagami-m, $\eta - \mu$, $\kappa - \mu$ and $\alpha - \mu$ fading channels will be provided. The performance metrics considered are probability of error, channel capacity and outage probability.

Chapter 6 will present about the space time block codes (STBC), which give diversity gain over SISO systems. We will first discuss about the STBC design criteria, generation and detection. Then we will provide the performance of STBC over Nakagami-m, $\eta - \mu$, $\kappa - \mu$ and $\alpha - \mu$ fading channels.

MIMO is already employed in fourth generation (4G) mobile and it will be fully utilized in fifth generation (5G) mobile. Chapter 7 will start with the introduction of multiuser MIMO. Three big pillars of 5G, viz. massive MIMO, millimeter wave (mmWave) and small cell networks, will be briefly discussed. Next, device-to-device (D2D) communications for internet of things (IoT) will be presented. Finally we will touch upon the basics of large scale MIMO systems.

We hope that this book will be useful as a reference for MIMO wireless researchers.

Acknowledgments

Both Brijesh Kumbhani and Rakhesh Singh Kshetrimayum are grateful to Senior Editor (Dr. G. Singh) and Editorial Assistant (Ms. Mauli Sharma) for all that they have done to make this book project successful.

Brijesh Kumbhani is thankful to all the colleagues and staffs of the Department of Electrical Engineering, IIT Ropar and his friends at IIT Guwahati. He is also thankful to the colleagues at IIIT Kota where he began this book project. He is thankful to all his teachers (at JNV Kodinar, DDU Nadiad and IIT Guwahati) who made it possible for him to gain the knowledge and cross all the hurdles in the path of his career with their continuous guidance and encouragements. He extends his thankfulness to his parents, siblings and Dr. Dipti for their continuous support in making this project a success.

Rakhesh Singh Kshetrimayum is grateful to all the colleagues and staffs of the Department of EEE, IIT Guwahati and his students who have worked in MIMO wireless communications for creating such a stimulating environment conducive to research work. He is also inspired by NTU and IIT Bombay class fellows who are doing excessively well in MIMO wireless research. He is thankful to his wife, children and family members for their encouragement and making this life meaningful and wonderful.

1

Introduction to MIMO Systems

CONTENTS

Marconi is regarded as the father of wireless communications. But there are some articles that regard J. C. Bose to be the inventor of the solid-state wireless receiver that was used by Marconi in his experiments [1, 10, 98]. Since then, wireless communication technology has been advancing with the advancements of semiconductor devices and highly capable computers. The demand is also increasing at an exponential rate. With commercialization of third generation (3G) wireless technology, the communication systems are no longer limited to mobile phone calls. Wireless systems have expanded their horizons to internet data which generate many folds of data traffic than the voice traffic alone. This led the researchers to find solutions and techniques that enable wireless systems to carry more data per unit bandwidth, i.e., better spectral efficiency. It can be achieved by using higher order modulation schemes but higher order modulation schemes are found to degrade the bit error rate (BER) performance of the communication systems because of the reduced separation of constellation points. Multiple-input multiple-output (MIMO) wireless systems come as a rescue. In wireless communication systems, we define MIMO systems as those systems in which the transmitter and the receiver terminals contain multiple antennas. In this chapter, we describe the need to go for MIMO systems. The following sections will summarize how the communication paradigm shifted from single antenna systems to MIMO systems. It is aimed to make the readers familiar with different channel models being used for the analysis of MIMO systems. And finally, various variants of MIMO systems will be discussed in two broad categories, as open loop MIMO systems and closed loop MIMO systems.

1.1 Evolution of MIMO systems

In this section, we will discuss the paradigm shift from analog communication systems to digital communication systems and from single-input single-output (SISO) systems to MIMO systems. The concept of wireless communication is not modern. There are evidences of passing messages through various sounds and fire alarms even in the era before the sapiens. It may be considered as the beginning of wireless communication (it was not electronic though). It was the carrier less wireless communication. The sequence of different sounds carried different messages. However, it did not make a possibility for people at distance to talk. This was given an electrical form by Morse in the name of Morse code in which each English character was coded using a sequence of dots and dashes. This was a transmission of messages in digital form but without any need to use a career (pulsed communication over wire line). However, the beginning of wireless communications in the form of electromagnetic signals started with analog communication systems using amplitude modulation (AM) and frequency modulation (FM). In fact, FM was the globally accepted modulation scheme for first generation (1G) wireless technology [107]. The second generation (2G) wireless technology was the result of germinating digital era. In 2G, the service providers used digital modulation schemes such as binary phase shift keying (BPSK), Gaussian minimum shift keying (GMSK) and differential phase shift keying (DPSK), and multiple access technologies like time division multiple access (TDMA) and code division multiple access (CDMA) which were not possible to use with the analog modulation schemes. Up to 2G, the voice traffic over wireless channels had been the main focus. It introduced the short message service (SMS) service though. Internet over wireless links begins with further advancement of 2G wireless technology. With updated softwares of 2G wireless systems, it introduced data services using technologies like general packet radio service (GPRS) and enhanced data rates for global evolution (EDGE). Further, to fulfill the increasing demands (mainly of internet data speed) from the exponentially growing consumers, researchers standardized wideband CDMA with wider bands and higher data rates which we commonly call 3G. By the time of the beginning of 3G, it was clear that further improvements in the data rates and performance were no more possible with single antenna systems. Use of multiple antennas at the receivers and different diversity combining schemes improved the BER performance. MIMO systems have already been used in fourth generation (4G) technologies and WiMAX. In long term evolution (LTE), up to eight antennas have been allowed at the transmitter and receiver. Researchers have moved into large MIMO systems where transmitter and receivers are considered to have 100s and 1000s of antennas [19]. A brief insight to the benefits and shortcomings of various modulation schemes and diversity techniques that motivated to shift the paradigm toward MIMO systems is given further.

At this point of discussion about the evolution of MIMO systems, we focused on the following requirements for better quality of service.

- Higher spectral efficiency, i.e., transmission of more number of bits per channel use per unit channel bandwidth
- Minimum possible BER

1.1.1 Higher order modulation schemes

Of the requirements mentioned above, again coming to SISO systems, higher spectral efficiency may be achieved with the help of M-ary modulation schemes. In M-ary modulation schemes, multiple bits are transmitted in every symbol. So, M-ary modulation schemes can effectively improve the spectral efficiency. But, to achieve gain in spectral efficiency of M-ary modulation schemes, we need to pay in terms of power efficiency to achieve the required BER. The power efficiency and BER are dependent on each other. It can be observed in the constellation diagram in Figure 1.1. It shows the constellation diagram for 4-(Quadrature Amplitude Modulation)QAM and 16-QAM keeping energy per bit the same for both cases. It can be observed that moving from 4-QAM to 16-QAM (from lower order modulation scheme to higher order modulation scheme) the constellation points come closer to each other. Nearer constellation points result in higher probability of error as observed in Figure 1.2. The higher probability of error can be explained as follows. The constellation points get dislocated from their original position due to any added noise in the signal while propagating and processing. It is possible that the point received has smaller distance from the other point than the actual transmitted constellation point. If the distances between constellation points are less, the probability of the received point becoming nearer to the point other than the actual transmitted point increases. So, to keep the BER of higher order modulation scheme that same as of the lower order modulation scheme, the

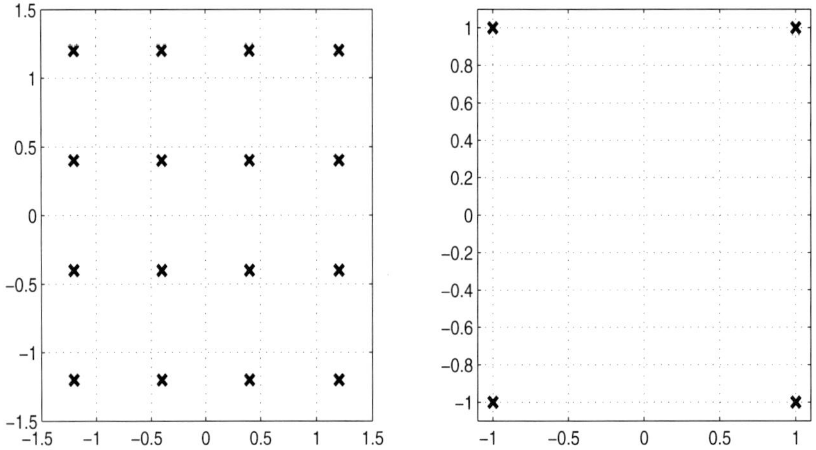

FIGURE 1.1
QAM constellation.

distance between constellation points must be made equal to that of the lower order modulation scheme. It can be achieved by transmitting more energy per symbol which results in degraded power efficiency of higher order modulation scheme if the bit rate is kept the same.

[It is important to note here that many authors consider equal symbol energy for the comparison of distances among constellation points/power efficiency and BER of various modulation schemes. However, considering equal energy per transmitted bit is a more generic way to compare these metrics, as bit is the basic unit of digital information and it remains the same for any modulation scheme.]

From constellation diagrams for 4-QAM and 16-QAM, it can be observed that the nearest constellation points are $2E_b$ apart for 4-QAM, while for 16-QAM they are approximately $0.8E_b$ apart. This results in almost 4dB loss in terms of signal-to-noise ratio (SNR) in 16-QAM as compared to 4-QAM as observed in Figure 1.2.

By this time, it is clear that M-ary modulation schemes can achieve better spectral efficiencies as the larger constellations are selected. But it degrades BER performance. Thus, M-ary modulation schemes introduce a trade-off between spectral efficiency and BER performance of the system.

Degradation in BER with higher order modulation schemes is already explained. The other factor that is responsible for degradation in BER performance is the environment. Multiple mobile and immobile obstacles in the path between transmitter and receiver are responsible for multiple copies reaching

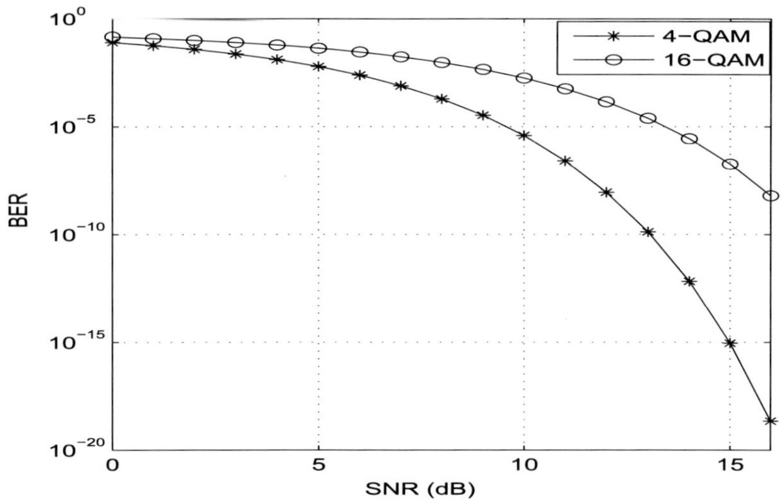

FIGURE 1.2
BER comparison for 4-QAM and 16-QAM over additive white Gaussian noise (AWGN) channels.

at the receiver with random strength and different phases. This results in variation of received power in temporal as well as spatial domain. This phenomenon is known as fading and the environment is referred to as scattering environment.

Now, the problem of improvement in BER performance of wireless communication systems opened up when the M-ary modulation schemes had been implemented in the scattering environment. To deal with the problem without increasing the transmitted power, various diversity techniques may be used. Diversity techniques use the fact that the signals received from different paths in the scattering environment may be of different strengths at different points in space and/or time, i.e., receiving multiple copies of the same signal and extracting the transmitted data stream from the strongest signal can help to improve the BER performance without demanding extra power/energy at the transmitter.

1.1.2 Diversity techniques

Fading causes severe degradation in the performances of wireless communication systems, the most important being erroneously received data (compare BER curves of Figure 1.2 without fading and Figure 1.3 with fading). The most commonly studied methods to overcome the effect of fading are the diversity combining schemes such as spatial diversity scheme, temporal diversity scheme, spectral diversity scheme, polarization diversity scheme, etc. of

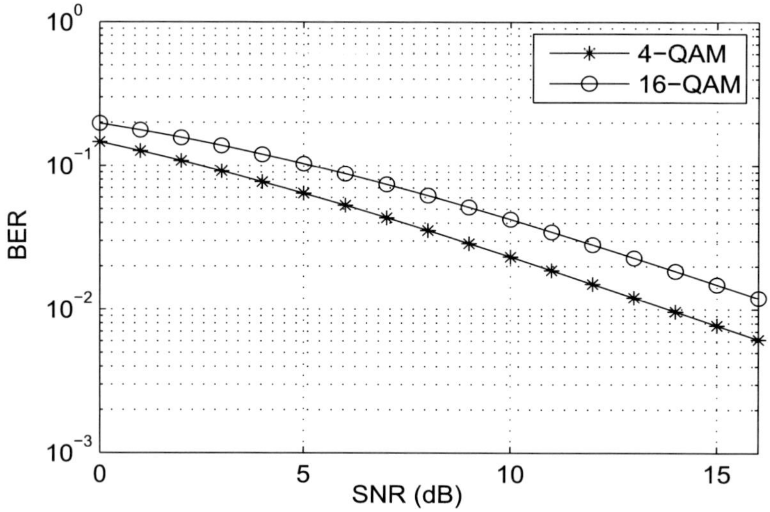

FIGURE 1.3
BER comparison for 4-QAM and 16-QAM over Rayleigh fading channels.

which spatial diversity techniques are the most commonly used and highly studied diversity techniques. Spatial diversity techniques take advantage of spatial variation in the signal strength due to multipath components and combine them accordingly using multiple antennas at the receiver. Thus, diversity systems can be categorized as single-input multiple-output (SIMO) systems. Implementing a diversity system requires no extra frequency band (except in frequency diversity), no added latency (except in time diversity), etc. The commonly used spatial diversity combining schemes are: selection combining (SC), equal gain combining (EGC) and maximal ratio combining (MRC). Each of these combining techniques is discussed in brief as follows.

1.1.2.1 Selection diversity

It is the simplest of all spatial diversity schemes. It takes advantage of the spatial variation in received power due to which the signal may have different strengths at different points in space. So, it is highly likely that the signal has sufficient strength at one of the antennas at the receiver. Keeping this fact into consideration, signal received at the antenna having the highest strength is passed to the detector and decoded to recover the transmitted signal.

1.1.2.2 Equal gain combining

EGC also takes advantage of different strengths at different locations. But unlike selection diversity scheme, it uses all the received signals including the weaker signals. Before passing the signal to the detector, received signals from

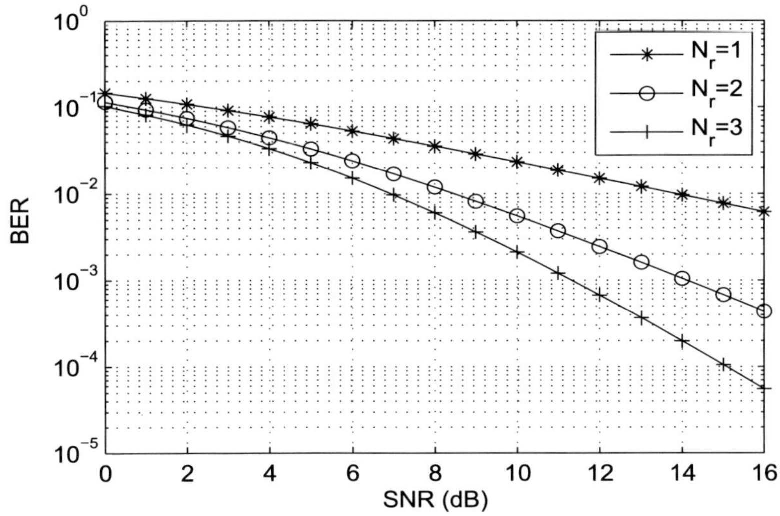

FIGURE 1.4
BER comparison for 4-QAM over Rayleigh fading channels with different branches at receiver

all the receiver antennas are co-phased and added. The resultant signals are then used to recover the transmitted message. EGC gives better performance compared to SC because it uses less strong signals also in addition to the strongest signal.

1.1.2.3 Maximal ratio combining

This is a more complex diversity combining scheme to implement as compared to EGC and SC. It has been inspired from EGC to further improve the system performance. In addition to co-phasing the received signals, each signal is amplified such that after combining the SNR gets maximized. With this added complexity, MRC gives the best error performance among all the spatial diversity combining schemes. So, sometimes, it is also referred to as optimal combining scheme, i.e., the combining scheme that gives optimum error performance. MRC has been a popular and widely explored technique for its optimality. However, this optimality is achieved at the cost of added implementation complexity.

Overall, all the diversity techniques take advantage of the fact that if multiple copies of signals are available at the receiver, the best one can be identified and processed. This not only improves the BER performance but also improves the rate at which BER reduces with increase in the SNR. The effect of number of branches on BER performance and the improved BER with more number of branches at the receiver can be observed in Figure 1.4.

At this point, readers may feel that BER performance can be improved using more and more number of branches at the receiver and this takes care of the degradation in the BER due to higher order modulation techniques. However, if we observe Figure 1.4 carefully, the improvement in the BER for each additional branch keeps on reducing as we go on increasing the number of branches at the receiver. Again, number of branches at receiver cannot be increased indefinitely. So, even after using diversity techniques, it is not possible to deal with the requirement of the spectral efficiency. This opened the doors for using multiple antennas at the transmitter so that information can be transmitted simultaneously. Multiple antennas at the transmitter transmitting parallel data streams using the same channel. This enhances the data transmission rate of the channel and hence improves the spectral efficiency. Such communication systems using multiple antennas both at transmitter and receiver are known as MIMO systems [43, 93, 131]. MIMO systems get benefit from the scattering environment. In fact, scattering environment is the requirement for MIMO systems to function well. The requirement of rich scattering is explained in detail in the next section. Further in this chapter, we will discuss various aspects of MIMO systems and their variants.

Note: Scattering environment actually degrades performance of wireless communication systems. But, MIMO systems use the same scattering phenomena to enhance the data rates and hence the spectral efficiency.

1.2 MIMO system and channel models

For the sake of introduction to MIMO channel models, let's begin with the process of signal transmission from a transmitter to a receiver (SISO case). A signal transmitted through wireless media from transmitter undergoes physical processes like reflection, refraction, diffraction and scattering on its way to the receiver in channel [107]. In process, the received signal is a group of multiple copies of the same signal received through multiple paths after experiencing multiple physical processes. Most of the time, the delays of all the received multipath components are negligible as compared to the bit/symbol duration. All the multipath components have undergone some physical process. So, they arrive at the receiver with different magnitude and different phase. In general, all the received multipath components can be decomposed into in-phase (real) and quadrature phase (imaginary) components. Thus, the received signal is a sum of in-phase components and quadrature phase components of the multipath copies received. All these components have random magnitude and phase. So, real and imaginary components may be assumed as random variables with either a positive or a negative sign. In practice, when the number of multipath components is sufficiently large with no dominant component, both the resultant components (real and imaginary) may be assumed to follow Gaussian distribution by applying the central limit theorem [91]. The envelopes of such channels are known to be Rayleigh distributed, which will be discussed in the next chapter. In such a case, the received signal is given as

$$r = hx + n \tag{1.1}$$

where $h = h_r + jh_i$ is the channel coefficient which is the impulse response of the link between transmitter and receiver, x is the transmitted bit/symbol, n is additive noise and h_r and h_i are real and imaginary components of the channel coefficient. Consider the MIMO system model shown in Figure 1.5. It contains N_t antennas at transmitter and N_r antennas at receiver. Throughout the text, we refer to such MIMO systems as $N_t \times N_r$ MIMO systems. For an $N_t \times N_r$ MIMO system, there will be a total of $N_t N_r$ links from transmitter to the receiver. Hence, the channel coefficient would no longer be a single element as in the SISO case. Now, the channel can be represented as a matrix of dimension $N_r \times N_t$. And the received signal would be a vector of dimension $N_r \times 1$. It can be given as

$$\mathbf{r} = \mathbf{H}\mathbf{x} + \mathbf{n} \tag{1.2}$$

where $\mathbf{H}$ is the channel matrix containing $N_r N_t$ complex elements, $\mathbf{x}$ is the transmitted vector of dimension $N_t \times 1$ and $\mathbf{n}$ is the additive noise vector of dimension $N_r \times 1$, i.e.

$$\begin{bmatrix} r_1 \\ r_2 \\ \vdots \\ r_{N_r} \end{bmatrix} = \begin{bmatrix} h_{11} & h_{12} & \cdots & h_{1N_t} \\ h_{21} & h_{22} & \cdots & h_{2N_t} \\ \vdots & \vdots & \ddots & \vdots \\ h_{N_r1} & h_{N_r2} & \cdots & h_{N_rN_t} \end{bmatrix} \begin{bmatrix} x_1 \\ x_2 \\ \vdots \\ x_{N_t} \end{bmatrix} + \begin{bmatrix} n_1 \\ n_2 \\ \vdots \\ n_{N_r} \end{bmatrix} \tag{1.3}$$

where r_i is the signal received at the ith receiving antenna, x_i is the symbol transmitted through the ith transmitting antenna and h_{ij} is the complex channel coefficient of the wireless link between i^{th} receive antenna and j^{th} transmit antenna. Channel modeling is an essential requirement to analyze various performance metrics of wireless communication systems. In most of the cases, it is assumed that the received signal is a collection of many multipath components generated as a result of reflections/diffraction/scattering from various obstacles in the path between transmitter and receiver. As a result, the real and imaginary part of the channel can be modeled as Gaussian distributed. The magnitude of gain of such channels follows Rayleigh distribution. Therefore, such channels are known as Rayleigh fading channels. Various generalized channel models are discussed in the next chapter.

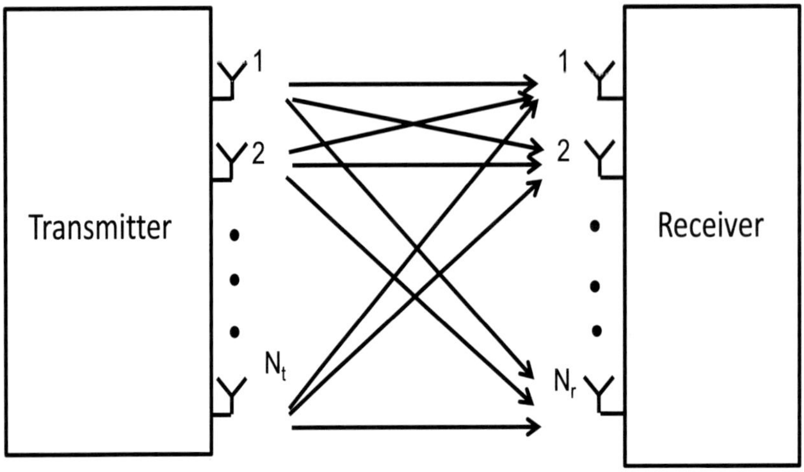

FIGURE 1.5
MIMO system model.

MIMO systems can be classified as open loop MIMO systems or closed loop MIMO systems based on the scheme used for transmission of information. Some transmission schemes need a feedback link from receiver to transmitter to provide information about channel states or some parameter of channel matrix. Such MIMO systems requiring feedback link are known as closed loop MIMO systems, while those MIMO systems which do not require any information at the transmitter back from receiver are known as open loop MIMO systems.

1.3 Open loop MIMO systems

Transmitter transmits a pilot sequence of binary data which is already known at the receiver. With the help of pilot transmission, the channel matrix is estimated at the receiver. In open loop MIMO systems, this information about channel matrix is used and required only at the receiver to decode the signal which is received after the pilot sequence is over. Various MIMO systems that use open loop configuration are spatial multiplexing, spatial modulation, space time block codes, space time block coded spatial modulation, etc. In the subsequent part of this section, we will familiarize with these variants of MIMO systems.

1.3.1 Spatial multiplexing

In spatial multiplexing, the incoming data stream is converted from serial to parallel for transmission. The parallel streams of information obtained after this conversion are transmitted simultaneously from multiple antennas available at the transmitter [43,93,131]. This has two main impacts on the practical aspects of wireless communication systems.

- Bandwidth requirement would reduce if the data are transmitted simultaneously using the same rate at which it is generated. This is more useful for real time data transmission.

- The effective data rate becomes proportional to the number of transmitting antenna. This makes overall communication faster. This becomes useful when already stored data are to be transmitted (more towards the applications of internet or for multimedia transmission).

When compared to systems with single antenna at the transmitter, this makes the transmission rate as many times as the number of antennas available at the transmitter [128].

Multipath/scattering and fading had been considered to be impairment until invention of MIMO systems. The following example demonstrates the positive aspects of rich scattering environment for MIMO systems. Consider a 2×2 MIMO system. It is used to transmit information using the BPSK modulation scheme. Assume that there is no multipath fading. In conditions without fading, all the channel coefficients are unity. Thus, the channel matrix is of size 2×2 with all entries equal to '1', i.e., $\mathbf{H} = \begin{bmatrix} 1 & 1 \\ 1 & 1 \end{bmatrix}$. Let, the groups of two incoming bits at a time be '0, 0', '0, 1', '1, 0' or '1, 1'. Using BPSK they will be respectively mapped as $\mathbf{x_1} = \begin{bmatrix} -1 \\ -1 \end{bmatrix}$, $\mathbf{x_2} = \begin{bmatrix} -1 \\ 1 \end{bmatrix}$, $\mathbf{x_3} = \begin{bmatrix} 1 \\ -1 \end{bmatrix}$, $\mathbf{x_4} = \begin{bmatrix} 1 \\ 1 \end{bmatrix}$, which will now onward be called "transmitted codewords". Both of the BPSK modulated bits will be transmitted simultaneously from the two antennas at the transmitter. Received signal at both the receive antennas in absence of noise for each of the cases may be given by $\mathbf{r_i} = \mathbf{Hx_i}$, which can respectively be represented as $\mathbf{r_1} = \begin{bmatrix} -2 \\ -2 \end{bmatrix}$, $\mathbf{r_2} = \begin{bmatrix} 0 \\ 0 \end{bmatrix}$, $\mathbf{r_3} = \begin{bmatrix} 0 \\ 0 \end{bmatrix}$, $\mathbf{r_4} = \begin{bmatrix} 2 \\ 2 \end{bmatrix}$.

It is clear from the above example that the receiver can decode the symbols correctly only in the case when the transmitted bits/symbols from each of the transmitter antennas are same. In the other two cases, the received vectors are the same for the two cases. So, a correct decision is not possible even in the absence of noise. However, in noisy conditions there is possibility of error in all the cases.

Now, consider wireless transmission through a rich scattering environment. This introduces randomness in the channel coefficients. Now, the channel matrix can be represented as $\mathbf{H} = \begin{bmatrix} h_{11} & h_{12} \\ h_{21} & h_{22} \end{bmatrix}$. Correspondingly, the received vectors under no noise conditions for each possible codeword at the transmitter become $\mathbf{r_1} = \begin{bmatrix} -h_{11} - h_{12} \\ -h_{21} - h_{22} \end{bmatrix}$, $\mathbf{r_2} = \begin{bmatrix} -h_{11} + h_{12} \\ -h_{21} + h_{22} \end{bmatrix}$, $\mathbf{r_3} = \begin{bmatrix} +h_{11} - h_{12} \\ +h_{21} - h_{22} \end{bmatrix}$, $\mathbf{r_4} = \begin{bmatrix} +h_{11} + h_{12} \\ +h_{21} + h_{22} \end{bmatrix}$.

For detection at the receiver using the maximum likelihood (ML) rule which gives optimum performance in terms of BER/symbol error rate (SER), the detected symbol is obtained as

$$i_d = \arg\min_{j} |\mathbf{r} - \mathbf{H}\mathbf{x_j}|^2 \tag{1.4}$$

where i_d is the index of the detected vector when received vector is $\mathbf{r}$. So, the detected vector at the receiver becomes $\mathbf{x_{i_d}}$.

The complete process for detection denoted mathematically in (1.4) can be described as follows.

1. For each received vector, evaluate $d_j = |\mathbf{r} - \mathbf{H}\mathbf{x_j}|^2$, i.e., the squared Euclidean distance of the received vector as if $\mathbf{x_j}$ were transmitted.
2. Among each d_j for a received vector, find j for which d_j is minimum.
3. Using j obtained in the previous step, identify the corresponding symbol, $\mathbf{x_j}$, as the detected symbol.

From (1.4), it is clear that if $\mathbf{x_j}$ happens to be the same as the transmitted symbol, d_j is zero under noiseless conditions and $d_j = |\mathbf{n}|^2$ when the system is noisy.

By this time, we know the following facts about spatial multiplexing

- All the antennas at transmitter are used to transmit the data streams in parallel.
- Transmission rate enhances linearly with number of transmitter antennas.
- The diversity order depends only on the number of antennas at the receiver and independent of antennas at the transmitter (Explanation to this will be provided in later chapters).

1.3.2 Space time block codes

Spatial multiplexing is known for improving the capacity and the spectral efficiency in the rich scattering environment. However, it does not provide any diversity gain at the transmitter. The diversity gain of spatial multiplexing systems depends only on the number of antennas available at the receiver. In 1998, Alamouti proposed a transmission scheme for MIMO systems to obtain transmit diversity at the same time as a simple decoding scheme for MIMO wireless systems [2]. However, it explores spatial and temporal aspects of diversity schemes and achieves full diversity order. The scheme is known as space time block coding (STBC). As the name suggests, an appropriate coding scheme is used at the transmitter to encode the information before it is transmitted. The coding scheme uses a block of incoming data bits and transmits it across the transmitting antennas which are separated in space. Hence, it is named the space time block code (STBC). In STBC systems, information is transmitted in blocks (say blocks of 'r' symbols each, $r = N_t$) over time slots (say 'n' time slots for one block, $n \geq r$). An STBC symbol can be mathematically expressed as

$$\mathbf{X} = \begin{bmatrix} s_{11} & s_{12} & \cdots & s_{1n} \\ s_{21} & s_{22} & \cdots & s_{2n} \\ \vdots & \vdots & \ddots & \vdots \\ s_{r1} & s_{r2} & \cdots & s_{rn} \end{bmatrix}$$

where s_{ij} is one of the symbols or its transformed version. The transformation here may be negation, complex conjugate or negation of complex conjugate according to the designed code.

Based on the number of symbols per block and time slots over which the block is transmitted, the code rate is defined as the ratio, $R_c = r/n$. The maximum code rate that can be achieved in STBC systems is 1, i.e., it does not achieve a gain in spectral efficiency. An STBC system that can achieve maximum code rate is known as *full rate* STBC system. However, this scheme is more interesting as far as diversity order and the receiver complexity are taken into consideration.

To understand the STBC transmission scheme, let's start with Alamouti STBC scheme for 2×1 multiple-input single-output (MISO) system. In this case, the channel matrix becomes a row vector and may be considered as $\mathbf{h} = \begin{bmatrix} h_{11} & h_{12} \end{bmatrix}$. Two transmit antennas are available. They transmit digitally modulated symbols simultaneously. Two symbols are transmitted in two time slots that makes the Alamouti scheme a full rate STBC system. Let two consecutive digitally modulated symbols from the incoming data bits at any time be s_1 and s_2. These symbols are transmitted as $\mathbf{x_1} = \begin{bmatrix} s_1 \\ s_2 \end{bmatrix}$ and $\mathbf{x_2} = \begin{bmatrix} -s_2^* \\ s_1^* \end{bmatrix}$ in first and second time slots respectively. So, in general, when s_1 and

s_2 are the digitally modulated symbols to be transmitted, the STBC symbol transmitted may be given as

$$\mathbf{X} = \begin{bmatrix} \mathbf{x_1} & \mathbf{x_2} \end{bmatrix} = \begin{bmatrix} s_1 & -s_2^* \\ s_2 & s_1^* \end{bmatrix} \tag{1.5}$$

The signal received in respective time slots of the transmission, r_1 and r_2, may be represented as

$$r_1 = h_{11}s_1 + h_{12}s_2 + n_1 \tag{1.6}$$
$$r_2 = -h_{11}s_2^* + h_{12}s_1^* + n_2 \tag{1.7}$$

To recover the transmitted symbols, it is assumed that the channel state information is available at the receiver. For detection of s_1 and s_2, the following computations are done.

$$\widehat{s}_1 = h_{11}^* r_1 + h_{12} r_2^* \tag{1.8}$$
$$\widehat{s}_2 = -h_{11} r_2^* + h_{12}^* r_1 \tag{1.9}$$

The simplification of the above operation results in the following relations:

$$\widehat{s}_1 = \left(\| h_{11} \|^2 + \| h_{12} \|^2\right) s_1 + h_{11}^* n_1 + h_{12} n_2^* \tag{1.10}$$
$$\widehat{s}_2 = \left(\| h_{11} \|^2 + \| h_{12} \|^2\right) s_2 - h_{11} n_2^* + h_{12}^* n_1 \tag{1.11}$$

The above expressions draw a very nice observation. The observations can be summarized as follows.

- After two time slots of transmission, s_1 and s_2 can be detected separately.
- s_1 and s_2 can be detected using the operations that involve only scalar multiplications and additions of complex numbers. This simplifies the detection process.
- The existence of the term $\left(\| h_{11} \|^2 + \| h_{12} \|^2\right)$ indicate that there is advantage of transmit diversity, i.e., the higher value of either of $\| h_{11} \|^2$ or $\| h_{12} \|^2$ would result in lower probability of erroneous detection of s_1 and s_2.

Two receiving antennas

Again, we consider the Alamouti STBC scheme for 2×2 MIMO system. In this case, the channel matrix becomes $\mathbf{H} = \begin{bmatrix} h_{11} & h_{12} \\ h_{21} & h_{22} \end{bmatrix}$ of which first row corresponds to the links from transmitter antennas to the first receiving antenna and second row corresponds to the links from transmitter antennas to the second receiving antenna. The signals received by the first antenna in respective time slots of the transmission, r_{11} and r_{12}, may be represented as

$$r_{11} = h_{11}s_1 + h_{12}s_2 + n_{11} \tag{1.12}$$

$$r_{12} = -h_{11}s_2^* + h_{12}s_1^* + n_{12} \tag{1.13}$$

and the signals received by the second antenna in respective time slots of the transmission, r_{21} and r_{22}, may be represented as

$$r_{21} = h_{21}s_1 + h_{22}s_2 + n_{21} \tag{1.14}$$

$$r_{22} = -h_{21}s_2^* + h_{22}s_1^* + n_{22} \tag{1.15}$$

For detection of s_1 and s_2, after applying the method discussed for the single receiver antenna separately on the signals received on the first and second receiver antenna, we have

$$\widehat{s}_{11} = \left(\| h_{11} \|^2 + \| h_{12} \|^2\right) s_1 + h_{11}^* n_{11} + h_{12} n_{12}^* \tag{1.16}$$

$$\widehat{s}_{12} = \left(\| h_{11} \|^2 + \| h_{12} \|^2\right) s_2 - h_{11} n_{12}^* + h_{12}^* n_{11} \tag{1.17}$$

$$\widehat{s}_{21} = \left(\| h_{21} \|^2 + \| h_{22} \|^2\right) s_1 + h_{21}^* n_{21} + h_{22} n_{22}^* \tag{1.18}$$

$$\widehat{s}_{22} = \left(\| h_{21} \|^2 + \| h_{22} \|^2\right) s_2 - h_{21} n_{22}^* + h_{22}^* n_{21} \tag{1.19}$$

From the above expressions, s_1 and s_2 can be detected as follows.

$$\begin{aligned}\widehat{s}_1 = &\left(\| h_{11} \|^2 + \| h_{12} \|^2 + \| h_{21} \|^2 + \| h_{22} \|^2\right) s_1 \\ &+ h_{11}^* n_{11} + h_{12} n_{12}^* + h_{21}^* n_{21} + h_{22} n_{22}^*\end{aligned} \tag{1.20}$$

$$\begin{aligned}\widehat{s}_2 = &\left(\| h_{11} \|^2 + \| h_{12} \|^2 + \| h_{21} \|^2 + \| h_{22} \|^2\right) s_2 \\ &- h_{11} n_{12}^* + h_{12}^* n_{11} - h_{21} n_{22}^* + h_{22}^* n_{21}\end{aligned} \tag{1.21}$$

Again, from the above expressions, it can be observed that the symbols s_1 and s_2 are separated and both the expressions contain $\left(\| h_{11} \|^2 + \| h_{12} \|^2 + \| h_{21} \|^2 + \| h_{22} \|^2\right)$. This means among four channels any single link being strong reduces the probability of error in the detection at the receiver. Thus, the system is observed to offer full diversity order of $N_t N_r$ which is 4 in the present system under consideration. In the examples presented here, we considered two transmit antennas, which is the requirement for Alamouti STBC. However, many other space time codes use a different number of antennas at the transmitter.

So, the key points about STBC can be summarized as follows.

- STBC scheme achieves full diversity order.
- STBC does not achieve gain in terms of spectral efficiency.
- The detection process for STBC symbols is computationally simple as it involves only scalar operations on complex numbers.

1.3.3 Spatial modulation

Unlike spatial multiplexing and STBC, in spatial modulation (SM), the information is not transmitted simultaneously from multiple antennas at the transmitter. This simplifies the hardware and also rectifies the problems like inter antenna interference and transmit antenna synchronization.

Moreover, in SM, all the information bits are not transmitted physically. A part of information bit sequence is mapped to transmit antenna indices (spatial constellation) and the antenna corresponding to mapped index is used to transmit information bits mapped to signal constellation. A system model for SM MIMO is given in Figure 1.6. This SM MIMO system model considers an $N_t \times N_r$ MIMO system having N_t antennas at transmitter side and N_r antennas at the receiver side. For M-ary modulation scheme, each block of $\log_2(N_t)+\log_2(M)$ bits of information is mapped by the SM mapper to signal constellation and spatial constellations. $\log_2(N_t)$ bits of each block are mapped to points in spatial constellations which select the antenna and $\log_2(M)$ bits are mapped in the signal constellation and modulated by symbol modulator using a suitable digital modulation scheme. The digitally modulated signal constellation point is then transmitted from the antenna to which the spatial constellation point is mapped. Thus, using SM, we effectively get spectral efficiency gain of $\log_2(N_t)$ bits/s/Hz without costing any extra bandwidth or power. Only one antenna depending on the incoming data bits is active at a time overcoming the problems of inter channel interference and the requirement of inter antenna synchronization. These are advantages of SM over (Vertical-Bell Laboratories Layered Space-Time) V-BLAST and MIMO orthogonal frequency division multiplexing (OFDM) systems. At the receiver, transmit antenna index and transmitted symbol are estimated separately in suboptimal scheme [76, 77]. SM de-mapper appends the estimated transmit antenna and detected symbol accordingly to regenerate the transmitted information bits. Table 1.1 shows how the bits are mapped in spatial domain and the signal domain for spectral efficiency of 3 bits/s/Hz using BPSK and 4QAM schemes with 4 and 2 transmitter antennas respectively. The transmitted vector will have all zero elements except the one which will transmit the digitally modulated data bits as shown below.

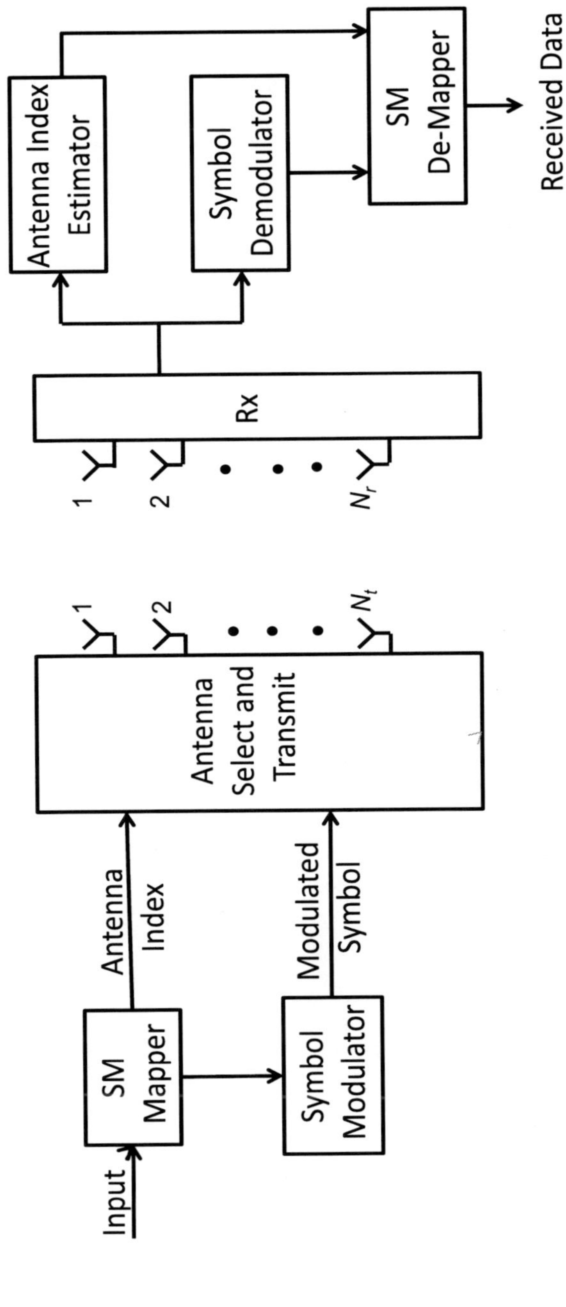

FIGURE 1.6
System model of spatial modulation (SM) MIMO systems.

TABLE 1.1
SM mapping table for 3 bits/s/Hz with BPSK and QAM modulation schemes

Input Bits	$N_t = 2, M = 4$ (QAM)		$N_t = 4, M = 2$ (BPSK)	
	Antenna Number	Transmit Symbol	Antenna Number	Transmit Symbol
000	1	+1+j	1	-1
001	1	+1-j	1	+1
010	1	-1-j	2	-1
011	1	-1+j	2	+1
100	2	+1+j	3	-1
101	2	+1-j	3	+1
110	2	-1-j	4	-1
111	2	-1+j	4	+1

$$\mathbf{x} = \begin{bmatrix} 0 & 0 & \cdots & \underset{k^{th}\ position}{\underset{\uparrow}{x_q}} & 0 & \cdots & 0 \end{bmatrix}^T \tag{1.22}$$

where $\mathbf{x}$ is the transmitted vector and x_q is the digitally modulated symbol that is to be transmitted from the kth antenna.

Sub-optimal detection

At receiver, overall detection needs two processes: one for transmit antenna index estimation and the other for estimation of transmitted symbol. Using ML criteria, antenna index can be estimated as [53]

$$\hat{j} = \arg\max_{j} \ |\boldsymbol{h}_j^H \boldsymbol{y}| \tag{1.23}$$

where $\hat{j}$ is the estimated transmit antenna index, $\boldsymbol{h}_j$ is the jth column of channel matrix $\boldsymbol{H}$ and $\boldsymbol{y}$ is the received signal vector. In the above equation, $(\cdot)^H$ denotes complex conjugate transpose. Detection of the transmitted symbol is the same as the detection process for the MRC diversity technique. It can be done using the ML rule given by

$$x_{\hat{q}} = \arg\min_{q} \parallel \boldsymbol{h}_{\hat{j}} x_q \parallel^2 - 2Re\{\boldsymbol{h}_j^H \boldsymbol{y} x_q^*\} \tag{1.24}$$

where $x_{\hat{q}}$ is the detected symbol from signal constellation space and x_q is the transmitted symbol. The error in any of the detection processes affects SER performance of the SM system.

Optimal detection

In optimal detection, the antenna index and digitally modulated symbol are detected jointly. It is also known as joint detection. This technique is optimal in terms of BER performance. It is joint ML detection scheme [53]. The optimality is achieved at the cost of computational complexity. The optimal detection rule using ML detection may be given by [53]

TABLE 1.2
Comparison of spatial multiplexing, STBC and SM

	Spectral Efficiency	Diversity Order
Spatial multiplexing	linearly related to N_t	N_r
STBC	No gain	Full
SM	Gain by $\log_2(N_t)$	N_r

$$\{\hat{j}, x_{\hat{q}}\} = \arg\min_{j,q} \| \boldsymbol{h}_j x_q \|^2 - 2Re\{\boldsymbol{h}_j^H \boldsymbol{y} x_q^*\} \tag{1.25}$$

There have been other detection methods proposed by researchers across the globe. However, in this book we emphasize mainly the suboptimal and optimal methods of detection for their simplicity. Readers interested to know about other methods of detection may follow [65, 144, 145] and references therein.

So, to compare the systems we discussed until now (spatial multiplexing, STBC and SM), we consider spectral efficiency and diversity order of each system.

Spatial multiplexing: It transmits parallel data streams from all available antennas at the transmitter. So, gain in the spectral efficiency is linearly related to number of transmit antennas, N_t, and the diversity order is the same as number of receiver antennas, N_r.

STBC: Recalling the discussions on STBC transmission, a symbol is transmitted over multiple time slots and hence there is no gain in spectral efficiency while these systems achieve full diversity order, i.e., N_tN_r.

SM: SM systems achieve gain of $\log_2(N_t)$ in the spectral efficiency while the diversity order remains the same as that in spatial multiplexing N_r.

This comparison is summarized in Table 1.2.

In the above comparison, we did not discuss about the hardware complexity and the computational complexity. Let's take the case of spatial multiplexing and STBC. In both these schemes, all available antennas at the transmitter are used to transmit a symbol all the time. So, each antenna requires separate radio frequency (RF) chain. This makes the system bulky. While in SM, single antenna is transmitting a symbol at a time. This requires single RF chain. Thus, SM hardware is simple and less bulky. This is desirable for mobile devices. Coming to computational complexity, we have already shown that STBC detection is simpler because of the separation of symbols transmitted during multiple time slots. If we look at the detection process of spatial multiplexing, the required computations increase exponentially with increasing number of antennas and/or the order of modulation scheme. The computational complexity of SM also has an exponential relation to the number of antennas and/or the order of modulation scheme. But the advantage of SM is that it requires vector operations while spatial multiplexing involves matrix operations. Thus, SM has less computational complexity in detection of symbols at the receiver.

1.4 Closed loop MIMO systems

The kinds of MIMO systems we already discussed use different strategies at the transmitter but they do not require the information about channel conditions at the transmitter. The closed loop MIMO systems, on the other hand, use different transmission strategies at the transmitter using the information about channel conditions, i.e., channel state information (CSI) is required at the transmitter. In general, the information about channel state is estimated at the receiver using a pilot sequence. Pilot sequence is a sequence of data bits which is already known at the receiver. Transmitter transmits this data sequence and the CSI is estimated at the receiver based on the already known sequence. This is known as pilot based channel estimation. But this makes CSI available only at the receiver. In closed loop MIMO systems, CSI is also required at the transmitter. This needs a feedback link from receiver to transmitter so that the information regarding channel conditions estimated at the receiver may be conveyed to the transmitter. Once CSI is available at the transmitter, it is used to choose suitable transmission strategy. Different transmission strategies are discussed as we proceed further in this section.

1.4.1 Power allocation

Now, we are considering closed loop MIMO systems. So, now onward, it is assumed that the transmitter knows CSI completely. Power allocation strategy extracts the information that which channels are good and which are not, and accordingly, power is allocated to the channels. Transmission through good channels is done with high power to get the most benefits from them. No transmission or very less power is allocated to the channels in bad condition. In this way, the overall capacity of the system is maximized so that the spectral efficiency is improved. This strategy is also referred to as *water filling algorithm* in MIMO communications. This method uses the concept of singular value decomposition (SVD) to convert the MIMO channels into parallel SISO channels, the number of parallel SISO channels being equal to the rank (R_H) of channel matrix, $\mathbf{H}$. As we have discussed in the earlier sections, when a set of symbols is transmitted simultaneously from multiple antennas

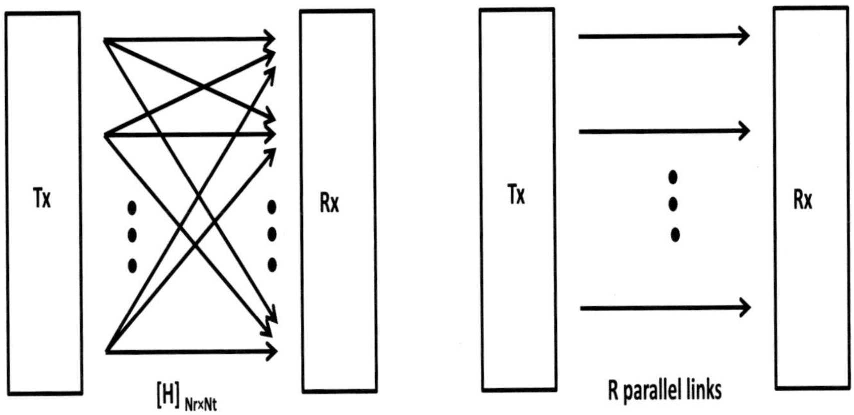

FIGURE 1.7
MIMO systems before and after power allocation.

at the transmitter, these symbols being transmitted at the same frequency interfere with each other. As a result, the signal received at each receiver antenna is a linear combination of these transmitted symbols. Thus, detecting the actual symbols transmitted from these interfered signals is a complex process. It demands high computational complexity. In this section, we will see that SVD can be helpful in actually getting rid of the interference among the symbols that are transmitted simultaneously from multiple antennas at the transmitter. After applying SVD and further processing, the conventional MIMO system becomes similar to collection of non-interfering SISO systems as shown in Figure 1.7. SVD is helpful not only in decomposing the interfering paths into parallel non-interfering channels but also it helps to identify the quality of those parallel channels. Based on the quality of channels, the power is allocated such that maximum data rate can be achieved.

The analysis that follows is available with different flavors in [94] and [52]. According to SVD, any matrix can be decomposed as a product of a unitary matrix, a diagonal matrix and another unitary matrix. In the case of MIMO systems, the channel matrix, $\mathbf{H}$, of dimensions $N_r \times N_t$ is factorized using SVD as follows.

$$\mathbf{H} = \mathbf{U}\mathbf{\Sigma}\mathbf{V}^H \tag{1.26}$$

where $\mathbf{U}$ is a unitary matrix of dimension $N_r \times N_r$, $\mathbf{\Sigma}$ is a diagonal matrix of dimension $N_r \times N_t$ and $\mathbf{V}$ is another unitary matrix of dimension $N_t \times N_t$. It should be noted here that the diagonal elements of $\mathbf{\Sigma}$ are the singular values of the channel matrix $\mathbf{H}$ which are square roots of the eigenvalues. So, the number of non-zero diagonal elements cannot exceed the rank of the channel matrix, $\mathbf{H}$, as the number of non-zero eigenvalues is the rank of the matrix.

At transmitter, instead of transmitting the symbol vector $\mathbf{x}$ directly it is first multiplied with the unitary matrix $\mathbf{V}$ that is obtained as a result of factorization of $\mathbf{H}$ using SVD. So, using (1.2) the received vector can be given by

$$\mathbf{r} = \mathbf{HVx} + \mathbf{n} \tag{1.27}$$

Now, in the above expression, representing the channel matrix as the SVD form as in equation (1.26), the received vector can be given by

$$\mathbf{r} = \mathbf{U\Sigma V^H Vx} + \mathbf{n} \tag{1.28}$$

Using the property of unitary matrices that $\mathbf{V}^H\mathbf{V} = \mathbf{I}$, the above expression can be further simplified as

$$\mathbf{r} = \mathbf{U\Sigma x} + \mathbf{n} \tag{1.29}$$

Then, some transformation is done at the receiver. It is the multiplication of the received vector given in the above expression by $\mathbf{U}^H$. This results in the following simplified expressions.

$$\mathbf{y} = \mathbf{U}^H\mathbf{r} \tag{1.30}$$
$$\mathbf{y} = \mathbf{U}^H\mathbf{U\Sigma x} + \mathbf{U}^H\mathbf{n} \tag{1.31}$$
$$\mathbf{y} = \mathbf{\Sigma x} + \mathbf{U}^H\mathbf{n} \tag{1.32}$$

where $\mathbf{y}$ is the vector obtained after transformation of received vector. Now, from equation (1.32), it is clear that all the transformations at the transmitter and receiver remove all the interferences that may happen in MIMO channels even if the antennas available at transmitter are simultaneously involved in transmission. This is because of the fact that the signal part in the final expression of $\mathbf{y}$ contains the product of symbol vector that is transmitted and the diagonal matrix obtained after SVD of $\mathbf{H}$. So, each element of the vector $\mathbf{y}$ can be represented as follows.

$$y_i = \sqrt{\lambda_i}x_i + \left[\mathbf{U}^H\mathbf{n}\right]_i \tag{1.33}$$

where y_i is the ith element of vector y, $\sqrt{\lambda_i}$ is the ith largest singular value of $\mathbf{H}$, x_i is the ith element of x and $\left[\mathbf{U}^H\mathbf{n}\right]_i$ is the ith element of vector $\mathbf{U}^H\mathbf{n}$. So, signal part in each element, y_i, is the product of a singular value of $\mathbf{H}$ and one symbol from the vector $\mathbf{x}$. And the noise part is $\left[\mathbf{U}^H\mathbf{n}\right]_i$.

Now, since the MIMO channels have been virtually decomposed as parallel SISO channels through the process of transformations at transmitter and receiver, the total channel capacity can be given as the sum of the individual channel capacity of the decomposed SISO channels. The channel capacity for $N_t = N_r$ full rank MIMO channels can be given by the following expression.

$$C = \sum_{i=0}^{N_r} \log_2 (1 + \gamma_i) \tag{1.34}$$

where C is the channel capacity and $\gamma_i = \frac{\lambda_i E_{x_i}}{N_0}$ is the SNR of the ith branch in the decomposed SISO channels, with E_{x_i} being the energy of the symbol transmitted through i^{th} branch/subchannel and N_0 as the noise power spectral density for the additive white gaussian noise (AWGN).

In such conditions, the channel capacity can be maximized by allocating unequal power to different branches/subchannels with a constraint that $\sum_{i=0}^{N_r} E_{x_i} = E_t$. This constraint is to ensure that the same energy is used on every instant of transmission. This maximization problem can be mathematically represented using the expression that follows:

$$C_{max} = \max_{\sum_{i=0}^{N_r} E_{x_i} = E_t} \sum_{i=0}^{N_r} \log_2 (1 + \gamma_i) \tag{1.35}$$

Solving the above maximization problem gives the power allocation strategy that must be followed to achieve the maximum capacity when the channel is known at the transmitter.

So, to summarize the power allocation strategy:

- Using SVD of the channel matrix both at the transmitter and the receiver, it is made possible to decompose the interfering MIMO channels into parallel and non-interfering SISO channels.
- The overall capacity of MIMO channel for such systems can be given as the sum of capacity of individual SISO channels obtained after decomposing the MIMO channels.
- Solving the optimization problem for power allocation as given by the expression of equation (1.35), the capacity can be maximized.

1.4.2 Transmit antenna selection

Transmit antenna selection (TAS) is another strategy that can be followed at the transmitter. In TAS, only one antenna is selected at the transmitter to transmit the information. So, the advantage of MIMO systems that there is parallel transmission of information no longer holds in TAS systems. But TAS systems have other advantages as compared to MIMO systems which will be clear after discussions of this section. In these systems, one antenna is selected at the transmitter based on the channel conditions such that the power of the

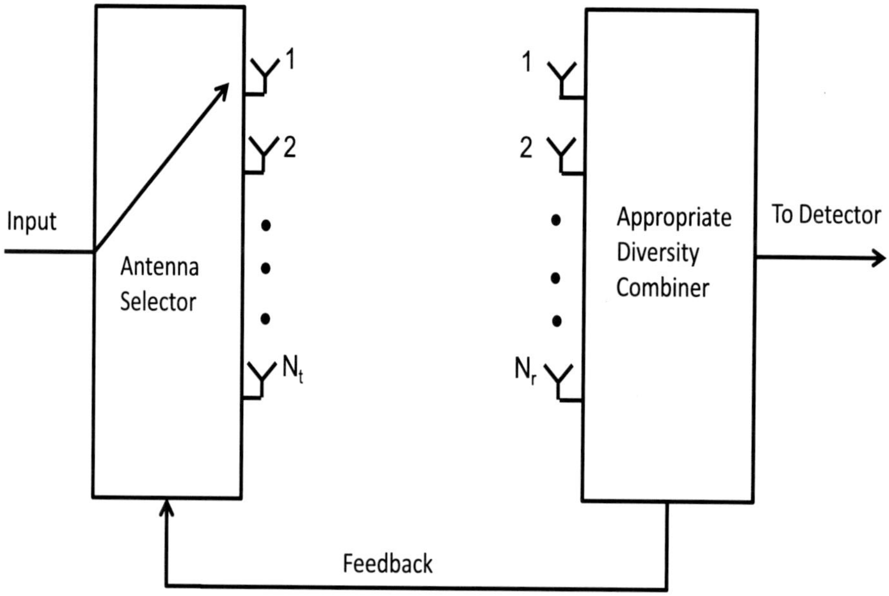

FIGURE 1.8
System model of TAS MIMO systems [15].

useful signal component in the received signal gets maximized. This maximizes the SNR and hence improves the error performance of the system. Not only the performance but since the instantaneous SNR is improved, this also improves the capacity of the system. This allows higher order modulation schemes to be used with sufficiently low error rates in the received symbols. Moreover, TAS systems use a single antenna at a time among all available antennas at the transmitter. So, it simplifies the hardware at transmitter requiring a single RF chain for single operating antenna as in the case of SM for open loop MIMO systems. Again, there are TAS systems that select more than one antenna and transmit information simultaneously on those selected antennas. But, we are limiting our discussion to single antenna selection at the transmitter.

Different criteria for TAS explored in the literature are received SNR, channel capacity and Euclidian distance [29, 105]. However, channel capacity and Euclidean distance based antenna selection algorithms are more suitable when more than one antenna are to be selected. As we have limited our discussion to single antenna selection, the antenna is selected at the transmitter which maximizes the received instantaneous SNR at the receiver. However, it does not require the estimation or complete knowledge of channel state information at the transmitter. The transmit antenna which maximizes the received SNR is estimated at the receiver and the information is made available to transmitter through a very low rate feedback link. Different diversity combining techniques can be used at the receiver with TAS of which MRC and SC are

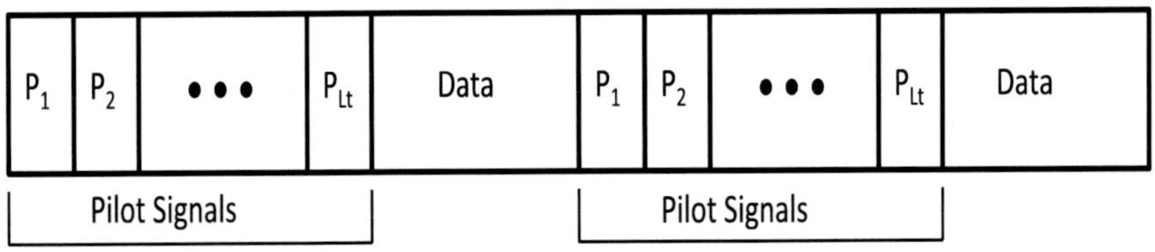

FIGURE 1.9
Frame structure of TAS MIMO systems [15].

most commonly explored in the literature because of optimality of MRC and simplicity of SC. The block diagram of the TAS system with spatial diversity combining at the receiver is shown in Figure 1.8. It consists of a feedback link from receiver to transmitter so that receiver can send information about the antenna that maximizes the received instantaneous SNR. According to this feedback information, the antenna selector connects to corresponding antenna and sends appropriately modulated data. Note that the antenna selector block is an RF switch and hence it does not require the complete set of RF chain for each antenna at the transmitter. To select an antenna at the transmitter, a set of pilot sequences is transmitted from each transmitter antenna one by one and for each transmission receiver compares the antenna selection criteria as per diversity scheme to be implemented and intimates the transmitter about the antenna maximizing SNR as in Figure 1.9. The size of data in this frame depends upon the coherence time of the channel. The pilot signals need to be transmitted for every instant of the channel change so as to guarantee that the best antenna remains selected all the time. These pilot data are not an overhead to the actual data to be transmitted. Pilot transmission is required in any way for channel estimation. The selection of the antenna that maximizes the instantaneous received SNR can be explained in the following manner.

We already know that the MIMO channel matrix can be given as

$$\mathbf{H} = \begin{bmatrix} h_{11} & h_{12} & \cdots & h_{1N_t} \\ h_{21} & h_{22} & \cdots & h_{2N_t} \\ \vdots & \vdots & \ddots & \vdots \\ h_{N_r1} & h_{N_r2} & \cdots & h_{N_rN_t} \end{bmatrix} \tag{1.36}$$

In this channel matrix, an element of each row of $\mathbf{H}$, say i^{th} row (a row vector), corresponds to the channel coefficients of the links between i^{th} receiver antenna and a transmitter antenna. Similarly, an element of jth column of $\mathbf{H}$ represents the channel coefficients of the links between j^{th} transmitter antenna and a receiver antenna. So, if an antenna is to be selected at the transmitter, the information of each column vector of the channel matrix $\mathbf{H}$ needs to be used. The antenna selection criterion for both (MRC and SC) combining schemes are discussed in the following sections.

1.4.2.1 Selection combining at the receiver

SC is the simplest diversity combining scheme. In SC, only the strongest received signal among all the received copies is processed further to detect the information. When SC is used with TAS, only the antennas corresponding to the link which gives maximum received SNR are activated. The link which gives the highest received SNR is determined by

$$I_{SC} = \underset{\substack{1 \le i \le N_t \\ 1 \le j \le N_r}}{\arg\max} \{\gamma_{j,i}\} \tag{1.37}$$

where I_{SC} denotes the link which gives maximum instantaneous SNR at the receiver and $\gamma_{j,i}$ is the instantaneous SNR of the link between i^{th} transmitting antenna and j^{th} receiver antenna. Only the antennas that correspond to the best link are made active at a time in this case. We refer to such systems as (N_t,1;N_r,1) joint transmit and receive antenna selection systems. The advantage of SC at the receiver is that it requires only one RF chain at the receiver and hence it is easy to implement. The systems with SC at the receiver and TAS at the transmitter are further referred to as TAS/SC systems in this book. In TAS/SC systems, originally there are multiple antennas at both the transmitter and the receiver which makes it a MIMO system. But, an effectively single antenna is active at both the transmitter and receiver at an instant. So, TAS/SC systems are a special kind of MIMO system that is converted into an SISO system while in operation.

1.4.2.2 MRC at the receiver

In this case, it is assumed that the number of RF chains at receiver is same as the number of receiver antennas. MRC is used at the receiver. The resulting received SNR for MRC combining scheme is given by $\gamma_t = \sum_{n=1}^{N_r} \gamma_n$, where γ_n is the instantaneous SNR of the nth branch. The transmitting antenna that maximizes SNR at the receiver can be determined by

$$I_{MRC} = \underset{1 \le i \le N_t}{\arg\max} \left\{ \gamma_{t,i} = \sum_{j=1}^{N_r} \gamma_{j,i} \right\} \tag{1.38}$$

where $\gamma_{t,i}$ denote total received instantaneous SNR when i^{th} transmitting antenna is selected and I_{MRC} denotes antenna index which corresponds to the transmitting antenna that maximizes the received SNR. We refer to such systems as (N_t,1;N_r) TAS/MRC systems. MRC requires as many RF chains at the receiver as the number of receiving antennas. With this hardware complexity, it promises optimum performance in terms of probability of error among all the diversity combining schemes. As TAS/SC systems are the form of MIMO systems that are converted into SISO systems while in operation, similarly, TAS/MRC systems are the form of MIMO systems that are converted into SIMO systems while in operation. In both the cases, TAS/SC and TAS/MRC, a single antenna is engaged in transmission at a time. This has inherent advantage that the system does not have any self-interference, unlike the case of systems which use multiple antennas simultaneously for transmission of information.

The features of TAS can be summarized as follows.

- For single antenna selection systems, transmitter needs to have a single RF chain which makes the hardware simple.
- TAS does not involve any inter antenna interference as single antenna is transmitting at an instant. So, detection at the receiver is computationally simple.
- TAS with SC at the receiver can further simplify the hardware, as in SC receiver also requires single RF chain.
- TAS systems do not require complete CSI to be conveyed to the transmitter. It only requires the information about the best antenna. This can be fulfilled by a very low rate feedback link which is not very difficult to implement.
- TAS systems can achieve full diversity order as in STBC. This will be discussed in detail in coming chapters.

1.4.3 Transmit antenna selection based spatial modulation

We already discussed the techniques of TAS and SM separately in Section 1.4.2 and in Section 1.3.3 respectively. Basically, TAS is a kind of closed loop MIMO system and SM is a kind of open loop MIMO system. In this section, we are defining a new kind of MIMO system which is a combination of both the TAS and SM. These are named as transmit antenna selection based spatial modulation (TAS-SM) [139]. This gets advantage of both the TAS in terms of diversity order and SM in terms of spectral efficiency. Again, at times it happens that one particular channel is highly faded and if SM is used for transmission, a large amount of data gets lost or erroneously detected at the receiver. In such situations, TAS-SM can be better option in which only the best/usable channels are in use after antenna selection [60,99,105]. Now, unlike in SM where we select an antenna for transmission based on the sequence of incoming bits, in TAS-SM, first some antennas are selected, say N_s based on some antenna selection algorithms and then use SM on the selected N_s antennas. Again, unlike in TAS, that we have considered in Section 1.4.2, more than one antenna is required to be selected at the transmitter in TAS-SM MIMO systems.

2

Generalized Fading Channels

CONTENTS

2.1 Introduction

A signal passing through wireless media undergoes various physical phenomena due to presence of different kinds of objects in the vicinity. These objects in the outdoor mainly consist of mobile and immobile obstacles of different physical properties. Various signal parameters like phase and amplitude get modified after incidence on such obstacles depending on the nature and size of the obstacle and the frequency of the signal. As a result the received signal is a collection of multiple components generated due to various processes like reflection, diffraction, scattering, etc. This results in momentary change in the received signal amplitude and power levels. These changes in the received power level become more severe for the situation in which transmitter and/or receiver and/or the surrounding objects are moving. Moreover, the nature of surrounding objects also affects the fluctuation in the received power level significantly. This variation is random in nature and can be modeled via various probabilistic distributions that may vary depending on the situation. Such variation in the received power over time and position is known as fading. The exact analysis of such variations is very difficult and even impossible in many cases. However, statistical modeling of received power or amplitude via various distributions can be used to simplify the analysis. The statistical modeling may vary depending on the nature of fading. Fading can be classified as frequency flat fading and frequency selective fading in the frequency domain. The statistical modeling for both kinds of fading is different. In this book, we would limit the discussion only to frequency flat fading.

Further, in this chapter, we will introduce different types of fading in brief. The types of fading and their relation with the rate of change of channel characteristics will be discussed concisely. It will be followed by discussion on some commonly known statistical models for fading distributions with their applicability over various wireless environments. Finally, we will discuss about generalized fading distributions. It includes details of the physical models that may be considered and modeled, and various statistical properties that may be useful in the analysis of wireless communication systems. Special cases of the generalized fading distributions will also be provided. This chapter also deals with the methods to generate fading channel coefficients that follow generalized fading distributions. These channel coefficients are especially useful while simulating wireless communication systems.

2.2 Introduction to fading channels

The fading characteristics depend mainly on the channel conditions. The two parameters that are used to classify the fading channels are coherence bandwidth and coherence time of the channel. Coherence time and coherence bandwidth are dependent on the Doppler shift and the delay spread respectively of the multipath components. It is well known that Doppler shift occurs when there is relative motion between transmitter and receiver while communicating. While moving, the transmitter and the receiver might be coming closer or going farther apart from each other. Accordingly, the Doppler shift might be positive or negative. This virtually increases or decreases the frequency of the signal. The relation between Doppler spread and *coherence time* can be given as

$$T_c \approx \frac{1}{f_d} \tag{2.1}$$

where T_c is the coherence time of the channel and f_d is the maximum Doppler spread. In general, *coherence time* is defined as the period of time for which the impulse response of the channel remains unchanged.

Coherence bandwidth is the other characteristic of fading channels. It can be defined as the band of frequencies over which the frequency response of the channel is constant or flat. Coherence bandwidth of the channel is related to the maximum delay spread of the multipath components. The relation can be given as follows.

$$f_c \approx \frac{1}{\tau_d} \tag{2.2}$$

where f_c is the coherence bandwidth and τ_d is the maximum delay spread.

2.2.1 Fast and slow fading

The channels are classified as fast fading channels or slow fading channels depending on the coherence time of the channel and the symbol duration of the signal. If the symbol duration is smaller than the coherence time, the channel is classified as slow fading channel. Otherwise, it falls in the category of fast fading channel. If the rate of change of impulse response of the channel is faster than the symbol duration then the channel is fast fading channel.

2.2.2 Frequency flat and frequency selective fading

The channel is classified as flat fading channel if the coherence bandwidth of the channel is greater than the signal bandwidth. In such types of channels, all the frequency components of the signal are affected equally. The effect may be of attenuation or phase variation. In case the signal bandwidth exceeds the coherence bandwidth, different frequency components of the signal would be affected differently while passing through the channel. Such channels are known as frequency selective fading channels.

2.3 Commonly used fading distributions

Considering a single-input single-output (SISO) system, the received symbol under the slow and frequency flat fading conditions can be given by

$$r = hs + n \tag{2.3}$$

where r is the received signal, h is the channel coefficient due to fading, s is the transmitted symbol and n is the noise. In this book we consider the noise to be additive white gaussian noise (AWGN). The noise is assumed to have one sided power spectral density of N_0 and independent of both the symbols and the fading amplitudes. In the above expression, the effect of fading is ascertained by h. h is a complex coefficient representing the phase and amplitude alterations in the signal due to fading. The amplitude represents the attenuation in the signal while it traverses from transmitter to receiver. While modeling fading by statistical distributions, distribution of phase and amplitude of the received signal needs to be considered. In most of the processes, phase can be assumed to be uniformly distributed and it is assumed that there is perfect phase estimation at the receiver. However, when the communication systems perform maximal ratio combining (MRC) at the receiver, phase shifts of different components of the received signal actually vanish as MRC itself involves the step of phase cancelation or co-phasing of all the received signals at different antennas. Eventually, the phase shift is canceled assuming perfect phase estimation. Similarly, in selection combining (SC) the phase shift of different components is canceled at the receiver while detection. But, there are cases where the phase shifts of different components need to be considered separately in addition to amplitude of the received signal. Apart from phase and amplitude of the received signal, signal-to-noise ratio (SNR) is an important parameter which also needs the statistical modeling. Average SNR is the independent variable in most of the performance metrics for wireless communication systems. Once the distribution of amplitude is modeled, it is easy to obtain the statistical properties of the SNR. SNR and amplitude are related in a similar manner as the relation of amplitude and power. So, the methods of random variable transformation can be applied to find the distribution of SNR from the distribution of amplitude. Now, we will briefly discuss some commonly used fading distributions. The detailed description of these descriptions can also be found in [45, 101, 107, 116, 118].

2.3.1 Rayleigh fading

The fading can be modeled as Rayleigh distributed when there is no line of sight (LOS) component present in the received signal. LOS signal may also be characterized by a component that contains considerably higher power than collective power of all the remaining components. The probability density function (PDF) of the fading amplitude can be given by [44, 116, 118]

$$p_X(x) = \frac{2x}{\Omega} e^{-\frac{x^2}{\Omega}}, \qquad x \geq 0 \tag{2.4}$$

where $\Omega = E(X^2)$ is known as fading power.

When the amplitude is Rayleigh distributed, instantaneous SNR follows exponential distribution. It can be given by

$$p_\gamma(\gamma) = \frac{1}{\overline{\gamma}} e^{-\frac{\gamma}{\overline{\gamma}}}, \qquad \gamma \geq 0 \tag{2.5}$$

where γ is the instantaneous SNR and $\overline{\gamma}$ is the average SNR.

2.3.2 Hoyt fading

Hoyt distribution is used to model the fading channels when it is more severe than the Rayleigh fading conditions. It is also known as Nakagami-q fading. The PDF of Hoyt/Nakagami-q faded amplitude can be given by [83, 92, 116]

$$p_X(x) = \frac{1+q^2}{\Omega} x e^{-\frac{(1+q^2)^2 x^2}{4q^2\Omega}} I_0\left(\frac{1-q^4}{4q^2\Omega} x^2\right), \qquad x \geq 0 \tag{2.6}$$

where q is the fading parameter for Hoyt distribution and I_ν is the Bessel function of first kind and ν^{th} order. In the present case, $\nu = 0$, i.e., Hoyt distribution involves Bessel function of the first kind and order 0. The value of q ranges from 0 to 1.

Using the distribution of Hoyt faded amplitude and methods of transformation of random variables, the PDF of Hoyt faded instantaneous SNR can be obtained as

$$p_\gamma(\gamma) = \frac{1+q^2}{2q\overline{\gamma}} e^{-\frac{(1+q^2)^2 \gamma}{4q^2\overline{\gamma}}} I_0\left(\frac{1-q^4}{4q^2\overline{\gamma}} \gamma\right), \qquad \gamma \geq 0 \tag{2.7}$$

The value of fading parameter, q shows the severity level of fading. $q = 0$ represents the highest severity that can be modeled by Hoyt distribution and $q = 1$ represents the lowest. It shall be noted that $q = 1$ is the special case of Hoyt fading which is equivalent to Rayleigh fading and $q = 0$ is the special case which is equivalent to one sided Gaussian fading.

2.3.3 Rician fading

Earlier two models were suitable for modeling the fading distributions in a non-LOS environment. Rician fading is more suitable for the LOS scenario where the collection of multipath components contains a dominant component that has high strength as compared to other components. Rice distribution is also known as Nakagami-n distribution [109]. The PDF of Rice-distributed amplitude can be given by

$$p_X(x) = \frac{2(1+K)}{\Omega} x e^{-K-\frac{(1+K)x^2}{\Omega}} I_0\left(2x\sqrt{\frac{K(1+K)}{\Omega}}\right), \qquad x \geq 0 \quad (2.8)$$

where K is the Rice K factor. It is the fading parameter of Rice distribution. Rice distribution and Nakagami-n distributed differ only in the definition of distribution parameters, K and n. They are related as $K = n^2$. K is defined as the ratio of power of the dominant component to the total power of the scattered components. Its value ranges from 0 to ∞ depending on the environmental conditions.

The PDF of the SNR for which the amplitude is modeled as Rician distributed can be given by

$$p_\gamma(\gamma) = \frac{(1+K)}{\overline{\gamma}} e^{-K-\frac{(1+K)\gamma}{\overline{\gamma}}} I_0\left(2\sqrt{\frac{K(1+K)\gamma}{\overline{\gamma}}}\right), \qquad \gamma \geq 0 \qquad (2.9)$$

The range of values of Rician fading parameter K which can model the channel conditions varies from no fading ($K = \infty$) to Rayleigh fading ($K = 0$). So, Rician distribution can be used to model the fading scenarios when there is LOS and less severity level of fading.

2.4 Generalized fading distributions

In the previous section, we discussed about some commonly used distributions to model fading statistics of the received signal. As indicated by the title of this book, our focus is to discuss generalized fading distributions in detail. In this section, we will make the readers familiar with various distributions that can be used to model fading in a generalized scenario. These distributions are categorized as generalized fading distributions because these are parametric distributions and varying the values of parameters can model a large range of fading environments. However, many commonly used fading distributions are also parametric and can model a range of fading environments but generalized fading distributions are capable of modeling a much wider range of scenarios. This also includes the non-homogeneous fading environment in which the reflecting surfaces and scatterers posses different properties. As a result, the delay spread may be very large and the received signal needs to be assumed as a collection of clusters of multipath components. Next, we discuss various generalized fading distributions.

2.4.1 Nakagami-m fading

Nakagami-m distribution is used to model fading scenario in various conditions (mostly in a non-LOS environment). It is widely explored by the researchers because of the simplicity of analysis. Nakagami-m distribution is also commonly used to model the fading statistics of the received signal but it can model fading under generalized conditions up to a certain level. So, in this book, we classified Nakagami-m fading also as a generalized fading model.

2.4.1.1 Physical model

In this part of the discussion, we aim to describe the physical realization of Nakagami-m distribution. It is essential to understand the actual physical modeling of wireless channels and to get an idea about the applicability of Nakagami-m distribution in statistical modeling of fading channels. In Nakagami-m fading model the received signal is modeled as a collection of clusters. Each cluster has a number of scattered multipath components. The delay spread of different clusters is relatively larger than the delay spread of multipath components within a cluster. Every cluster is assumed to have the same power. In such a model, the envelope X of the fading signal can be represented as

$$X^2 = \sum_{i=0}^{n} \left(I_i^2 + Q_i^2\right) \tag{2.10}$$

where n is the number of clusters in the received signal, and I_i and Q_i are respectively the in-phase and quadrature phase component of the resultant signal of the ith cluster. Note that I_i and Q_i are mutually independent random processes with zero mean and equal variance. Both I_i and Q_i, being resultant of multipath components of a cluster, may be assumed to be Gaussian distributed with $E(I_i) = E(Q_i) = 0$ and $E(I_i^2) = E(Q_i^2) = \sigma^2$.

So, the fading amplitude can be expressed as $X^2 = \sum_{i=0}^{n} \left(R_i^2\right)$ with $R_i^2 = I_i^2 + Q_i^2$. From the fact that I_i and Q_i are Gaussian distributed, it is to be noted here that each R_i^2 is exponentially distributed. As far as distributions of SNR are concerned, it is the same as the square of amplitude distribution. So, it is valid to say that the PDF of X^2 is the same as that of SNR. The SNR in this case is the sum of mutually independent Gamma distributed random variables. The PDF of SNR follows Gamma distribution and can be given as

$$p_\gamma(\gamma) = \frac{m^m \gamma^{m-1}}{\overline{\gamma}^m \Gamma(m)} e^{\frac{-m\gamma}{\overline{\gamma}}}, \qquad \gamma \geq 0 \tag{2.11}$$

where $\Gamma(\cdot)$ is the Gamma function and m is the fading parameter given by [83]

$$m = \frac{E(X^2)^2}{E(X^2 - E(X)^2)}, \qquad m \geq \frac{1}{2} \tag{2.12}$$

Using equation (2.11), the PDF of amplitude follows central Chi-square distribution and may be given by

$$p_X(x) = \frac{2m^m x^{2m-1}}{\Omega^{m-1}\Gamma(m)} e^{\frac{-mx^2}{\overline{\gamma}}}, \qquad x \geq 0 \tag{2.13}$$

Next, we will see different functions that are useful in performance evaluation of wireless communication systems.

2.4.1.2 CDF and MGF of Nakagami-m fading

Cumulative distribution function (CDF) and moment generating function (MGF) are two properties of random variables which are most essential in evaluating the performance metrics of a wireless communication system. In the following sections of this chapter, we will discuss the metrics and methods for evaluation of the metrics which are used as performance analysis tools for wireless communication systems.

CDF of Nakagami-m distributed SNR can be given by

$$P_\gamma(y) = \int_0^y \frac{m^m \gamma^{m-1}}{\overline{\gamma}^m \Gamma(m)} e^{\frac{-m\gamma}{\overline{\gamma}}} d\gamma \tag{2.14}$$

$$= \frac{\gamma\left(m, \frac{my}{\overline{\gamma}}\right)}{\Gamma(m)} \tag{2.15}$$

where $\gamma(\cdot,\cdot)$ is the lower incomplete Gamma function defined as $\gamma(p,y) = \int_0^y x^{p-1}e^{-x}dx$ [46].

For integer values of m, the above equation can be represented as finite sum series given below.

$$P_\gamma(y) = 1 - e^{-\frac{my}{\overline{\gamma}}} \sum_{k=0}^{m-1} \frac{\left(\frac{my}{\overline{\gamma}}\right)^k}{k!} \tag{2.16}$$

MGF of Nakagami-m distributed SNR can be given by

$$M_\gamma(s) = \int_0^\infty p_\gamma(\gamma) e^{-s\gamma} d\gamma \tag{2.17}$$

$$= \left(\frac{m}{m - \overline{\gamma}s}\right)^m \tag{2.18}$$

2.4.1.3 Special cases

If we observe equation (2.10), it is clear that a Nakagami-m faded signal is composed of various Gaussian distributed random variables. The parameter m is defined such that its different values can model different fading environments. If we consider one cluster, the situation corresponds to $m = 1$ and the signal described by equation (2.10) can be represented as

$$X = \sqrt{I^2 + Q^2} \tag{2.19}$$

with I and Q being Gaussian distributed with zero mean and equal variance. This is by definition a Rayleigh distributed random variable. Thus, $m = 1$ in Nakagami-m fading model represents the special case which is equivalent to Rayleigh fading.

Another case is $m = \frac{1}{2}$. This means half a cluster which is physically not possible. But it can be described taking into consideration the previous case in which $m = 1$ represents the square root sum of two squared Gaussian distributed random variables. So, in the case of $m = \frac{1}{2}$, it can be thought of as the single Gaussian distributed random variable. Since the PDF of Nakagami-m distribution is restricted only to positive values, this case represents one sided Gaussian fading.

As a generalized fading model, Nakagami-m distribution can also model Hoyt (Nakagami-q) and Rice (Nakagami-n) distributions as explained in [89, 116]. If we revisit the discussion about Hoyt fading in Section 2.3.2, it can model the fading environments ranging between severe fading as high as one sided Gaussian to moderately severe as Rayleigh fading. Nakagami-m distribution includes these two cases in the range $m = \frac{1}{2}$ to $m = 1$; thus, the range of values $q = 0$ to $q = 1$ of fading parameter, q, in the case of Hoyt fading maps to the range of values $m = \frac{1}{2}$ to $m = 1$ of the fading parameter, m, in the case of Nakagami-m fading model. The relation for approximation between these parameters for representation of Hoyt fading as a special case of Nakagami-m fading can be given by

$$m = \frac{\left(1 + q^2\right)^2}{2\left(1 + 2q^4\right)}. \tag{2.20}$$

Proceeding in the same manner, revisiting the discussion about Rician fading in Section 2.3.3, the cases of severity that can be modeled by Rician distribution range from Rayleigh fading ($K = 0$) at one extreme to less severe fading like static channels with no fading ($K = \infty$). Nakagami-m distribution includes these two cases in the range $m = 1$ to $m = \infty$; thus, the range of values $K = 0$ to $K = \infty$ of fading parameter, K, in the case of Rician fading maps to the range of values $m = 1$ to $m = \infty$ of the fading parameter, m, in the case of Nakagami-m fading model. The relation for approximation between these parameters for representation of Rician fading as a special case

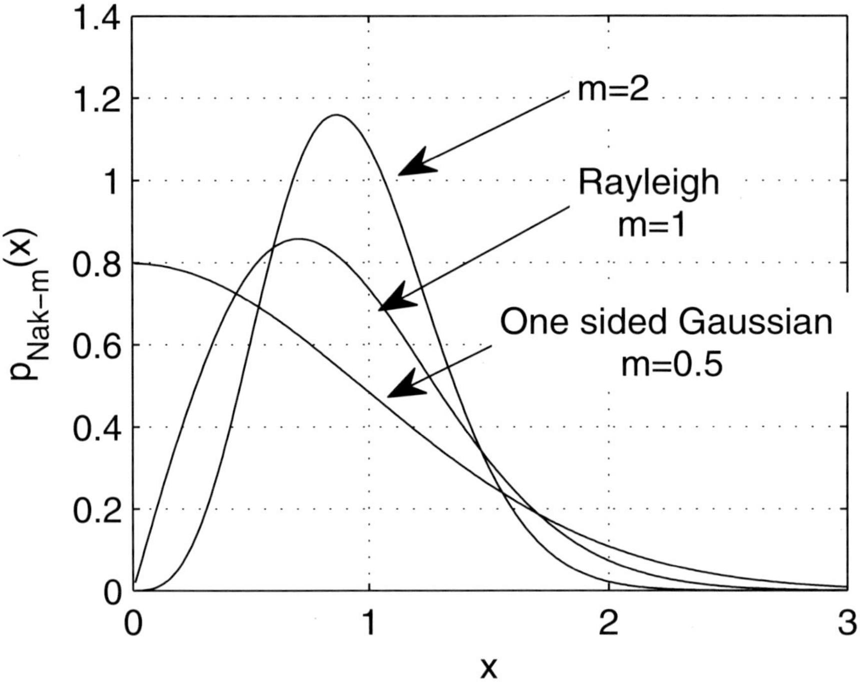

FIGURE 2.1
PDFs of some special cases of Nakagami-m distribution

of Nakagami-m fading can be given by

$$m = \frac{(1+K)^2}{1+2K}. \tag{2.21}$$

All the special cases of Nakagami-m fading distribution discussed above have been summarized in Table 2.1. The PDFs of some special cases of Nakagami-m distributed envelope corresponding to the fading parameter values are plotted in Figure 2.1.

2.4.2 $\eta - \mu$ fading

The proposal of $\eta - \mu$ fading was laid down by M. D. Yacoub in [132] as a generalized distribution to model different fading environments. $\eta - \mu$ distribution is also suitable to model non-LOS environments. After the proposal as a generalized fading distribution, $\eta - \mu$ fading distribution is popularly being used by many researchers for analysis of wireless communication systems. Like Nakagami-m fading, $\eta - \mu$ distribution also models a generalized fading

TABLE 2.1
Summary of special cases of Nakagami-m fading

Fading Distribution	Fading Parameter	Relation with Nakagami Fading
Rayleigh	-	$m = 1$
One sided Gaussian	-	$m = \frac{1}{2}$
Hoyt	q	$m = \frac{\left(1+q^2\right)^2}{2(1+2q^4)}$
Rician	K	$m = \frac{(1+K)^2}{1+2K}$

scenario which includes the non-homogeneous environment which is composed of the reflecting obstacles, scattering elements, etc., of different physical properties.

2.4.2.1 Physical model

Similar to Nakagami-m fading model, it is assumed that the multi-path components in the received signal are in the form of clusters and the clusters do not have any dominating or LOS component in $\eta - \mu$ distribution. Each cluster has a number of scattered multipath components. The delay spread of different clusters is relatively larger than the delay spread of multipath components within a cluster. Every cluster is assumed to have the same average power. However, the parameter η makes it different from Nakagami-m fading. η is defined as the ratio of power of the in-phase component to power of the quadrature phase component of the received signals. In such a model, the envelope X of the fading signal can be represented in the same way as that in Nakagami-m fading with different statistical parameters as described later, i.e.,

$$X^2 = \sum_{i=0}^{n} \left(I_i^2 + Q_i^2\right) \tag{2.22}$$

where n is the number of clusters in the received signal, I_i and Q_i are respectively the in-phase and quadrature phase component of the resultant signal of the ith cluster. Both I_i and Q_i, being resultant of multipath components of a cluster, may be assumed to be Gaussian distributed with zero mean, i.e., $E(I_i) = E(Q_i) = 0$. In $\eta - \mu$ fading, the variation from Nakagami-m fading is that the variance which is the same as the power content of I_i and Q_i is different. $E(I_i^2) = \sigma_{I_i}^2$ and $E(Q_i^2) = \sigma_{Q_i}^2$.

Again, as in the case of Nakagami-m fading model, the fading amplitude can be expressed as $X^2 = \sum_{i=0}^{n} \left(R_i^2\right)$ with $R_i^2 = I_i^2 + Q_i^2$. From the fact that I_i and Q_i are Gaussian distributed but with different variance, it is to be noted here that R_i^2 is no longer exponentially distributed. In this case, the PDF of

$\eta-\mu$ faded amplitude may be given by [132, 136]

$$p_X(x) = \frac{4\sqrt{\pi}\mu^{\mu+\frac{1}{2}}h^{\mu}}{\Gamma(\mu)H^{\mu-\frac{1}{2}}\Omega^{\mu+\frac{1}{2}}}x^{2\mu}e^{-\frac{2\mu h x^2}{\Omega}}I_{\mu-\frac{1}{2}}\left(\frac{2\mu H x^2}{\Omega}\right), \qquad x \geq 0 \tag{2.23}$$

where $\mu > 0$ is a fading parameter directly related to number of clusters, n, $\Gamma(\cdot)$ is the Gamma function, $I_\nu(\cdot)$ is the modified Bessel function of first kind and order ν, $\mu = \frac{n}{2}$ but it constrains the values of μ to be discrete on the account of discrete values of n. To make the parameter, μ, to take continuous values, it is alternatively defined as $\mu = \frac{1}{2V(X^2)}\left[1 + \left(\frac{H}{h}\right)^2\right]$, $V(\cdot)$ denotes the variance operator, and H and h are the functions of fading parameter η which is the fading parameter defined in two ways and thus there are two formats for $\eta-\mu$ fading channels.

$\eta-\mu$ fading: Format 1

In this format, the in-phase component and quadrature phase component of the resultant signal in each cluster are assumed to be independent and with different powers. η is defined as the ratio of power of the in-phase component to the power of quadrature component, i.e., $\eta = \frac{\sigma_{I_i}^2}{\sigma_{Q_i}^2}$. It is also assumed that this ratio is constant for all the clusters in the received signal. In this case, the values of η range between 0 and ∞, and H and h, the functions of η, are defined as $H = \frac{\eta^{-1}-\eta}{4}$ and $h = \frac{2+\eta^{-1}+\eta}{4}$ respectively. It can be shown that the values of h and H are symmetrical around $\eta = 1$, i.e., the values of H and h for $0 < \eta \leq 1$ are the same for the range $1 \leq \eta < \infty$. This means the statistical property of amplitude of the received signal remains unchanged for both these ranges of η. Therefore, it suffices to consider either $0 < \eta \leq 1$ or $1 \leq \eta < \infty$. In this format, the ratio $\frac{H}{h}$ simplifies to $\frac{1-\eta}{1+\eta}$.

$\eta-\mu$ fading: Format 2

In this format, the in-phase component and quadrature phase component of the envelope of the resultant signal in each cluster are assumed to be correlated and with equal powers. In this case, η is defined as the correlation coefficient between the in-phase component and the quadrature component, i.e., $\eta = \frac{E(I_i,Q_i)}{E(I_i^2)}$ or $\eta = \frac{E(I_i,Q_i)}{E(Q_i^2)}$. It is also assumed that the correlation coefficient between the in-phase component and the quadrature component is the same for all the clusters in the received signal. In this case, the values of η range between -1 and <1; and H and h, the functions of η are defined as $H = \frac{\eta}{1-\eta^2}$ and $h = \frac{1}{1-\eta^2}$. It can be shown that the values of h and H are symmetrical around $\eta = 0$. Therefore, it suffices to consider either $0 \leq \eta < 1$ or $-1 < \eta \leq 0$. In this format, the ratio $\frac{H}{h}$ simplifies to η.

Relation between format 1 and format 2

Physically, both the formats are different but the distributions of format 1 and format 2 match with each other for certain values of fading parameters. The parameter μ is defined in the same way for both the formats. So, to relate format 1 and format 2, the relation between η of format 1 and format 2 (though defined differently) needs to be established. This relation can be given as follows [30] by equating the ratio $\frac{H}{h}$ of both formats discussed earlier.

$$\eta_{format2} = \frac{1 - \eta_{format1}}{1 + \eta_{format1}} \tag{2.24}$$

2.4.2.2 $\eta - \mu$ fading: PDF of SNR

As far as distributions are concerned, SNR is the same as the square of amplitude distribution. So, it is valid to say that the PDF of X^2 is the same as that of SNR. The PDF of $\eta - \mu$ distributed instantaneous SNR can be given as [34, 132, 136]

$$p_\gamma(\gamma) = \frac{2\sqrt{\pi}\mu^{\mu+\frac{1}{2}}h^\mu}{\Gamma(\mu)\, H^{\mu-\frac{1}{2}}\bar{\gamma}^{\mu+\frac{1}{2}}}\gamma^{\mu-\frac{1}{2}}e^{\frac{-2\mu h\gamma}{\bar{\gamma}}} I_{\mu-\frac{1}{2}}\left(\frac{2\mu H\gamma}{\bar{\gamma}}\right), \qquad \gamma \geq 0 \tag{2.25}$$

where $\mu > 0$ is a fading parameter defined as $\mu = \frac{1}{2V(P_{\eta-\mu})}\left[1 + \left(\frac{H}{h}\right)^2\right]$.

Next, we will see different functions that are useful in performance evaluation of wireless communication systems.

2.4.2.3 CDF and MGF of $\eta - \mu$ fading

CDF and MGF are two statistical functions of random variables (instantaneous SNR in our case) which are most essential in evaluating the performance metrics of a wireless communication system. In the following sections of this chapter, we will discuss the metrics and methods for evaluation of the metrics which are used as performance analysis tools for wireless communication systems.

CDF of $\eta - \mu$ distributed instantaneous SNR can be given by

$$P_\gamma(y) = \int_0^y p_\gamma(\gamma)\, d\gamma \tag{2.26}$$

$$= 1 - Y_\mu\left(\frac{H}{h}, \sqrt{\frac{2h\mu y}{\bar{\gamma}}}\right) \tag{2.27}$$

where $Y_\mu(x, y)$ is Yacoub's integral [136]. It is defined as

$$Y_\mu(x, y) = \frac{\sqrt{\pi}\left(1 - x^2\right)^\mu 2^{\frac{3}{2}-\mu}}{\Gamma(\mu) x^{\mu-\frac{1}{2}}} \int_y^\infty e^{-t^2} t^{2\mu} I_{\mu-\frac{1}{2}}\left(tx^2\right) dt \tag{2.28}$$

where $-1 < x < 1$ and $y \geq 0$. The solution to Yacoub's integral can be given by [81]

$$Y_\mu(x,y) = 1 - \frac{\left(1-x^2\right)^\mu y^{4\mu}}{\Gamma(1+2\mu)} \Phi_2^{(2)}\left(\mu,\mu;1+2\mu;-(1+x)y^2,-(1-x)y^2\right) \tag{2.29}$$

where $\Phi_2^{(2)}$ is the confluent Lauricella function. It can be further simplified for special cases. The simplified expressions for integer values of 2μ and even values of 2μ are obtained and reported in [81].

MGF of $\eta - \mu$ distributed instantaneous SNR can be given by [38]

$$M_\gamma(s) = \int_0^\infty p_\gamma(\gamma) e^{-s\gamma} d\gamma \tag{2.30}$$

$$= \frac{\left(4\mu^2 h\right)^\mu}{\left(2(h-H)\mu + s\overline{\gamma}\right)^\mu \left(2(h-H)\mu + s\overline{\gamma}\right)^\mu} \tag{2.31}$$

For detailed evaluation of CDF and MGF of $\eta - \mu$ distributed instantaneous SNR, the readers may refer to Appendices A.8 and A.9.

2.4.2.4 Special cases

Different values of fading parameters, η and μ, can represent different fading distributions. But, certain values of η and μ can model specific distributions which are well known and useful in modeling wireless fading channels. Though the parameter η is defined in two ways for format 1 and format 2, we already discussed that its value for one format can be transformed to another format using equation (2.24). So, special cases pertaining to one format can also be interpreted to another format. However, most special cases can be explained with format 1 as far as physical models are concerned. To discuss the special cases, we again refer the reader to the physical model and format 1. Limit the number of clusters in the received signal to 1 which represents $\mu = 0.5$. Then, the expression of equation (2.22) reduces to

$$X = \sqrt{I^2 + Q^2} \tag{2.32}$$

where I and Q are Gaussian distributed with zero mean. But, in this case the variances of I and Q are different and the difference is related by $\eta = \frac{\sigma_{I_i}^2}{\sigma_{Q_i}^2}$. With a slight restriction on the value of η, this can match to the definition of Rayleigh fading distribution. For the values $\eta = 1$ and $\mu = 0.5$, Rayleigh fading is represented as a special of $\eta - \mu$ fading distribution.

Again, restricting $\eta = 1$, another case is $\mu = \frac{1}{4}$. This means half a cluster which is physically not possible. But it can be described taking into consideration the previous case in which $\mu = 0.5$ represents square root sum of two squared Gaussian distributed random variables. So, in the case of $\mu = \frac{1}{4}$, it can be thought of as the single Gaussian distributed random variable. Since

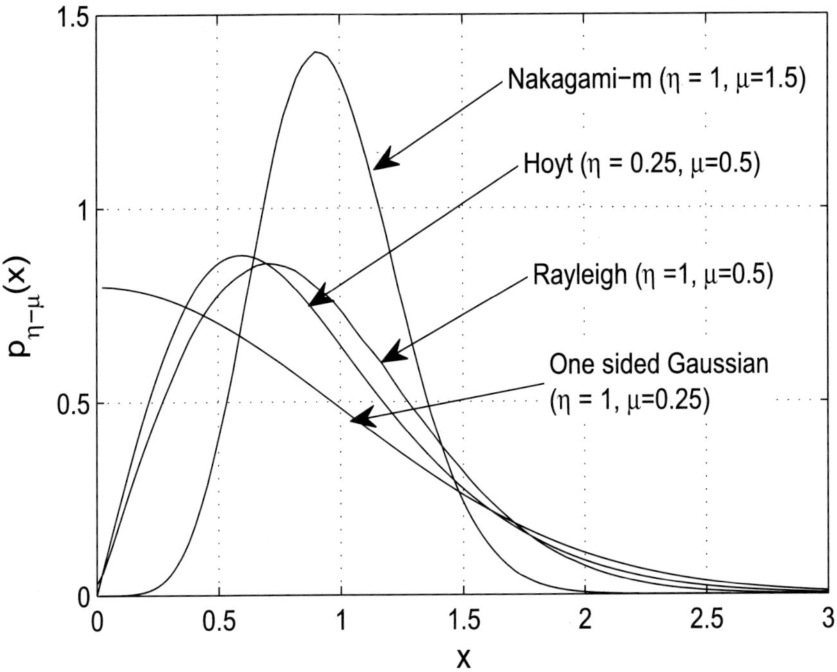

FIGURE 2.2
PDFs of some special cases of $\eta - \mu$ distribution.

the PDF of $\eta - \mu$ distributed random variable is restricted only to positive values, this case represents one sided Gaussian fading.

Again, we restrict number of clusters to 1, i.e., $\mu = 0.5$. η can take arbitrary value. In this case, the expression of equation (2.22) reduces to

$$X = \sqrt{X_1^2 + \eta X_2^2} \tag{2.33}$$

with X_1 and X_2 both being Gaussian distributed with zero mean and equal variance. However, the factor $\sqrt{\eta}$ as coefficient of X_2 manipulates the variance of the second term, i.e., the variance of $\sqrt{\eta}X_2$ as a whole would be different from that of X_2. This represents Hoyt fading. So, Hoyt fading can be represented as a special case of $\eta - \mu$ fading for the values $\eta = q^2$ and $\mu = 0.5$.

As was already mentioned while discussing the physical model of $\eta - \mu$ distribution, it differs from Nakagami-m fading model in only one parameter and that parameter is the variance of in-phase components and quadrature phase components of resultant of each cluster in the received signal. Otherwise, the expressions that model the amplitude of received signal in Nakagami-m fading, expression of equation (2.10), and $\eta - \mu$ fading, expression of equation

TABLE 2.2
Values of fading parameters to represent different fading distributions as special case of $\eta - \mu$ fading distribution

Distribution	Value of η	Value of μ
One sided Gaussian	1	0.25
Rayleigh	1	0.5
Nakagami-m	1	$m/2$
Nakagami-q (Hoyt)	q^2	0.5

(2.22), are exactly the same. So, restricting the value of η to 1, Nakagami-m fading model can be realized as a special case of $\eta - \mu$ fading distribution with the parameter, $m = 2\mu$. So, $\eta - \mu$ fading can model Nakagami-m distribution as a special case. This means that all the special cases of Nakagami-m fading discussed in Section 2.4.13 can also be modeled using $\eta - \mu$ distribution. This makes $\eta - \mu$ distribution more widely accepted as generalized distribution than Nakagami-m fading distribution.

The special cases of Nakagami-m fading distribution discussed above have been summarized in Table 2.2. The PDFs for some special cases of $\eta - \mu$ distributed normalized envelope corresponding to values of fading parameters are plotted and can be observed in Figure 2.2.

2.4.3 $\kappa - \mu$ fading

The proposal of $\kappa - \mu$ fading was laid down by M. D. Yacoub in [133] as a generalized distribution to model fading environment. Unlike Nakagami-m and $\eta - \mu$ fading models, $\kappa - \mu$ distribution is suitable to model LOS environments. After the proposal as a generalized fading distribution, $\kappa - \mu$ fading distribution is popularly used by the researchers for analysis of wireless communication systems. Like Nakagami-m fading and $\eta - \mu$ distributions, $\kappa - \mu$ distribution also can model a generalized fading scenario which includes the non-homogeneous environment which is composed of the reflecting obstacles, scattering elements, etc. of different physical properties.

2.4.3.1 Physical model

Similar to Nakagami-m and $\eta - \mu$ fading models, it is assumed that the multipath components in received signal are the form of clusters in $\kappa - \mu$ distribution. Each cluster has a number of scattered multipath components. The delay spread of different clusters is relatively larger than the delay spread of multipath components within a cluster. Every cluster is assumed to have the same average power. Unlike in $\eta - \mu$ fading and like Nakagami-m fading, it is assumed that the in-phase and quadrature phase components are independent and have equal powers in $\kappa - \mu$ fading. However, each cluster is assumed to have some dominant components considered to be LOS components. In such a model, the representation of envelope X of the fading signal is slightly different from that of Nakagami-m and/or $\eta - \mu$ fading. It can be given as

$$X^2 = \sum_{i=0}^{n} \left((I_i + p_i)^2 + (Q_i + q_i)^2 \right) \tag{2.34}$$

where n is the number of clusters in the received signal, and $(I_i + p_i)$ and $(Q_i + q_i)$ are respectively the in-phase and quadrature phase component of the resultant signal of the ith cluster. Both I_i and Q_i are mutually independent and Gaussian distributed with zero mean, i.e., $E(I_i) = E(Q_i) = 0$ and equal variance, i.e., $E(I_i^2) = E(Q_i^2) = \sigma^2$. p_i and q_i are the respective means of in-phase and quadrature components of the ith cluster in the received signal. The non-zero mean of in-phase and quadrature phase components reveals the presence of a dominant component in the clusters of the received signal.

Again, as in the case of Nakagami-m and $\eta - \mu$ fading model, the fading amplitude can be expressed as $X^2 = \sum_{i=0}^{n} \left(R_i^2\right)$ with $R_i^2 = (I_i + p_i)^2 + (Q_i + q_i)^2$. From the fact that I_i and Q_i are Gaussian distributed, it is to be noted here that R_i^2 follows non-central Chi-squared distribution. In this case, the PDF of $\kappa - \mu$ faded amplitude may be given by [133, 136]

$$p_X(x) = \frac{2\mu(1+\kappa)^{\frac{\mu+1}{2}}}{\kappa^{\frac{\mu-1}{2}} e^{\mu\kappa} \Omega^{\frac{\mu+1}{2}}} x^{\mu} e^{-\frac{\mu(1+\kappa)x^2}{\Omega}} I_{\mu-1}\left(2\mu x \sqrt{\frac{\kappa(1+\kappa)}{\Omega}}\right), \qquad x \geq 0 \tag{2.35}$$

where κ is the ratio of total power in the dominant (LOS) components to that of scattered components and $\mu > 0$ is a fading parameter directly related to number of clusters, n, in the received signal as $\mu = n$ with the constraint that values of μ must be discrete on the account of discrete values of n. To make the parameter μ to take continuous values, it is alternatively defined as $\mu = \frac{1}{V(X^2)} \frac{1+2\kappa}{(1+\kappa)^2}$.

SNR in the case of fading channels varies as the square of amplitude of the received signal. So, as far as distributions are concerned, PDF of SNR is the same as the square of amplitude distribution. PDF of SNR may be obtained from the PDF of amplitude in equation (2.35) using the techniques of transformation of random variables. The PDF of $\kappa - \mu$ distributed instantaneous SNR can thus be given as

$$p_\gamma(\gamma) = \frac{\mu(1+\kappa)^{\frac{\mu+1}{2}}}{\kappa^{\frac{\mu-1}{2}} e^{\mu\kappa} \overline{\gamma}^{\frac{\mu+1}{2}}} \gamma^{\frac{\mu-1}{2}} e^{-\frac{\mu(1+\kappa)\gamma}{\overline{\gamma}}} I_{\mu-1}\left(2\mu \sqrt{\frac{\kappa(1+\kappa)\gamma}{\overline{\gamma}}}\right), \qquad \gamma \geq 0 \tag{2.36}$$

where $\mu = \frac{1}{V(\gamma)} \frac{1+2\kappa}{(1+\kappa)^2}$.

2.4.3.2 CDF and MGF of $\kappa - \mu$ fading

CDF and MGF are two statistical functions of random variables (instantaneous SNR in our case) which are most essential in evaluating the performance metrics of a wireless communication system. In general, CDF is useful in analyzing the outage probability and MGF is useful for bit error rate (BER)/symbol error rate (SER) analysis of the wireless communication systems. In the following sections of this chapter, we will discuss the metrics and methods for evaluation of the metrics which are used as performance analysis tools for wireless communication systems.

CDF of $\kappa - \mu$ distributed instantaneous SNR can be given by

$$P_\gamma(y) = \int_0^y p_\gamma(\gamma)\, d\gamma \tag{2.37}$$

$$= 1 - Q_\mu\left(\sqrt{2\kappa\mu}, \sqrt{\frac{2\mu(1+\kappa)y}{\overline{\gamma}}}\right) \tag{2.38}$$

where $Q_\mu(\alpha, \beta)$ is generalized Marcum Q-function [46]. It is defined as

$$Q_\mu(\alpha, \beta) = \int_{\beta}^{\infty} t \left(\frac{t}{\alpha}\right)^{\mu-1} e^{-\frac{t^2+\alpha^2}{2}} I_{\mu-1}(\alpha t)\, dt \tag{2.39}$$

MGF of $\kappa - \mu$ distributed instantaneous SNR can be given by [38]

$$M_\gamma(s) = \int_0^\infty p_\gamma(\gamma) e^{-s\gamma} d\gamma \tag{2.40}$$

$$= \left(\frac{\mu(1+\kappa)}{\mu(1+\kappa) + s\overline{\gamma}}\right)^\mu e^{\left(\frac{\mu^2\kappa(1+\kappa)}{\mu(1+\kappa)+s\overline{\gamma}}\right) - \kappa\mu} \tag{2.41}$$

For detailed evaluation of CDF and MGF of $\kappa - \mu$ distributed instantaneous SNR, the readers may refer to Appendices A.5 and A.6. It shall be noted here that alternative expressions for MGF of $\kappa - \mu$ and $\eta - \mu$ distributions in the form of Meijer's G function are also reported [26]. But the expressions given in this book (originally derived in [38] by N. Y. Ermolova) are in the form of basic mathematical functions which are easy to evaluate, while one needs to use numerical techniques to evaluate Meijer's G functions.

2.4.3.3 Special cases

It was already discussed that $\kappa - \mu$ distribution can model generalized fading environments. This enables us to represent different fading channels by selecting different values of fading parameters, κ and μ. But, certain values of κ and μ can model specific distributions which are well known and useful in modeling wireless fading channels.

To discuss a special case, we again refer to the physical model. Limit the number of clusters in the received signal to 1 which represents $\mu = 1$. Then, the expression of equation (2.34) reduces to

$$X = \sqrt{(I+p)^2 + (Q+q)^2} \tag{2.42}$$

where I and Q are Gaussian distributed with zero mean and equal variance. But, in this case, the in-phase and quadrature phase components contain some dominant component, which the means of in-phase component and quadrature component are respectively p and q. This matches the definition of Rician fading distribution which is commonly used to model fading channel with LOS conditions, i.e., the value $\mu = 1$ represents Rician fading as a special of $\kappa - \mu$ fading distribution. In this case, κ represents the Rician K factor which is also defined as the ratio of total power in the dominant component to that of the scattered components.

We have already discussed that Rician fading distribution can model the fading scenarios ranging from as severe as Rayleigh fading to less severe fading including no fading at all. Similarly, in the expression of equation (2.42), if we consider that there is no dominant component, i.e., $p = q = 0$, it reduces to

$$X = \sqrt{I^2 + Q^2} \tag{2.43}$$

which is exactly the same as the amplitude of the Rayleigh fading model which is the square root of sum of two squared independent Gaussian distributed random variables. This case is obtained for no dominant component, i.e., $K = \kappa = 0$. So, Rayleigh fading can be represented as a special case of $\kappa - \mu$ fading with the values $\kappa = 0$ and $\mu = 1$.

Again, restricting the value of κ to 0, another case which may be considered is $\mu = \frac{1}{2}$. This means half a cluster with no dominant component, which is not possible to explain with the help of a physical model. But it can be described taking into consideration the previous case in which $\mu = 1$ represents square root sum of two squared Gaussian distributed random variables. So, in the case of $\mu = \frac{1}{2}$, it can be thought of as a single Gaussian distributed random variable. Since the PDF of $\kappa - \mu$ distributed random variable is restricted only to positive values, this case represents one sided Gaussian fading.

Again, consider that the received signal arrives in clusters but the clusters do not have any dominant component, i.e., put $p_i = q_i = 0$ in the physical model of $\kappa - \mu$ fading represented in equation (2.34). This reduces the expression to the following form:

$$X^2 = \sum_{i=0}^{n} \left(I_i^2 + Q_i^2\right) \tag{2.44}$$

with each I_i and Q_i being mutually independent and Gaussian distributed with zero mean and equal variance. This represents the physical model which is exactly the same as that of Nakagami-m fading. So, Nakagami-m fading

TABLE 2.3
Values of fading parameters to represent different fading distributions as special case of $\kappa - \mu$ fading distribution

Distribution	Value of κ	Value of μ
One sided Gaussian	0	0.5
Rayleigh	0	1
Nakagami-m	0	m
Rice	K	1

can be represented as a special case of $\kappa - \mu$ fading for the values $\kappa = 0$ and $\mu = m$. As described in the special cases of Nakagami-m fading, similar manipulations in the fading parameters of $\kappa - \mu$ distribution can represent all the special cases of Nakagami-m fading too.

So, $\kappa - \mu$ distribution includes one-sided Gaussian, Rayleigh, Rician and Nakagami-m fading distributions as its special cases. The values of $\kappa - \mu$ fading parameters for these special cases are listed in Table 2.3. The PDFs for special cases of $\kappa - \mu$ distributed normalized envelope are plotted as shown in Figure 2.3. It can be observed that the PDF curves shift toward right for non-zero values of κ when compared with the same value of μ (see Rayleigh and Rician case in Figure 2.3). This is because of a stronger LOS component for a larger value of κ.

2.4.4 $\alpha - \mu$ fading

The proposal of $\alpha - \mu$ distribution had been made with the name of Stacy distribution as a generalization to Gamma distribution [117, 135]. The initial proposal of Stacy distribution was to deal with the statistical problems, which was renamed later by M. D. Yacoub as $\alpha - \mu$ distribution [134, 135]. $\alpha - \mu$ distribution can be used to model fading channels in the environment characterized by non-homogeneous obstacles that may be nonlinear in nature. Like the other generalized fading models discussed earlier in this section, $\alpha - \mu$ fading can also model various fading distributions as a special case.

2.4.4.1 Physical model

As discussed in Nakagami-m, $\eta - \mu$ and $\kappa - \mu$ fading models, $\alpha - \mu$ fading also considers the received signal to be collection of clusters of multipath components. The clusters are assumed to have no dominant component. The delay spread of different clusters is relatively larger than the delay spread of multipath components within a cluster. Every cluster is assumed to have the same average power. This model differs from other models in the definition

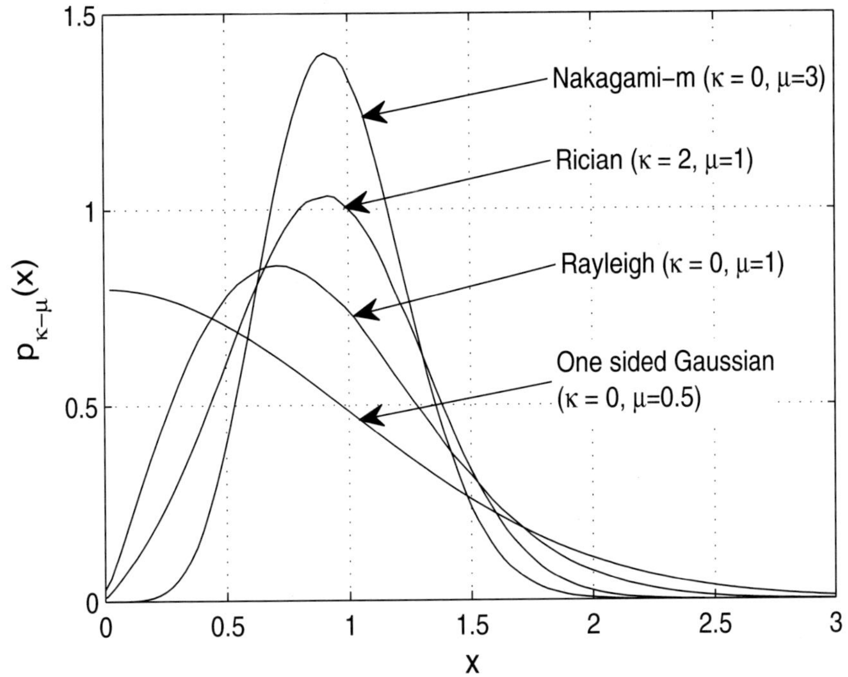

FIGURE 2.3
PDFs of some special cases of $\kappa - \mu$ distribution.

of power. In $\alpha - \mu$ distribution, fading amplitude is defined as the αth root of the received power in the faded signal. In all other fading models, fading amplitude is defined as the square root of the received power in the faded signal. The physical relation between resultants of received multipath clusters and fading amplitude for $\alpha - \mu$ distribution can be given as [135]

$$X^{\alpha} = \sum_{i=1}^{n} \left(I_i^2 + Q_i^2\right) \tag{2.45}$$

where n is the number of clusters, I_i and Q_i are the resultant in-phase and quadrature phase components of ith cluster in the received signal, α is the fading parameter that defines the relation that arises due to the non-linear nature of the objects in-between the power of the received signal and amplitude of the fading signal envelope. I_i and Q_i are mutually independent Gaussian distributed random variables with zero mean and equal variance, i.e., $E(I_i) = E(Q_i) = 0$ and $E(I_i^2) = E(Q_i^2) = \sigma^2$. Physically, $\alpha - \mu$ distribution is the same as Nakagami-m distribution with an extra parameter, α, which is introduced to take care of non-linearity in certain fading conditions.

In such a case, X^α follows Gamma distribution as it is represented as the sum of squared Gaussian random variables. The PDF of amplitude of the received faded signal, X, can be obtained using transformation of random variables. It can be given as

$$p_X(x) = \frac{\alpha\mu^\mu x^{\alpha\mu-1}}{\Gamma(\mu)\Omega_\alpha^{\alpha\mu}} e^{-\mu\left(\frac{x}{\Omega_\alpha}\right)^\alpha} \qquad \mu > 0,\ \alpha > 0,\ x \geq 0 \qquad (2.46)$$

where Ω_α is the αth root mean value of the fading envelope, X, defined as $\Omega_\alpha = \sqrt[\alpha]{E\left(X^\alpha\right)}$, and $\mu > 0$ is the fading parameter defined as the number of clusters which is discrete in nature. But similar to the same parameter of $\eta-\mu$ fading and $\kappa-\mu$ fading distributions, to make the values of μ continuous, alternatively, it is defined as [68]

$$\mu = \frac{E(X^\alpha)^2}{E(X^\alpha - E(X)^\alpha)}, \qquad \mu > 0 \qquad (2.47)$$

Again, the parameter of interest in most communication systems is the received SNR. In the case of $\alpha-\mu$ distribution, the PDF of received SNR can be given as

$$p_\gamma(\gamma) = \frac{\alpha\mu^\mu \gamma^{\frac{\alpha\mu}{2}-1}}{2\Gamma(\mu)\bar{\gamma}^{\frac{\alpha\mu}{2}}} e^{-\mu\left(\frac{x}{\bar{\gamma}}\right)^{\frac{\alpha}{2}}} \qquad \mu > 0,\ \alpha > 0,\ \gamma \geq 0 \qquad (2.48)$$

Various other functions related to the SNR that describe the statistical properties of the received SNR are discussed in the following parts of this section.

2.4.4.2 CDF and MGF of $\alpha-\mu$ fading

CDF and MGF are two statistical functions of random variables (instantaneous SNR in our case) which are most essential in evaluating the performance metrics of a wireless communication system. In general, CDF is useful in analyzing the outage probability and MGF is useful for BER/SER analysis of the wireless communication systems. In the following sections of this chapter, we will discuss the metrics and methods for evaluation of the metrics which are used as performance analysis tools for wireless communication systems.

For $\alpha-\mu$ distributed instantaneous SNR, CDF can be given in the form of lower incomplete Gamma function as

$$P_\gamma(y) = \int_0^y p_\gamma(\gamma)d\gamma \qquad (2.49)$$

$$= \frac{\gamma\left(\mu, \mu\left(\frac{y}{\bar{\gamma}}\right)^{\frac{\alpha}{2}}\right)}{\Gamma\left(\mu\right)} \qquad (2.50)$$

where $\gamma(\cdot,\cdot)$ is the lower incomplete Gamma function defined as $\gamma\left(p, y\right) = \int_0^y x^{p-1}e^{-x}dx$ [46].

MGF is useful in evaluating BER/SER performance of the wireless communication systems, which is the most important parameter. As far as the closed form MGF of $\alpha-\mu$ fading distributions is concerned, a form which is easy to evaluate like that in the case of Nakagami-m, $\eta-\mu$ and $\kappa-\mu$ fading distributions are not available for the general case. But an MGF in the form of Meijer G-function is available and can be given by [68]

$$M_\gamma(s)=\int_0^\infty p_\gamma(\gamma)e^{-s\gamma}d\gamma \tag{2.51}$$

$$=\frac{\alpha\mu^\mu}{2\overline{\gamma}^{\frac{\alpha\mu}{2}}}\frac{\sqrt{k}l^{\frac{\alpha\mu-1}{2}}}{(2\pi)^{\frac{l+k-2}{2}}s^{\frac{\alpha\mu}{2}}}G^{k,l}_{l,k}\left(\left(\frac{\mu}{\overline{\gamma}^{\frac{\alpha}{2}}}\right)^k\frac{l^l}{s^lk^k},\ \begin{matrix}P\left(l,1-\frac{\alpha\mu}{2}\right)\\ P(k,0)\end{matrix}\right) \tag{2.52}$$

where $P(a,b)$ is an array of the elements $b/a,(b+1)/a,...(b+a+1)/a$, and $G^{r,s}_{p,q}\left(x,\ \begin{matrix}P(r,s)\\ P(p,q)\end{matrix}\right)$ is a Meijer's G-function with p and r being integers [103].

2.4.4.3 Special cases

All the distributions under the umbrella of generalized fading distributions can model different fading conditions for different sets of values for fading parameters. These different values represent various physical conditions. Some of those conditions can be explained with the physical model of the distributions. We have already discussed such specialized physical models for Nakagami-m, $\eta-\mu$ and $\kappa-\mu$ fading distributions. However, generalized fading distributions can model many more fading conditions as complex as such that cannot even be explained with physical model description. In the following discussion, we will discuss about special cases of $\alpha-\mu$ fading distribution which can be used to model certain fading environments.

To begin with the special cases of $\alpha-\mu$ fading distribution, let us restrict the number of clusters in the received signal to one, i.e., $\mu=1$. In this case, the fading amplitude of the received signal described by equation (2.45) can be given by

$$X=\sqrt[\alpha]{I^2+Q^2} \tag{2.53}$$

with I and Q being the in-phase and quadrature phase components of the received signal. Here, I and Q are Gaussian distributed with equal variances, i.e., the power contained in in-phase and quadrature phase components is equal. The relation described by equation (2.53) can be characterized by Weibull distribution [89,91]. The PDF of Weibull distributed fading amplitude can be given by

$$p_X(x)=\frac{\alpha x^{\alpha-1}}{\Omega}e^{-\frac{x^\alpha}{\Omega}} \tag{2.54}$$

where $\Omega=E(x^\alpha)$. Note that α also indicates severity of fading in addition to the nonlinearity in the fading environment. Higher values of parameter α indicate less severe fading. So, $\mu=1$ represents Weibull distribution as a special case of $\alpha-\mu$ fading distribution with α being the fading parameter of Weibull distribution.

Now, considering $\alpha=1$ in the expression of the physical model described by the expression of equation (2.53), the amplitude of received signal can be modeled as

$$X = I^2 + Q^2 \tag{2.55}$$

with I and Q being Gaussian distributed with zero mean and equal variance. This represents exponential distribution. In such a case, the PDF of fading envelope can be given by

$$p_X(x) = \frac{1}{\Omega} e^{-\frac{x}{\Omega}} \tag{2.56}$$

So, considering $\alpha=1$ and $\mu=1$ as fading parameters of $\alpha-\mu$ distribution, exponential distribution can be represented as a special case of $\alpha-\mu$ fading distribution.

Again, considering the physical model described by the expression of equation (2.53) and $\alpha=2$, the amplitude of received signal can be represented as

$$X = \sqrt{I^2 + Q^2} \tag{2.57}$$

with I and Q being Gaussian distributed with zero mean and equal variance. This represents Rayleigh distribution which is already explained in Section 2.3.1, i.e., considering $\mu=1$ and $\alpha=2$, Rayleigh fading can be represented as a special case of $\alpha-\mu$ fading distribution.

Similar to explanations given for the case of $\mu=\frac{1}{2}$ and $\mu=\frac{1}{4}$ in the respective cases of $\kappa-\mu$ and $\eta-\mu$ fading distributions, considering $\alpha=2$ and $\mu=\frac{1}{2}$, one sided Gaussian distribution can be represented as a special case of $\alpha-\mu$ distribution.

We now go back to equation (2.45) and assume the value of $\alpha=2$. For this case, the received envelope of the faded signal can be represented as

$$X = \sqrt{\sum_{i=1}^{n} (I_i^2 + Q_i^2)} \tag{2.58}$$

with each I_i and Q_i being mutually independent and Gaussian distributed with zero mean and equal variance. This represents the physical model which is exactly the same as that of Nakagami-m fading. So, Nakagami-m fading can be represented as a special case of $\alpha-\mu$ fading for the values $\alpha=2$ and $\mu=m$. As described in the special cases of Nakagami-m fading, similar manipulations in the fading parameters of $\alpha-\mu$ distribution can represent all the special cases of Nakagami-m fading too.

So, $\alpha-\mu$ distribution includes Weibull, exponential, one-sided Gaussian, Rayleigh and Nakagami-m distributions as its special cases. The values of $\alpha-\mu$ fading parameters for these special cases are listed in Table 2.4. The PDFs for special cases of $\alpha-\mu$ distributed normalized envelope corresponding to values of fading parameters are plotted and can be observed in Figure 2.4.

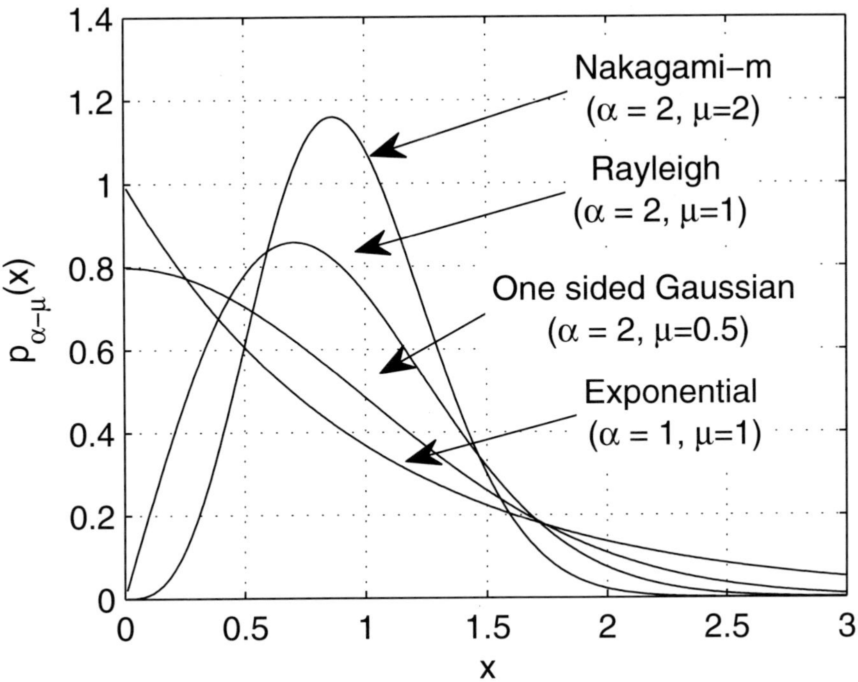

FIGURE 2.4
PDFs of some special cases of $\alpha - \mu$ distribution.

TABLE 2.4
Values of fading parameters to represent different fading distributions as special case of $\alpha - \mu$ fading distribution

Distribution	Value of α	Value of μ
One sided Gaussian	2	0.5
Rayleigh	2	1
Nakagami-m	2	m
Exponential	1	1

2.5 Generation of channel coefficients

Simulation of wireless communication systems involves realization of various processes at transmitter, receiver and channel using suitable software tools. The processes at transmitter include generation of bits, grouping the bits into symbols, coding and/or modulation. At receiver, the involved processes are demodulation, detection of symbols and getting the bitstream. Random noise and channel coefficients are to be generated to realize the wireless channel. The channel coefficients are assumed to be collective response of all the multipath components and hence they may be assumed to be a distributed according to any of the fading models we discussed earlier depending on the environment. In the following subsections, we will discuss methods to generate the channel coefficients in MATLAB® for various fading distribution models. MATLAB is the commonly used software tool for simulation of wireless communication systems.

2.5.1 Channel coefficients for commonly used fading distributions

The methods to generate channel coefficients for commonly used fading distributions are well known to the researchers and the academicians. However, to maintain the continuity, we are discussing them in brief. Here, we will show the ways to generate channel coefficients in MATLAB. In all the generated channel coefficients, we will assume that the phase of the received envelope is uniformly distributed in the range from 0 to 2π. The most frequently considered and the simplest fading model is Rayleigh fading. It has gained popularity for its simplicity in determining statistical parameters. Physically, Rayleigh distributed variable can be represented as the square root of sum of two independent squared Gaussian distributed random variables with zero mean and equal variance. So, a single sample of Rayleigh distributed channel coefficient, h, can be generated using the instruction that follows:

```
h=(1/sqrt(2))*(randn+j*randn);
```

where the factor $\frac{1}{\sqrt{2}}$ is to normalize the variance of the generated fading coefficients. Similarly, a matrix of channel coefficients can be generated using the following instruction in case the system under consideration has multiple antennas at transmitter and the receiver (multiple-input multiple-output (MIMO) system).

```
H=(1/sqrt(2))*(randn(Nr,Nt) +j*randn(Nr,Nt));
```

The above instruction generates channel matrix of dimension $Nr \times Nt$ which corresponds to the channel matrix of a MIMO system with N_t antennas at the transmitter and N_r antennas at the receiver.

Rayleigh distribution can model a non-LOS fading environment. But that is not always the case. To model the fading distribution when there exist LOS or dominant components in the received signal, Rician distribution comes to the rescue. Generation of Rician distributed fading channel coefficients is similar to that of Rayleigh fading. In Rician fading coefficient generation, the LOS component is characterized by a constant term in the in-phase component. The value of this constant term depends on the K factor which is a parameter of Rician distribution. So, a single sample of Rician distributed channel coefficient, h, can be generated using the MATLAB instruction that follows:

```
hi=sqrt(K/(K+1))+sqrt(1/(2*(K+1)))*randn;
hq=sqrt(1/(2*(K+1)))*randn;
h=hi+j*hq;
```

where the factors $\sqrt{\frac{K}{K+1}}$ and $\sqrt{\frac{1}{2(K+1)}}$ are used to indicate the strength of LOS/dominant component and to normalize the fading signal power. Similarly, the same set of MATLAB instructions can be considered to generate channel matrix for MIMO systems.

```
Hi=sqrt(K/(K+1))+sqrt(1/(2*(K+1)))*randn(Nr,Nt);
Hq=sqrt(1/(2*(K+1)))*randn(Nr,Nt);
H=Hi+j*Hq;
```

The coefficients that are generated in the channel matrix are identically distributed and mutually independent. In the above set of instructions for h and H, it can be observed that for the limiting case of $K \to \infty$, all the coefficient values become 1, i.e., there is no fading, and for the case of $K = 0$, it becomes equivalent to Rayleigh fading coefficients. Thus, referring to the discussion about Rician fading channels in Section 2.3.3, it is verified that Rician distribution can model fading environments ranging from no fading to Rayleigh fading.

Rician distribution for modeling the fading environments is suitable for the fading conditions that are less severe than Rayleigh fading. Next, we discuss the distribution that can model fading conditions that are more severe than Rayleigh fading. Such conditions can be modeled by Hoyt (Nakagami-q) distribution. Hoyt fading channel is characterized by absence of the dominant/LOS component unlike in Rician fading. The difference of Hoyt fading from Rayleigh fading is that it considers unequal powers in in-phase and quadrature phase components of the received signal. So, a single sample of Hoyt distributed channel coefficient, h, can be generated using the MATLAB instruction that follows:

```
h=(1/sqrt(1+q^2))*(randn+j*q*randn);
```

where q, the fading parameter, indicates the severity of fading. Similarly, a matrix of channel coefficients can be generated using the following instruction in case the system under consideration has multiple antennas at transmitter and the receiver (MIMO system).

```
H=(1/sqrt(1+q^2))*(randn(Nr,Nt) +j*q*randn(Nr,Nt));
```

In the next section we will discuss the methods of generating channel coefficients for generalized fading channels.

2.5.2 Channel coefficients for generalized fading distributions

Simulation of wireless communication systems over generalized fading channels is important because simulating them over generalized fading can produce results for any well-known fading scenario just by selecting suitable values of fading parameters. This eliminates the need of producing multiple software programs for different fading conditions. In this section, we will discuss the methods for obtaining channel coefficients of various generalized fading distributions that are discussed earlier. In most cases, we will use the relation of physical models that were defined to explain generalized fading distributions while generating the channel coefficient on MATLAB. The same methods will be used to generate the channel coefficients in the simulations of different variants of MIMO communication systems in the chapters to follow. We will now discuss the procedure to obtain channel coefficients one by one for various generalized fading distributions.

2.5.2.1 Generating Nakagami-m faded channel coefficients

Note that the received signal following Nakagami-m fading distribution is assumed to be a collection of multipath components that reach the receiver in form of clusters. The number of clusters is related to the fading parameter, m, as per the physical model discussed in Section 2.4.1. Primarily, Nakagami-m fading distribution is already shown to be suitable for medeling non-LOS fading conditions as per the physical model. However, there are some indirect manipulations of fading parameters that can model Rician fading as a special case of Nakagami-m distribution (already discussed in Section 2.4.1). Following the expression of physical model of equation (2.10), channel coefficients following Nakagami-m distribution can be generated by the following set of MATLAB instructions.

```
function h=nak_m(m)
n=0;
for i=1:2*m
n=n+randn^2;
end
n=n/(2*m);
phi=2*pi*rand;
h=sqrt(n)*cos(phi)+j*sqrt(n)*sin(phi);
```

This generates a single sample of channel coefficient (useful for SISO systems). We first generated the square variate of the Nakagami-m fading signal and then converted it to the complex envelope form using the uniformly distributed phase component. The same function can be modified to obtain independent and identically distributed channel coefficients with slight modifications as shown below.

```
function H=nak_m(m,Nr,Nt)
n=zeros(Nr,Nt);
for i=1:2*m
n=n+randn(Nr,Nt).^2;
end
n=n/(2*m);
phi=2*pi*rand(Nr,Nt);
H=(n.^0.5).*cos(phi)+j*(n.^0.5).*sin(phi);
```

The method to generate Nakagami-m distributed channel coefficients used the physical model. In the physical model, the fading parameter, m, is defined as the number of clusters of multipath components in the received signal. Alternatively, it is observed that the instantaneous SNR follows Gamma distribution. Hence, the coefficients of Nakagami-m distributed fading channels can also be derived from Gamma distributed random variable. To do that we need to relate the fading parameter, m, of Nakagami-m fading distribution model with the parameters of Gamma distribution. In MATLAB, Gamma distributed random variables can be generated using the function **gamrnd**. It has two parameters, shape parameter, A, and rate parameter, B, as input arguments. A and B are related to the Nakagami-m fading parameter m as A=m and B=$\overline{\gamma}/m$ respectively. Thus, a sample of Nakagami-m distributed fading channel coefficients can be generated using the following set of instructions.

```
n=gamrnd(m,a_SNR/m);
phi=2*pi*rand;
h=(n^0.5)*cos(phi)+j*(n^0.5)*sin(phi);
```

where a_SNR is the average SNR or variance of the generated samples of Gamma/Nakagami-m distributed random variable.

Similar to the other cases, channel coefficient matrix for MIMO systems can be generated using the following set of MATLAB instructions for independent and identically distributed fading links.

```
n=gamrnd(m,a_SNR/m,Nr,Nt);
phi=2*pi*rand(Nr,Nt);
H=(n.^0.5).*cos(phi)+j*(n.^0.5).*sin(phi);
```

2.5.2.2 Generating $\eta - \mu$ faded channel coefficients

In $\eta - \mu$ fading distribution, the received signal is assumed to be a collection of clusters of multipath components. The resultant of each cluster is in the form of Gaussian distributed in-phase and quadrature phase components. The difference from Nakagami-m distribution and $\eta - \mu$ distribution is that the variance of in-phase and quadrature phase component is different in $\eta - \mu$ distribution. Again, $\eta - \mu$ fading is found suitable to model a non-LOS environment and the fading scenarios ranging from as severe as one sided Gaussian to the case where there is no fading at all. To generate the fading channel coefficients that follow $\eta - \mu$ distribution, we refer to the physical model discussed in Section 2.4.2. A sample of $\eta - \mu$ distributed fading channel coefficient may be obtained using the following set of MATLAB instructions.

```
function h=eta_mu_channel(eta,mu);
coeff=sqrt(eta);
ni=0;
nq=0;
for j=1:2*mu

        norm1=randn;
        ni=ni+norm1.^2;
        norm2=coeff*randn;
        nq=nq+norm2.^2;
end
h_abs=(sqrt(ni+nq))/sqrt((2*mu*(1+eta)));
theta = 2*pi*rand;
h=h_abs*cos(theta)+sqrt(-1)*h_abs*sin(theta);
```

For MIMO systems, the MATLAB function can be modified as follows to generate the required channel coefficient matrix.

```
function x=eta_mu_channel(eta,mu,Nr,Nt);
coeff=sqrt(eta);
ni=0;
nq=0;
for j=1:2*mu

        norm1=randn(Nr,Nt);
        ni=ni+norm1.^2;
        norm2=coeff*randn(Nr,Nt);
        nq=nq+norm2.^2;

end
h_abs=(sqrt(ni+nq))/sqrt((2*mu*(1+eta)));
theta = 2*pi*rand(Nr,Nt);
x=h_abs.*cos(theta)+sqrt(-1)*h_abs.*sin(theta);
```

It shall be noted here that the procedure for generation of $\eta-\mu$ distributed channel coefficients in this book is using physical a model. The physical model relates number of clusters to the fading parameter, μ. Thus, this technique is valid for generation of fading coefficients with discrete values of μ. The restriction is that the values of μ have to be half integer, i.e., integer values of 2μ. For other values of μ, it is not possible to generate the channel coefficients using such a simple set of instructions. One needs to use statistical methods of transformation of a uniformly distributed random variable to the random variable that follows required distribution. Various methods of such transformation of uniformly distributed random variables have been discussed in detail in Section 7.3 of [127]. Among the methods that can transform a uniformly distributed random variable to the required distribution, the rejection method has been popularly used. In general, the samples of a random variable may be generated using the rejection method in the case that the PDF of the distribution is available in closed form. This method gives accurate results as reported in [23].

2.5.2.3 Generating $\kappa-\mu$ faded channel coefficients

Most of the characteristics of $\kappa-\mu$ fading distribution have been discussed in Section 2.4.3. To simulate a wireless communication system over $\kappa-\mu$ fading channels, it is necessary to generate channel coefficients that follow $\kappa-\mu$ distribution. To generate $\kappa-\mu$ distributed fading channel coefficients, we note that $\kappa-\mu$ fading model is well suited for LOS conditions. That means each cluster of the received signal contains some dominant component. So, we will again refer to the physical model of $\kappa-\mu$ represented by the expression of equation (2.34) in Section 2.4.3. It was shown that a $\kappa-\mu$ distributed random variable can be obtained from mutually independent Gaussian random variables with non-zero mean that represents the LOS component. The MATLAB function that follows can be used to generate a single sample that comes from $\kappa-\mu$ distributed random variables. The same can be used as a channel coefficient the of $\kappa-\mu$ fading model.

```
function x=kappa_mu_channel(kappa,mu);
m = sqrt( kappa/((kappa+1))) ;
s = sqrt( 1/(2*(kappa+1)) );
ni=0;
nq=0;
for j=1:2*mu
    if mod(j,2)==1
        norm1=m+s*randn;
        ni=ni+norm1.^2;
    else
        norm2=s*randn;
        nq=nq+norm2.^2;
    end
end
h_abs=sqrt(ni+nq)/sqrt(mu);
```

```
theta = 2*pi*rand;
x=h_abs*cos(theta)+sqrt(-1)*h_abs*sin(theta);
```

For MIMO systems, the MATLAB function can be modified as follows to generate the required channel coefficient matrix. The only difference for MIMO systems from the previous case is that now multiple samples need to be generated in the form of a matrix of size $N_r \times N_t$ which is accomplished in the following MATLAB function.

```
function x=kappa_mu_channel(kappa,mu,Nr,Nt);
m = sqrt( kappa/((kappa+1)) ) ;
s = sqrt( 1/(2*(kappa+1)) );
ni=0;
nq=0;
for j=1:2*mu
    if mod(j,2)==1
        norm1=m+s*randn(Nr,Nt);
        ni=ni+norm1.^2;
    else
        norm2=s*randn(Nr,Nt);
        nq=nq+norm2.^2;
    end
end
h_abs=sqrt(ni+nq)/sqrt(mu);
theta = 2*pi*rand(Nr,Nt);
x=h_abs.*cos(theta)+sqrt(-1)*h_abs.*sin(theta);
```

It shall be noted here that the procedure for generation of $\kappa-\mu$ distributed channel coefficients in this book is using a physical model. The physical model relates number of clusters to the fading parameter, μ. Thus, this technique is valid for generation of fading coefficients with discrete values of μ. The restriction is that the values of μ have to be integers. For other values of μ, it is not possible to generate the channel coefficients using such a simple set of instructions. One needs to use statistical methods of transformation of a uniformly distributed random variable to the random variable that follows required distribution. Various methods of such transformation of uniformly distributed random variables have been discussed in detail in Section 7.3 of [127]. In general, $\kappa-\mu$ distributed fading channel coefficients may be generated using the rejection method as reported in [23].

2.5.2.4 Generating $\alpha - \mu$ faded channel coefficients

It has already been discussed that $\alpha - \mu$ distribution has been a generalized version of Stacy distribution which was again proposed as a generalized form of Gamma distribution [135]. In addition, it was discussed that the square variate of Nakagami-m distributed variable is related to Gamma distribution in Section 2.4.1. Also, comparing the physical models of Nakagami-m fading distribution and $\alpha - \mu$ fading distribution respectively given by expression of equation (2.10) and expression of equation (2.45), the relation between

Nakagami-m distributed random variable and $\alpha-\mu$ distributed random variable can be given by

$$X_{Nak}^2 = X_{\alpha-\mu}^{\alpha} \tag{2.59}$$

where X_{Nak} is a Nakagami-m distributed random variable and $X_{\alpha-\mu}$ is an $\alpha-\mu$ distributed random variable.

Using the relation above between Nakagami-m distributed random variable and $\alpha-\mu$ distributed random variable, we can generate the channel coefficients that follow $\alpha-\mu$ distribution using the set of MATLAB instructions that generated Nakagami-m fading coefficients and manipulating them as required to obtain $\alpha-\mu$ distributed fading channel coefficients. The exact set of MATLAB instructions to generate $\alpha-\mu$ fading channel matrix coefficients for MIMO systems is given below.

```
function H=alpha_mu_channel(alpha,mu,Nr,Nt)
a_inv=1/alpha;
n=zeros(Nr,Nt);
for i=1:2*mu
n=n+randn(Nr,Nt)^2;
end
n=n/(2*mu);
phi=2*pi*rand(Nr,Nt);
H=(n.^a_inv).exp(j*phi);
```

Similarly, as discussed in Section 2.4.1 and Section 2.5.2.1, Nakagami-m distributed random variables are directly related to Gamma distributed random variables. We can use this fact and generate $\alpha-\mu$ distributed fading channel coefficient using the MATLAB instructions that follow.

```
n=gamrnd(mu,a_SNR/mu,Nr,Nt);
a_inv=1/alpha;
phi=2*pi*rand(Nr,Nt);
H=(n.^a_inv).*exp(j*phi);
```

The samples and matrices of fading channel coefficients generated in this chapter would be frequently used in the subsequent chapters of this book. The following chapters are going to consider performance analysis of various variants of MIMO systems discussed in Chapter 1 over generalized fading channels discussed in this chapter. In all the cases, we are going to consider that the channel envelopes follow particular distribution and the phase of received signal follows uniform distribution over the range 0 to 2π radians.

3

Spatial Multiplexing

CONTENTS

3.1 Introduction

In Chapter 1, we introduced spatial multiplexing. It has been mentioned that it can achieve spectral efficiencies which vary linearly with the number of antennas available at the transmitter. This is made possible by transmitting parallel streams of information from multiple transmit antennas. The improvement in the spectral efficiency and hence the transmission rate comes at the cost of other complexities like hardware requirement, computational complexity at the receiver, requirement of inter antenna synchronization, high levels of inter antenna interference, etc. It is a well established fact that due to inter antenna interference, the symbols transmitted by multiple antennas cannot be detected correctly at the receiver in the absence of scattering. Unlike other wireless communication systems, multiple-input multiple-output (MIMO) systems take advantage of the randomness introduced by scattering and multipath components to improve the spectral efficiency.

MIMO systems basically serve two purposes: providing multiplexing gain by parallel transmission of independent streams of data and providing diversity gain by receiving multiple copies of independently faded data streams. So, MIMO systems are capable of achieving multiplexing gains as well as diversity gains. But both these gains cannot increase to infinity. For an $N_t \times N_r$ MIMO system the maximum diversity gain that can be achieved is $N_t N_r$ and the maximum multiplexing gain that can be achieved is $\min(N_t, N_r)$. Both the gains cannot be maximum at an instant. So, there exists a trade-off between diversity gain and multiplexing gain.

In this chapter, we will discuss various performance metrics of MIMO systems with spatial multiplexing over various generalized fading channels. Basically, the metrics under consideration are error probability and channel capacity. We will see some insights to analytical expressions of these metrics and simulation results using MATLAB®.

3.2 Diversity multiplexing trade-off and the capacity of MIMO systems

MIMO systems are known for improving the data rates by parallel transmission of an independent stream of digitally modulated symbols. Obviously, this augments to the total transmission rate. At the same time, MIMO systems offer diversity gain. The diversity gain is achieved as a result of multiple antennas at the receiver when multiplexing is used for transmission. In many cases, diversity is achieved at the cost of multiplexing. Diversity gain and multiplexing gain both cannot increase simultaneously. To achieve one, we need to compromise the other. So, there is a trade-off between diversity gain and multiplexing gain [146]. MIMO systems can be used only for achieving diversity. In that case, all the antennas at the transmitter are used to transmit the same symbol from the incoming bit/symbol sequence. In the case of MIMO systems used only for multiplexing, incoming stream of information is converted into parallel bit/symbol sequences and each of these parallel sequences are transmitted simultaneously from multiple antennas available at the transmitter. For multiplexing, it can be shown that 8×8 MIMO systems can achieve capacity as high as 40 times than that of single-input single-output (SISO) systems under certain conditions [41]. For the case of multiplexing, the channel capacity per unit channel bandwidth for MIMO systems at high signal-to-noise ratio (SNR) conditions can be given as [41, 146]

$$C \approx min(N_t, N_r) \log_2 (SNR) \tag{3.1}$$

The multiplexing gain is limited by the minimum of the number of antennas at transmitter and the number of antennas at the receiver. This is because it is considered that there are non-interfering parallel links in multiplexing. This establishes that number of parallel links cannot be more than the minimum number of antennas on either side. The spatial multiplexing gain [146] can be defined as

$$r = \lim_{SNR \to \infty} \frac{C}{\log_2(SNR)} \tag{3.2}$$

where r is the spatial multiplexing gain. This shows that for any MIMO system, the maximum spatial multiplexing gain is $\min(N_t, N_r)$.

In the same way, [146] defines the diversity gain in the following way:

$$d = \lim_{SNR \to \infty} -\frac{\log(BER)}{\log(SNR)} \tag{3.3}$$

where d is the diversity gain. In other words, the diversity order of a communication system is said to be d if the bit error rate (BER) of the system varies as inverse of SNR^d in the high SNR regime. Here, we also recall that diversity order is alternatively defined as the number of independent paths from transmitter to receiver. In the case of receiver diversity systems such as selection combining or maximal ratio combining, we get the same diversity order as the number of antennas available at the receiver. In the case of MIMO systems, the maximum number of independent paths from transmitter to receiver is the same as the product of number of antennas at transmitter and at receiver. So, the maximum diversity order that can be achieved by any MIMO system is $N_t N_r$.

So, the diversity multiplexing trade-off can be given as [146]

$$d_{opt} = (N_t - r)\,(N_r - r) \tag{3.4}$$

where d_{opt} is the optimum diversity gain when the system is used to achieve multiplexing gain of r. As stated earlier, the trade-off given in the above expression is considering the case that multiplexing gain is obtained assuming that the parallel links are non-interfering links, i.e., an antenna at transmitter side is connected to single antenna of receiver side or there is one-to-one link among the antennas at transmitter and receiver. At the same time, while calculating the diversity gain, it considers that a link is available from an antenna at the transmitter to every antenna of the receiver. However, with fully interfering MIMO channels, the channel capacity per unit bandwidth can be given by [43, 80]

$$C_{MIMO} = \log_2\left(\det\left(\mathbf{I_{N_r}} + \frac{\overline{\gamma}}{N_t}\mathbf{H}\mathbf{H^H}\right)\right) \tag{3.5}$$

where $\mathbf{I_{N_r}}$ is an identity matrix of dimension $N_r \times N_r$, and $\overline{\gamma}$ is the average SNR per receiver antenna. Again, for optimum use of resources in this case, a subset of receiver antennas can be selected for maximum channel capacity without compromising on the number of parallel data streams that can be transmitted simultaneously. This virtually means that the channel capacity of MIMO systems can scale linearly with the number of transmitting antennas and we will also see in the next section that the diversity order depends on the number of antennas at the receiver. Note that the expression of MIMO channel capacity, equation (3.5), is valid only when spatial multiplexing is employed at the transmitter and the channels are fully interfering.

Spatial multiplexing is seen as a potential technology for improving transmission rates. But for the improved transmission rates, cost has to be paid in terms of hardware and computational complexity. Parallel transmission from multiple antennas requires separate radio frequency (RF) chain for each antenna at transmitter. These RF chains consist of complex hardware comprised of A/D converters, modulators, RF amplifiers, etc. This makes the system bulky (at least for the user handset). Although, receiver diversity techniques also demand such RF chains in multiple number, mainly receiver diversity techniques are implemented at the base stations which are not mobile in nature. So, bulky hardware can be afforded at the receiver. In addition to this, spatial multiplexing systems need inter antenna synchronization.

3.3 Spatial multiplexing

In the previous section, we discussed about two main parameters which can be improved using MIMO systems. It is also discussed that if the multiplexing in MIMO systems consider non-interfering channels, the multiplexing gain that can be achieved limits to $\min(N_t, N_r)$ and no gain can be achieved in the diversity order as compared to SISO systems. But, in practice, the existence of non-interfering MIMO channels is an idealistic scenario. All the practical MIMO channels are interfering in nature, i.e., a symbol/signal transmitted from an antenna at the transmitter is surely going to suffer interference from the symbols/signals being transmitted by all other antennas at the transmitter. This fact that the practical MIMO channels are interfering in nature is helpful and can be used in a way to achieve diversity even while using multiple antennas of transmitter for parallel transmission of information, i.e., multiplexing. Interfering channels or, in fact, the channel model that is under consideration to achieve the diversity order of N_tN_r is helpful in explaining how MIMO systems can be useful in achieving diversity and multiplexing simultaneously, provided there exists a rich scattering environment. In such a channel model, each transmitter antenna has a link to all the receiver antennas. So, the system as a whole may be considered as a parallel combination of an N_t number of $1 \times N_r$ receiver diversity systems, i.e., each such system can achieve diversity order of N_r separately. Now, these $1 \times N_r$ systems are combined such that the set of receiver antennas remain the same. Each receiver antenna receives signals from all the transmitter antennas. So, received signal at each antenna is a combination of the signals being transmitted by all the transmitter antennas. Thus, each receiver antenna receives a signal that is interfered by all the transmitted symbols from each transmitter antenna at the corresponding time instant. But also, each receiver antenna receives one copy of each symbol from every transmitter antenna. Thus at the receiver, number of copies of the same signal/symbol is the same as that of the antennas at the receiver. This means that the diversity order offered by such a system is N_r. We have already discussed that each antenna at the transmitter is transmitting an independent sequence of symbols. This means the multiplexing gain is of the order of the number of antennas at the transmitter. Again, revisiting the expression of equation (1.2), the relation between transmitted symbols and received vector can be expressed as follows.

$$\mathbf{r} = \mathbf{H}\mathbf{x} + \mathbf{n} \tag{3.6}$$

$$\begin{bmatrix} r_1 \\ r_2 \\ \vdots \\ r_{N_r} \end{bmatrix} = \begin{bmatrix} h_{11}x_1 + h_{12}x_2 + \cdots + h_{1N_t}x_{N_t} \\ h_{21}x_1 + h_{22}x_2 + \cdots + h_{2N_t}x_{N_t} \\ \vdots \\ h_{N_r1}x_1 + h_{N_r2}x_2 + \cdots + h_{N_rN_t}x_{N_t} \end{bmatrix} + \begin{bmatrix} n_1 \\ n_2 \\ \vdots \\ n_{N_r} \end{bmatrix} \tag{3.7}$$

From the above expression, it is observed that the useful signal content in the receiver antennas is a linear combination of transmitted symbols at one particular instant. This gives rise to the need of antenna synchronization at transmitter side. If the antennas are not transmitting the symbols simultaneously, an overlap of multiple symbols need to be considered at every receiver antenna which would make the system highly complex and introduce large numbers of detection errors. So, in general when there is simultaneous transmission from multiple antennas, the transmitter antennas have to be synchronous in transmission.

The following set of MATLAB instructions can be used for the generation of a spatially multiplexed symbol and obtaining the vector received at the receiver after passing through the noisy fading channel (the fading coefficients can be generated using the MATLAB instructions discussed in Chapter 2 for various fading models):

```
x=round((M-1)*rand(Nt,1));
x_q = pskmod(x,M)/sqrt(Nt*Eavg);     % digital modulation
y =  H*x_q;  % faded signal
r = y+10^(-EbNo(c)/20)*n; % Received noisy signal
```

The above MATLAB instructions are for M-ary phase shift keying (PSK) modulation scheme where M is the modulation order, $Eavg$ is the average energy per symbol of the constellation points in the modulation scheme, Nt is the number of transmitter antennas, $EbNo$ is the SNR in dB, r is the vector of signals received at each receiving antenna and H is the channel matrix.

For detection of symbol vector at the receiver, maximum likelihood (ML) detection is found to be optimum and we will follow the same for detection in spatial multiplexing systems. ML detection for MIMO systems is based on the exhaustive search algorithm [6]. The ML detection rule for spatially multiplexed systems can be given by

$$\hat{\mathbf{x}} = \arg\min_{\mathbf{x}\in\mathbf{X}} \|\mathbf{r} - \mathbf{H}\mathbf{x}\|^2 \tag{3.8}$$

where $\hat{\mathbf{x}}$ is the detected symbol vector, $\mathbf{X}$ is the set of all possible symbol vectors that may be transmitted, $\mathbf{r}$ is the received vector and $\mathbf{H}$ is the $N_r \times N_t$ channel matrix.

Note that the ML detection of a spatially multiplexed symbol represented by the rule in expression of equation (3.8) needs the set of all possible spatially multiplexed symbols. This set can be obtained using the MATLAB instructions that follow:

```
lut = pskmod([0:M-1],M); % Look up table
Eavg = sum(abs(lut).^2)/M; % Average energy of constellation
N_lut = lut/sqrt(Eavg); % Normalized look up table

TA2=zeros(Nt,M^Nt); % Initialization of the set of all possible
                    %symbols
for m=1:M^Nt
    for n=1:Nt
        for p=1:Nt
            TA2(p,m)= N_lut(mod(floor((m-1)/(M^(Nt-p))),M)+1)..
            ../sqrt(Nt);
        end
    end
end
```

Again, the above instructions are used for M-ary PSK modulation schemes. For other modulation schemes, the first instruction requires appropriate change. However, other instructions would remain the same. In the above instructions, TA2 is the variable which is a matrix that contains all possible spatial multiplexing codes at the transmitter as column vectors. The set of all possible vectors of spatial multiplexing symbols is used for the detection of symbols at the receiver. Note that the look-up table having entries of all possible spatial multiplexed symbols is kept ready at the receiver as a part of the system design. The entries of the same table may be used once the receiver senses some reception. Detection of the received symbol can be done using the following set of MATLAB instructions.

```
%% Detection of spatially multiplexed symbol
    d = repmat(r,1,M^Nt)-H*TA2;
    nor = sum(abs(d).^2);
    [mi,ind] = min(nor);
    hat_x_q = TA2(:,ind);
```

The MATLAB instructions above can be combinely used for simulations of spatial multiplexing systems. The complete MATLAB script file for symbol error rate (SER) performance of spatial multiplexing systems is available at the end of this book in Appendix B. Simulation of a spatial multiplexing system on MATLAB was performed to study the effect of different numbers of antennas at transmitter and receiver. The results of the simulations can be observed in Figure 3.1. More number of transmitter antennas ($N_t = 4$ in

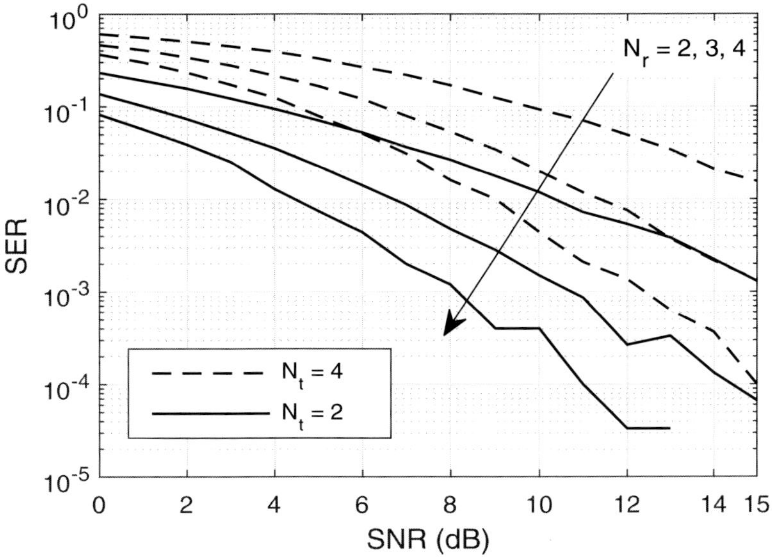

FIGURE 3.1
BER of spatial multiplexing systems for 4-QAM Rayleigh fading channels for different number of transmitter ant receiver antennas.

our case) degrades the performance but this degradation improves the spectral efficiency by a factor of 2 as compared to the case that $N_t = 2$, which has better SER performance for the spatial multiplexing system under consideration.

3.3.1 Detection complexity of spatially multiplexed systems

Apart from performance metrics of wireless communication systems, detection complexity is an important parameter. Detection complexity is directly related to the computational power required at the receiver and battery life of mobile equipment. More computation implies more power requirement. In mobile devices, battery is the only source of this required power. So, it is always preferred to have minimum detection complexity for mobile devices used in wireless communication systems. In the following discussions, we will study the complexity of the detection process of spatially multiplexed MIMO systems. The relation of decoding complexity with the order of modulation scheme, and number of antennas at transmitter and receiver will also be discussed. In this book, detection complexity is assumed to be of the order of the number of complex operations required to detect a symbol vector from the received signal vector.

MIMO systems involve matrix operations. So, it is obvious that the computational complexity is going to be more than that of SISO systems and diversity systems. From the expression of equation (3.8), it is clear that the detection complexity is directly related to the channel matrix size, i.e., the number of transmitter antennas and the number of receiver antennas. It is also related to the total number of all possible symbol vectors which is related to the order of modulation scheme. We first establish the number of complex operations required to calculate the squared norm $\|\mathbf{r} - \mathbf{H}\mathbf{x}\|^2$ for single value of $\mathbf{x}$. This involves multiplication of matrix $\mathbf{H}$ with vector $\mathbf{x}$. This step consumes $N_t N_r$ number of complex multiplications and $N_r(N_t - 1)$ number of complex additions.

The calculation of norm is required to be done for the same number of times as the number of all possible symbol vectors that may be transmitted. For an $N_t \times N_r$ MIMO system, the transmitted symbol vector is of dimension N_t. For a vector of dimension N_t with the possibility that each element can come only from any of the M possible entries, the number of all possible signal vectors N_{sv} for M-ary modulation scheme is given as

$$N_{sv} = 2^{N_t \log_2(M)} = M^{N_t} \tag{3.9}$$

From the order of modulation, M, number of bits per symbol can be given as $\log_2(M)$. Overall detection complexity is mainly governed by the number of complex multiplications. So, the complexity can be given by

$$N_{mult} = N_t N_r M^{N_t} \tag{3.10}$$

The detection complexity varies linearly with the number of receiver antennas. It varies exponentially with the number of transmitter antennas and modulation order. Mainly, MIMO systems are found to improve the spectral efficiency by parallel transmission of independent sequence of information symbols. But, the number of parallel data sequence (which is the same as number of transmitting antennas) is responsible for exponential increase in the detection complexity. So, for example, the detection complexity of an 8×8 MIMO system with 8-QAM modulation scheme is of the order 8^8 and it increases to 16^8 (by a factor of 2^8) for 16-QAM modulation scheme with an improvement of 8 bits/s/Hz in spectral efficiency.

Next, we will discuss the statistics of generalized fading channel models for spatial multiplexing systems. It would involve discussions on the methodology to analyze different performance metrics of wireless communication systems.

3.4 Nakagami-m fading channel model for spatially multiplexed systems

It is already discussed that MIMO systems perform well in rich scattering environments. The Nakagami-m fading can model a wide range of fading scenarios. We will use this characteristic of Nakagami-m fading model to study performance of spatial multiplexing MIMO systems over a wide range of fading environments. In Nakagami-m fading MIMO channels considered in this section, the channel matrix is the set of independent and identically distributed channel coefficients. Each channel coefficient has a resultant in-phase component and a resultant quadrature phase component. All these components and coefficients are assumed to be mutually independent of each other. From equation (3.7), the received signal at k^{th} receiver antenna can be given as

$$r_i = [h_{i1}x_1 + h_{i2}x_2 + \cdots + h_{iN_t}x_{N_t}] + n_i \tag{3.11}$$

So, each r_i is a linear combination of N_t number of Nakagami-m faded channel coefficients. Again considering the physical model of Nakagami-m fading in expression of equation (2.10) of Section 2.4.1, it can be shown that each complex element of channel coefficient matrix, $\mathbf{H}$, can be represented as follows.

$$h_{pq}|_m = h_{I_{pq}}|_{m/2} + jh_{Q_{pq}}|_{m/2} \tag{3.12}$$

The above expression says that a Nakagami-m distributed complex envelope can be decomposed into its real part (in-phase component) and imaginary part (quadrature phase component). Each in-phase component and quadrature phase component is Nakagami-m distributed envelope but the fading parameter for both these components is $m/2$, i.e., half the fading parameter value of the complex envelope. Though physical model restricts the values of fading parameter, m, to be discrete and half integer, a relation of Nakagami-m distributed random variables has been established with Gamma distributed random variables so that generation of Nakagami-m distributed channel coefficients can be easily done on suitable software platforms for any positive real value of m. The relation of Gamma distributed random variables and Nakagami-m distributed random variables not only makes the simulation of communication systems simpler but it also simplifies the analysis of wireless communication systems while modeling them for various performance metrics.

From the expressions of equation (3.11) and equation (3.12), signal received at i^{th} antenna, r_i for $i \in [1, 2, \cdots, N_r]$ can be represented as the weighted sum of the fading envelopes with the fading parameter, which is half the Nakagami-m fading parameter for the fading channel model. Again, the weights depend on the type and order of the modulation scheme. For spatially multiplexed systems, if the symbols are coming from a complex constellation, the symbol transmitted from j^{th} antenna for $j \in [1, 2, \cdots, N_t]$ will have real and imaginary components given by $x_j = x_{jI} + jx_{jQ}$. Thus, the jth term in r_i can be represented as

$$\begin{aligned} h_{ij}x_j &= (h_{ijI} + jh_{ijQ})(x_{jI} + jx_{jQ}) && (3.13)\\ &= (h_{ijI}x_{jI} - h_{ijQ}x_{jQ}) + j(h_{ijQ}x_{jI} + h_{ijI}x_{jQ}) && (3.14) \end{aligned}$$

This can be further helpful in the analysis of error performance of spatially multiplexed MIMO systems. In spatially multiplexed MIMO systems, the transmitted symbol is a vector of signals coming from constellation of digitally modulated signals. It consists of N_t elements. Let a transmitted symbol be denoted by $\mathbf{x}$. In case any other symbol $\mathbf{x}'$ is received on the transmission of $\mathbf{x}$, it is said that the symbol is received in error. Here, let us define pairwise error probability as the probability that a symbol $\mathbf{x}$ is transmitted and another symbol $\mathbf{x}'$ is detected. It is given by

$$P(\mathbf{x} \to \mathbf{x}') = Q\left(\sqrt{\frac{\|\mathbf{H}\boldsymbol{\Delta}\mathbf{x}\|_F^2}{2N_0}}\right) \tag{3.15}$$

where $\boldsymbol{\Delta}\mathbf{x} = \mathbf{x} - \mathbf{x}'$ is the difference vector of the transmitted symbol and the detected symbol. Note that the pairwise error probability is for a given channel matrix, $\mathbf{H}$. To find average pairwise error probability, we need to average the conditional pairwise error probability over the probability density function (PDF) of $\|\mathbf{H}\boldsymbol{\Delta}\mathbf{x}\|_F$. Note that $\mathbf{H}\boldsymbol{\Delta}\mathbf{x}$ is a column vector having N_r elements. The ith element of $\mathbf{H}\boldsymbol{\Delta}\mathbf{x}$ can be given as

$$\Delta r_i = [h_{i1}\Delta x_1 + h_{i2}\Delta x_2 + \cdots + h_{iN_t}\Delta x_{N_t}] \tag{3.16}$$

Thus, the Frobenius norm in the expression of pairwise error probability can be given by

$$\|\mathbf{H}\boldsymbol{\Delta}\mathbf{x}\|_F^2 = \sum_{i=1}^{N_r} |\Delta r_i|^2 \tag{3.17}$$

Now, to average the pairwise error probability conditioned on given channel matrix, the PDF of $\|\mathbf{H}\boldsymbol{\Delta}\mathbf{x}\|_F^2$ is required. From the above expression, for independent and identically distributed (i.i.d.) channel matrix, it is clear that every Δr_i is independent for all values of i. Thus, PDF of $\|\mathbf{H}\boldsymbol{\Delta}\mathbf{x}\|_F^2$ can be obtained by convolving the individual PDFs of $|\Delta r_i|^2$ or moment generating function (MGF) of $\|\mathbf{H}\boldsymbol{\Delta}\mathbf{x}\|_F^2$ can be obtained by multiplying the individual MGFs of $|\Delta r_i|^2$ for all the values of $i \in [1, 2, \cdots, N_r]$. Again, the PDF of each $|\Delta r_i|^2$ need to be obtained. It can be obtained using the principle of random variable transformations and using the methods to obtain PDFs for functions of random variables. The PDF of each $|\Delta r_i|^2$ can be further obtained by using equation (3.16). So, each Δr_i is a summation of N_t complex random variables each of which will be a channel coefficient scaled by a factor dependent on Δx_j. Thus, each $|\Delta r_i|^2$ is a summation of $2N_t$ random variables including in-phase and quadrature phase components. This process of obtaining PDF for evaluating average pairwise error probability of spatially multiplexed MIMO systems is straightforward for i.i.d. Rayleigh fading channels in which the in-phase and quadrature phase components follow Gaussian distribution. Further, without going into the detailed derivation of an expression for the PDF of $\|\mathbf{H}\boldsymbol{\Delta}\mathbf{x}\|_F^2$, we will discuss the behavior of spatial multiplexing MIMO systems for different values of fading parameters.

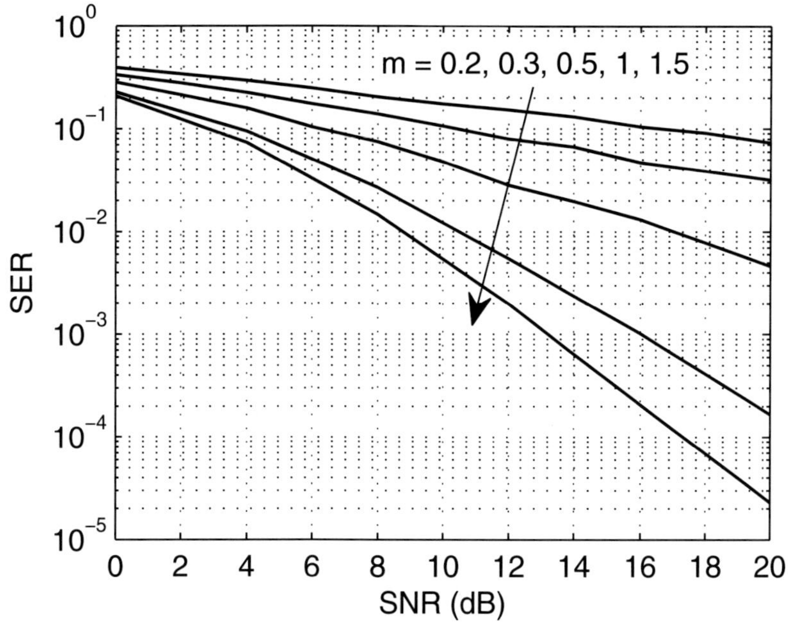

FIGURE 3.2
SER of spatial multiplexing systems for binary phase shift keying (BPSK) modulation over Nakagami-m fading channels for different values of m.

The results of MATLAB simulation of 2×2 MIMO systems with spatial multiplexing over Nakagami-m fading channels are shown in Figure 3.2. The results are shown for different values of fading parameter m. An improvement is observed in both the SER performance and the rate at which SER reduces with increasing SNR as m increases. Note that as per discussion in Chapter 2, the increasing values of fading parameter, m, indicates reduced fading severity. This results in the improved performance with higher values of m. The results are shown for $m < 2$. For this case, the improvement in fading conditions is the reason behind the improvement of SER. We will see that similar improvements are not observed by further increasing the value of fading parameter, m.

The cases of higher fading parameter values $m > 2$ are also discussed for spatial multiplexing systems. To discuss the performance of spatial multiplexing systems over Nakagami-m fading channels for the fading parameter values such that $m > 2$, we do it in two parts to reach to some fruitful observation. Part one is the performance for lower SNR regions (SNR <10 dB) and the second part is the performance for higher SNR regions (SNR >10 dB). The results for both cases are plotted and shown in Figure 3.3 and Figure 3.4 respectively. First, let's observe the results for the low SNR case in Figure 3.3.

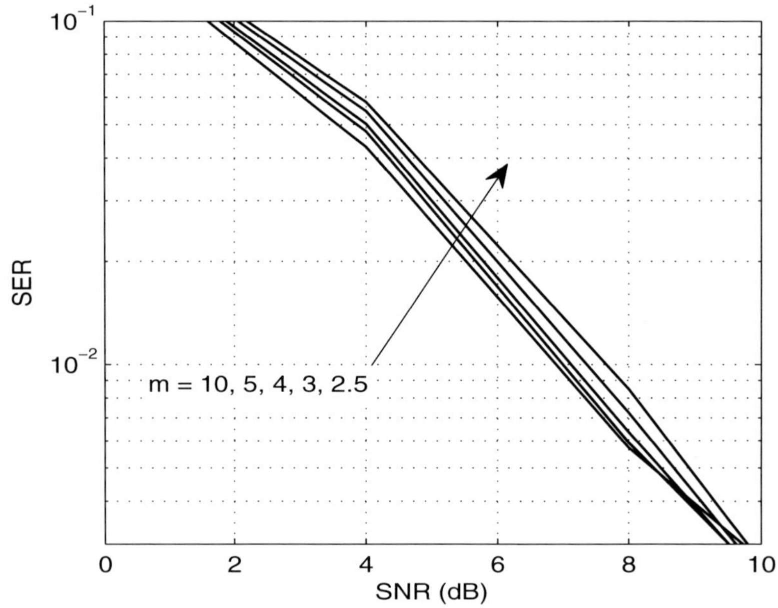

FIGURE 3.3
SER of spatial multiplexing systems for BPSK modulation over Nakagami-m fading channels for $m > 2.5$ and lower values of SNR.

It is observed that the SER performance is improved on increasing the value of fading parameter, m. The interesting thing to observe in this case is the region of SNR between 8 dB and 10 dB. In this region it is observed that the SER curves for different values of m start crossing each other. Finally, after this crossing happens, the performance is observed as shown in Figure 3.4 for the same case for SNR values greater than 10 dB. In this case, it is observed that the SER performance shows degradation with increased values of m. This degradation is observed because of the fact that reduced fading severity is described by greater values of m in Nakagami-m fading channel models, i.e., higher values of m indicate the fading channels are more like additive white Gaussian noise (AWGN) channels ($m \to \infty$ models no fading case). Note that rich scattering environment is a primary requirement for MIMO systems to perform better. In the case of low SNR region, noise is the dominating factor and this is the reason for better performance observed for higher values of m for SNR $<$10 dB.

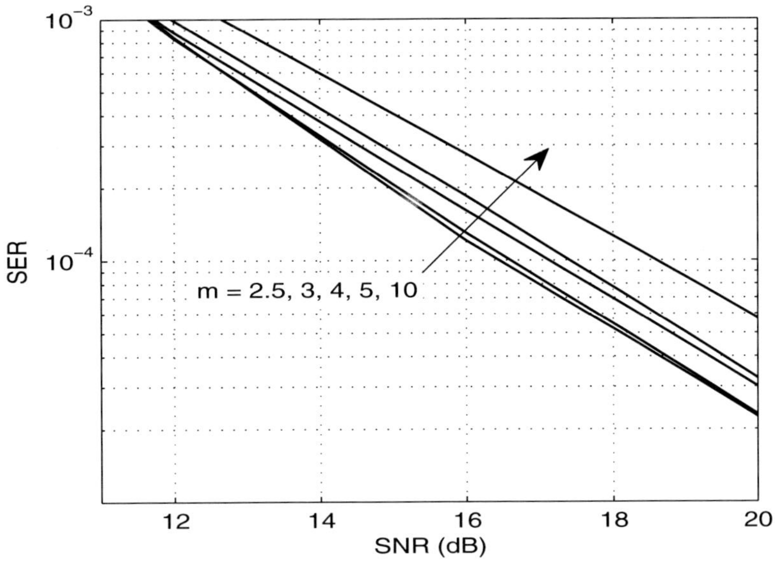

FIGURE 3.4
SER of spatial multiplexing systems for BPSK modulation over Nakagami-m fading channels for $m > 2.5$ and higher values of SNR.

3.5 Spatial multiplexing over $\eta - \mu$ fading channels

In the previous chapter, we discussed the importance of $\eta - \mu$ fading distribution model. It is more general than Nakagami-m fading model. In fact, Nakagami-m fading is a special case of $\eta - \mu$ fading distribution. Again, as discussed in the previous section, the received signal at each antenna would be a linear combination of N_t number of mutually independent $\eta - \mu$ faded channel coefficients. For analysis of wireless communication systems, we may need PDF of the resultant signal that is received at each antenna from which the statistics of total SNR may be easily obtained. So, in the case of $\eta - \mu$ distributed channel coefficients with uniformly distributed phase, the real and imaginary parts can be related to the complex envelope as follows.

$$h_{pq}|_{\eta-\mu} = h_I|_{\eta-\mu/2} + jh_Q|_{\eta-\mu/2} \tag{3.18}$$

For further analysis, a relation of $\eta - \mu$ distributed random variable may be established with Gamma distributed random variables. From the above expression and the physical model of $\eta - \mu$ distributed fading channels, it is clear that the in-phase and quadrature phase components of the fading coefficients can be separately approximated by Gamma distributed random

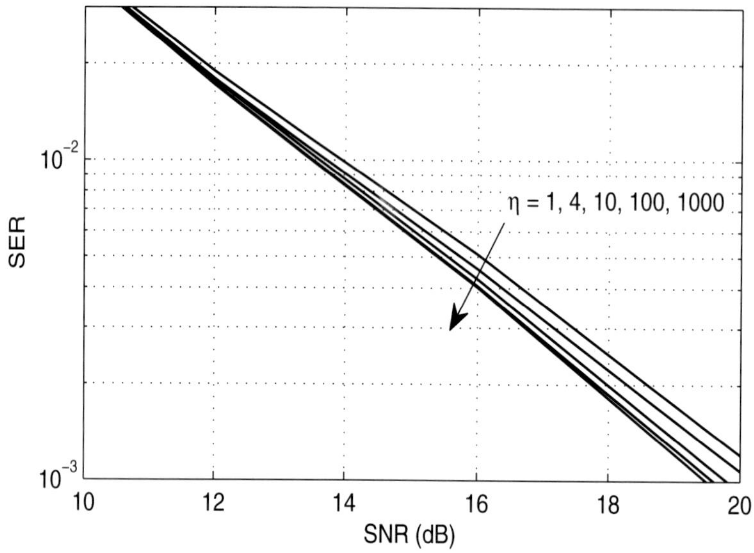

FIGURE 3.5
SER of spatial multiplexing systems with QPSK modulation over η-μ fading channels for various values of η and $\mu = 1$.

variables [97]. This can be done by matching the first two moments of random variables to obtain the shape parameter and the scale parameter [59]. This relation can make all the analyses very simple. However, this relation may not be true for all values of η and μ.

In a similar way to the case of Nakagami-m fading channels, the spatially multiplexed MIMO system has been modeled and simulated in MATLAB for various fading parameter values. In the case of $\eta - \mu$ fading channels, there are two fading parameters which are used to characterize a fading environment. The results for various fading conditions have been obtained from the simulations. Figure 3.5 shows SER results of spatially multiplexed MIMO systems over $\eta - \mu$ fading channels. For generating these results, quadrature phase shift keying (QPSK) modulation scheme is considered and the values of fading parameter η are varied from 1 to 1000. For this case, the value of $\mu = 1$ is kept unchanged. It can be observed that for a fixed value of μ, the error performance of a spatially multiplexed MIMO system improves when the value of η is increased from 1. Similar trend of improvement can be observed on decreasing the values of η from 1. This is because $\eta = 1$ represents the point of symmetry for $\eta - \mu$ distribution. It comes from the definition of η, i.e., η is the ratio of powers contained in the in-phase components of the received multipath clusters to that of the quadrature phase components. Thus, the cases $\eta > 1$ and $\eta < 1$ can be alternatively considered by the values η and $\frac{1}{\eta}$.

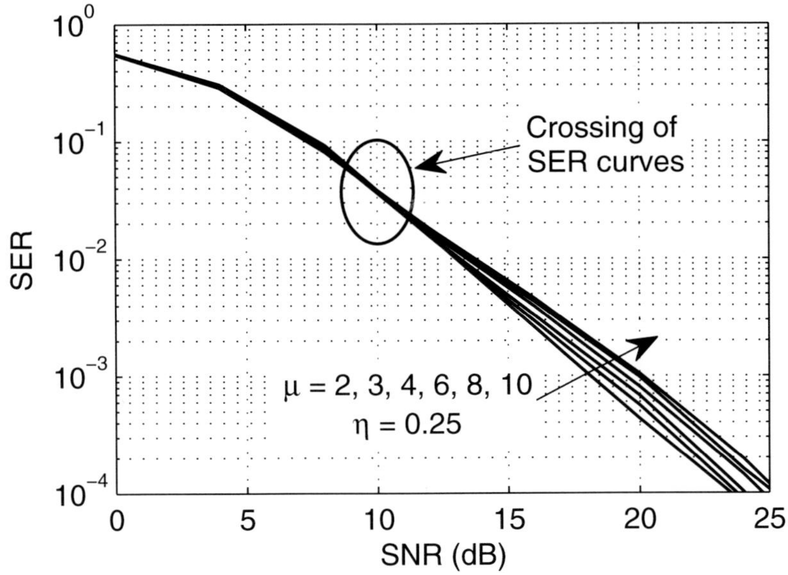

FIGURE 3.6
SER of spatial multiplexing systems with QPSK modulation over η-μ fading channels for various values of μ and $\eta = 0.25$.

The improvement of the SER results is on the account of the fact that higher values of η for $\eta > 1$ indicate more severe fading conditions. However, the complete range of values of η from $\eta = 1$ to $\eta = 0$ is equivalent to modeling the range of fading channel conditions that vary from μ to $\mu/2$ (for uniform phase coefficients). From the physical model, $\eta = 0$ means no power in quadrature phase component and the number of clusters gets automatically reduced to half the number that exists for $\eta = 1$. The distribution of phase in both cases no longer remains the same. However, this explanation is good for presenting an analogy.

The error performance is also studied for different values of μ and keeping $\eta = 0.25$ fixed. The results of SER of spatial multiplexing systems over η-μ fading channels with QPSK modulation scheme and 2×2 MIMO systems are shown in Figure 3.6. To study the effect of variation in the values of μ on the SER performance, in this case, the SER results are obtained for various values of μ ranging from 2 to 10. It is observed that the SER performance shows slight degradation when the values of μ are increased. Again, to explain this behavior of SER performance, note that rich scattering (severe fading) environment for MIMO systems for better performance is required. As the value of μ increases, the fading severity decreases as $\mu \to \infty$ is the special case which models non-fading environments. Thus, higher values of μ means less severe fading and degraded SER performance. Looking at Figure 3.6 closely, it can be observed that in the lower SNR conditions, the SER performance trend is different than that in the higher SNR conditions. It seems that the SER performance is the same for all μ values for SNR <10 dB. But, in actuality, this performance shows a similar trend as was discussed in the case of Nakagami-m fading channels. The SER performance in low SNR regions is dominated by the noise rather than the severity of fading. There is slight variation in the SER performance for different values of μ in a low SNR case. This observation is the same as in Figure 3.3 and Figure 3.4 for Nakagami-m fading channels. For the same reason, a region is marked by a circle to indicate the crossing points in SER curves in Figure 3.6.

3.6 Spatial multiplexing over $\kappa - \mu$ fading channels

So far, it is clear that a straightforward approach for obtaining analytical expressions for various performance metrics for spatially multiplexed MIMO systems over generalized fading channels does not exist. This is mainly because of the interference among the signals transmitted from multiple antennas. Analytical approaches to obtain expressions for the performance metrics for spatially multiplexed MIMO systems involve a complex process of finding PDF of the received effective SNR. This is because of multiple scaling and summation operations involved collectively at the transmitter, receiver and in the channel. However, if a reader is interested in obtaining an expression, it can be done in the ways discussed for Nakagami-m and $\eta - \mu$ fading channels in the previous sections of this chapter. Again, to get an idea of the PDF of in-phase and quadrature phase envelopes of $\kappa - \mu$ distributed fading envelope, let us refer to the physical model given by equation (2.34),

$$X^2 = \sum_{i=0}^{n} \left((I_i + p_i)^2 + (Q_i + q_i)^2 \right) \tag{3.19}$$

From the above expression, it can be observed that the square of in-phase and quadrature phase envelopes is a summation of squared Gaussian random variables with non-zero mean. So, square of both these envelopes can be assumed to follow non-central chi-squared distribution, each with $\mu/2$ degrees of freedom. Further, this can be used to obtain the PDF of the received signal at each antenna without noise which is required to average the conditional probability of error given by the following expression:

$$P\left(\mathbf{x_i} \rightarrow \mathbf{x_j}\right) = Q\left(\sqrt{\frac{\|\mathbf{H}\boldsymbol{\Delta}\mathbf{x}\|_F^2}{2N_0}}\right) \tag{3.20}$$

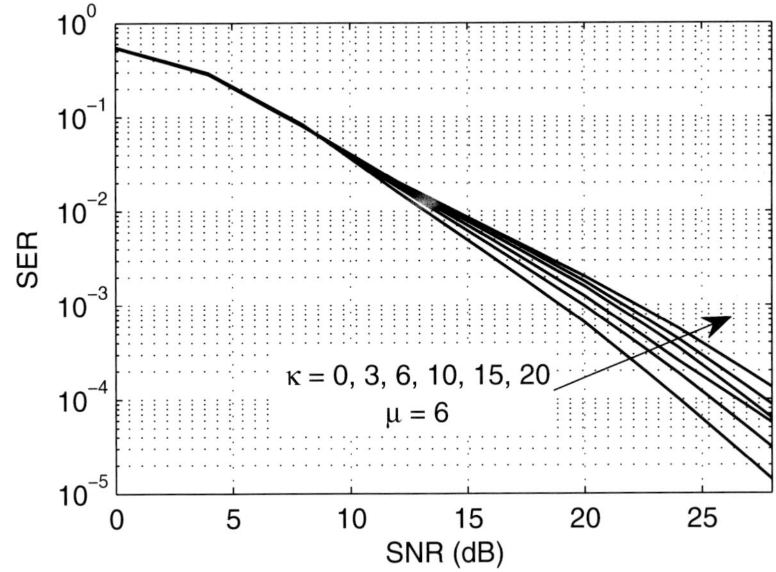

FIGURE 3.7
SER of spatial multiplexing systems with QPSK modulation over κ-μ fading channels for various values of κ and $\mu = 6$.

Averaging the above expression gives average pairwise error probability. Further, using union bound on the pairwise error probability of all possible combinations, SER can be obtained.

Now, we will discuss some of the simulation results for error performance of spatially multiplexed MIMO systems over $\kappa - \mu$ fading channels. The effect of fading parameter values on the SER performance will be discussed by studying the variation in SER performance with the variation of fading parameters.

The SER results from simulation of spatial multiplexing MIMO systems over $\kappa - \mu$ fading channels are plotted for various values of κ in Figure 3.7 with $\mu = 6$ and Figure 3.8 with $\mu = 2$. For these simulations 2×2 MIMO systems and the QPSK modulation scheme are considered. So, the achieved spectral efficiency is 4 bits/s/Hz by transmitting two symbols of two bits simultaneously from two antennas at the transmitter. In both figures, it is observed that the SER performance degrades with increasing values of κ. Increasing values of κ indicates stronger LOS components in the received signal clusters. This reduces the amount of scattering. Hence, the performance degradation is justified by reduced scattering with stronger LOS component for higher values of κ.

Note that in Figure 3.7 and Figure 3.8, the value of κ is varying from 0 to 20. There is an interesting observation while moving from low SNR regime to high SNR regime. For $\mu = 6$, the SER curves for low SNR conditions are

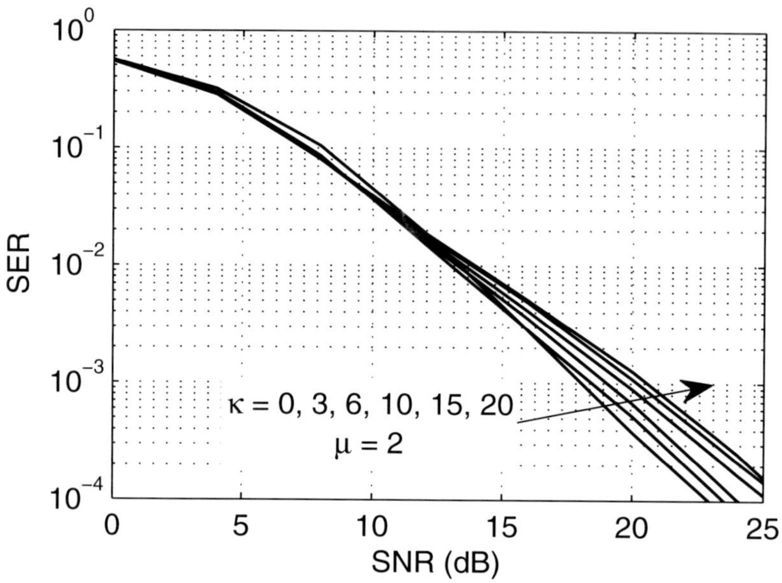

FIGURE 3.8
SER of spatial multiplexing systems with QPSK modulation over κ-μ fading channels for various values of κ and $\mu = 2$.

almost overlapping with very small variation, while we can observe a different trend for the case of $\mu = 2$. For the case of $\mu = 2$, the SER performance is better for higher values of κ for low SNR cases. In this case, the effect of noise dominates the degradation in performance because of reduction in the fading severity due to stronger LOS components with increasing values of κ. While in the case of $\mu = 6$, the reduction in severity of fading is not only because of stronger LOS component but also due to higher number of clusters in the received signal (larger value of μ accounts for less severe fading). Similarly, crossing of SER corves while moving from low SNR conditions to high SNR conditions is observed in the case of $\eta - \mu$ and Nakagami-m fading channels too. But, in the case of Nakagami-m and $\eta - \mu$ fading case, the observations were taken by varying the values of m and μ respectively.

To observe the variation of SER with different values of κ and μ in high SNR and low SNR regimes, the results are obtained and plotted in Figure 3.9 and Figure 3.10 for high SNR conditions (SNR $=$ 16 dB and 20 dB) and for low SNR conditions (SNR $=$ 4 dB and 8 dB) respectively. To obtain these SER results, we considered a 2×2 spatial multiplexing MIMO system with QPSK modulation scheme. The SER values are plotted by varying κ from 0 to 20 for two values of μ, $\mu = 2$ and $\mu = 6$ for both, the low SNR case and the high SNR case. In these figures, a degradation in the SER performance can be observed for higher values of μ and κ in general. However, in low

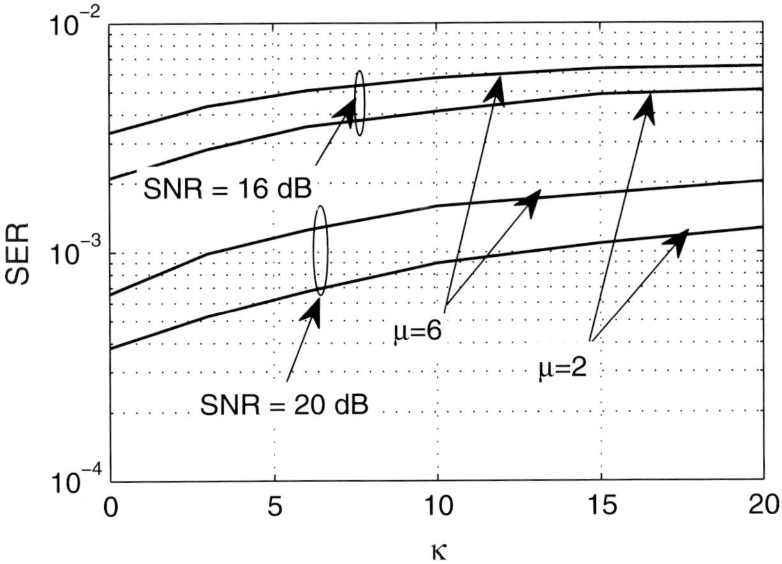

FIGURE 3.9
SER comparison of spatial multiplexing systems with QPSK modulation over κ-μ fading channels for $\mu = 6$ and $\mu = 2$ for high SNR conditions.

SNR conditions, the SER performance shows an improvement with increasing strength of LOS component. But, at high SNR conditions, an increase in SER is observed while moving from $\mu = 2$ to $\mu = 6$ for all values of κ from 0 to 20. And for a given value of μ and SNR, the SER is increasing with higher values of κ. For low SNR conditions, an improvement in SER is observed for initial increase in the values of κ for $\mu = 2$. Then, it becomes constant. In low SNR conditions, even an improvement in the SER performance is observed with stronger LOS component, although only for lower values of κ.

3.7 Spatial multiplexing over $\alpha - \mu$ fading channels

Our discussions in the previous sections of this chapter consider homogeneous and non-homogeneous fading environments. None of these fading distribution models including Nakagami-m, $\kappa - \mu$ and $\eta - \mu$ distributions consider non-linearities in the fading environments. $\alpha - \mu$ fading channel models take care of non-linearities in fading environments. The physical model of $\alpha - \mu$ fading channels relates the fading signal amplitude, X, by equation (2.45) discussed

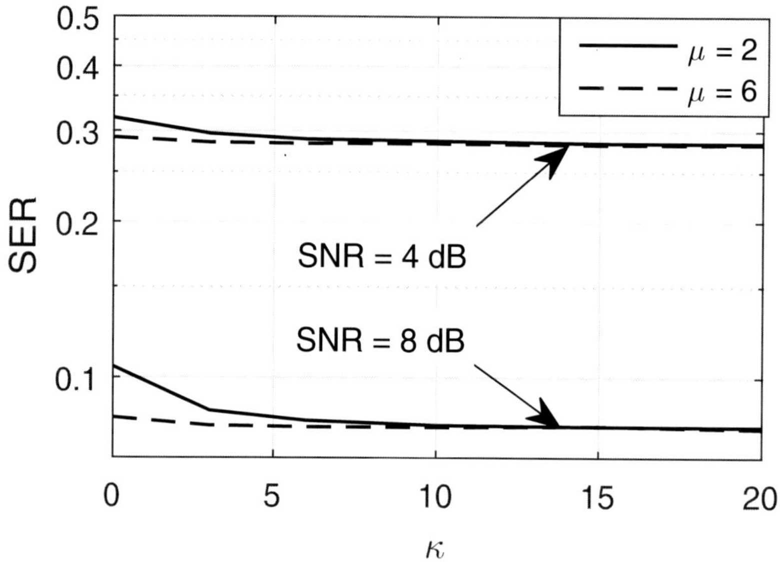

FIGURE 3.10
SER comparison of spatial multiplexing systems with QPSK modulation over κ-μ fading channels for $\mu = 6$ and $\mu = 2$ for low SNR conditions.

in Chapter 2 and given as follows.

$$X^{\alpha} = \sum_{i=1}^{n} \left(I_i^2 + Q_i^2\right) \tag{3.21}$$

The difference of powers on the right-hand side and the left-hand side of the above expression adds complexities in obtaining expressions for the PDFs of in-phase components and the quadrature components of the fading envelope. This had been less complex for the cases of Nakagami-m, $\kappa - \mu$ and $\eta - \mu$ fading channel models. Now, we will discuss the effect of non-linearities in the fading environment on the SER performance of spatial multiplexing MIMO systems.

To discuss the effects of fading parameter values on SER performance, a 2×2 MIMO system is considered with BPSK and QPSK modulation schemes. Simulations are performed for various cases which are discussed as follows.

The first case considers a 2×2 spatially multiplexed MIMO system with BPSK modulation over $\alpha - \mu$ fading channels. We begin the discussion with increasing values of α for a constant value of μ ($\mu = 2$ in this case). It is already discussed that the parameter α is introduced to model the non-linearities in the fading environment. The SER results are shown in Figure 3.11. In this case, we considered the variation of α in-between 0.5 and 1.7. An improvement in

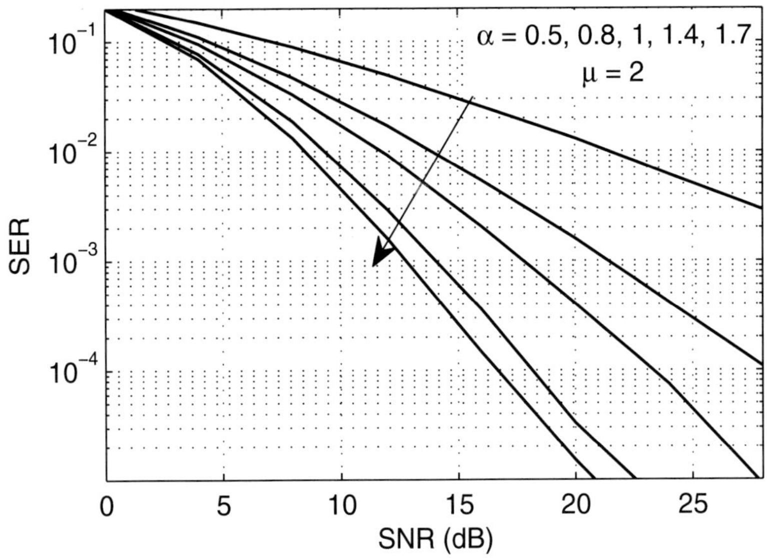

FIGURE 3.11
SER of spatial multiplexing systems with BPSK modulation over α-μ fading channels for $\mu = 2$.

the SER performance is observed with increasing values of α. As α approaches 2, the scattering environment approaches homogeneous [90].

Further, we consider variation of the parameter μ to observe its effect on the SER performance. In Figure 3.12, the SER results can be observed for different values of μ ranging from 2 to 20. Note that for this case, we considered a 2×2 spatially multiplexed MIMO system with QPSK modulation scheme. Figure 3.12 shows results for $\alpha = 2$. The observations in this case can be explained in the same manner as in the case of Nakagami-m fading channels. The case that $\alpha = 2$ is a special case of $\alpha - \mu$ fading scenario that models the fading environment as Nakagami-m fading distribution. Thus, for this case, the observations are the same as those in the Nakagami-m fading case for different values of m. A degradation in the SER is observed with higher values of μ. Again, a crossover in the SER curves is observed while moving from low SNR regions to high SNR regions, indicating dominance of noise over the fading environment for low SNR conditions.

Performance of spatially multiplexed MIMO systems over a non-linear fading environment is also studied for $\alpha = 3$. For this case also, we consider a 2×2 spatially multiplexed MIMO system with QPSK modulation scheme. The SER performance is plotted for different values of μ and $\alpha = 3$ in Figure 3.13. In this case also a crossover in SER curves is observed while moving from low SNR regions to high SNR regions. But, comparing the two figures (Figure

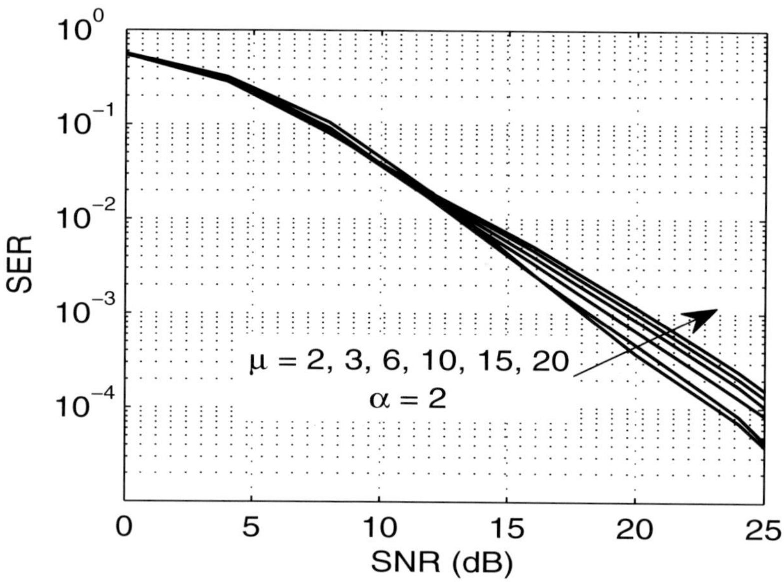

FIGURE 3.12
SER of spatial multiplexing systems with QPSK modulation over α-μ fading channels for $\alpha = 2$.

3.13 and Figure 3.12) minutely, it can be observed that the variation of SER for different values of μ is less for $\alpha = 3$ in low SNR conditions as compared to that for $\alpha = 2$.

Finally, the effect of variation in non-linearity of fading environment on SER performance is shown in Figure 3.14. Again, a 2×2 MIMO system with spatial multiplexing is considered. The SER curves are plotted for BPSK modulation with $\mu = 2$ and different values of α ranging from 0.5 to 8. From the figure, it is observed that SER performance shows an improvement with increasing values of α initially. Then, after reaching a minimum value, SER starts increasing with increased values of α. It is interesting to observe that the minima of SER with respect to α for different SNR values occur at different values of α. This is mainly due to different dominance levels of noise at different levels of non-linearities in the fading environments. To explain the behavior of SER with increasing values of α, consider the expression for envelope of the received signal over $\alpha - \mu$ fading channels. It is given by

$$X = \sqrt[\alpha]{\sum_{i=1}^{n} (I_i^2 + Q_i^2)} \tag{3.22}$$

In the above expression, we consider two extreme cases, $\alpha = 0$ and $\alpha = \infty$.

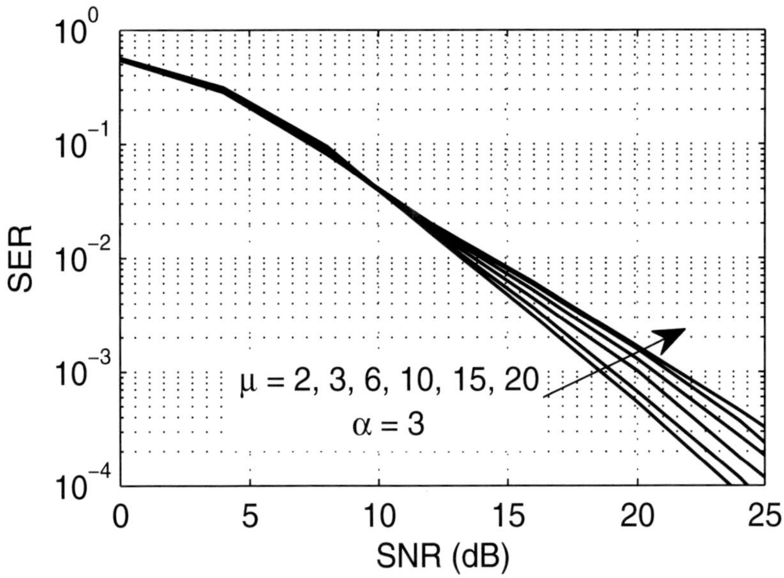

FIGURE 3.13
SER of spatial multiplexing systems with QPSK modulation over α-μ fading channels for $\alpha = 3$.

Note that both of them are hypothetical cases. But, it is helpful to consider them to explain the SER behavior with the non-linearity parameter, α.

- For $\alpha = 0$, the envelope can be described by the form $X = \sqrt[0]{R}$ or $X = R^{\infty}$. In this case, the envelope can have two values.
  - For $R > 1$, the envelope values will be ∞.
  - For $R < 1$, the envelope values will be 0.

  The effect of this on signal transmission is as follows. When $R > 1$, signal strength will be very high (equivalent to no fading and infinite channel gain), while no signal will reach the receiver if $R < 1$, i.e., only noise is received because the channel coefficient is 0.

- For $\alpha = \infty$, the envelope can be described by the form $X = \sqrt[\infty]{R}$ or $X = R^0$. This means for any value of R, the fading signal envelope will be unity. This case represents the condition equivalent to no fading. This means all the transmitted signals reach the receiver with equal strength.

So, in the case of $\alpha = 0$, the signals are received either with very high strength or no signal is received. In this case, it is obvious that the detection will be erroneous. In fact, both the cases, no signal and signals with high strength, are undesirable in MIMO systems with spatial multiplexing. This is

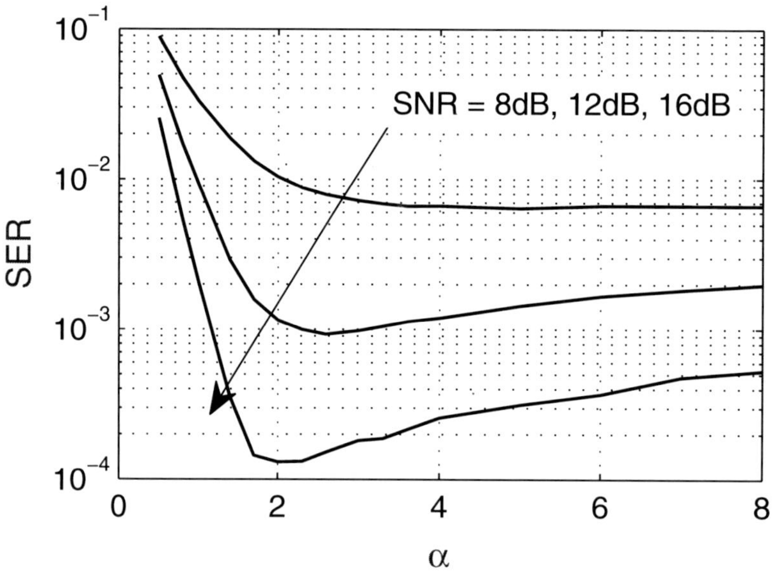

FIGURE 3.14
SER variation of spatial multiplexing systems with BPSK modulation over α-μ fading channels with the values of α.

because of the fact that detection of spatially multiplexed systems is effective in a rich scattering environment. This accounts for higher SER values for lower values of alpha as observed in Figure 3.14.

In the case of $\alpha = \infty$, the signals from all the transmitter antennas reach the receiver with equal amplitudes. In fact, all the channel coefficients become unity in this case representing a situation that there is no fading in the channel. This case is already discussed in Section 1.3 of Chapter 1. It has been already discussed that if all the channel coefficients are unity, there are chances of erroneous detection even in the case of noiseless systems. Thus, an increase in the SER values for higher values of α as observed in Figure 3.14 is justified.

4

Spatial Modulation

CONTENTS

4.1 Introduction

Spatial modulation (SM) systems have become popular mainly because of their reduced hardware and detection complexity. We have already discussed in Chapter 1 that, in SM systems, a single antenna is made active for transmission. This has advantages like requirement of single radio frequency (RF) chain, no need of inter-antenna synchronization, inherent avoidance of inter-antenna interference, etc. The hardware at the transmitter is simplified to a great extent due to employment of a single RF chain. RF chains are very bulky hardware which are comprised of amplifiers, (analog to digital converter/digital to analog converter)ADC/DACs, filters, modulators, etc. This reduction in hardware complexity is important specially for mobile devices. Even though (very large scale integration)VLSI technologies have been so advanced that most hardware is in the form of integrated circuits with considerably smaller sizes and weights but at the RF level, multiple RF chains make the hardware very bulky for mobile devices.

SM systems have been widely studied by several researchers across the globe. Especially with the advancement of multiple-input multiple-output

(MIMO) systems and the proposal of large MIMO systems [19], SM has been viewed as one of the promising technologies for implementing large MIMO systems with reduced hardware and computational complexity [64]. The computational complexity of SM systems is already compared with various existing technologies. It is claimed that SM systems are computationally efficient as far as detection is concerned and they perform better than other competitive techniques [77]. Performance of SM systems has already been studied over various fading channels including Nakagami-m fading channels and $\eta - \mu$ fading channels [5, 32, 33, 57, 75]. Further, various detection methods have also been proposed with reduced computational complexity in detection at the receiver [65, 100, 130, 144, 145].

In this chapter, we will discuss SM systems in detail. We will discuss about its spectral efficiency, diversity order and detection complexity. Further, performance of SM systems over various generalized fading channels will be discussed. In this book, we will mainly focus on the bit error rate/symbol error rate (BER/SER) of SM systems.

4.2 SM transmission reception schemes

Chapter 1 briefed about the transmission and reception schemes being used in SM. SM is a transmission scheme for MIMO systems that uses single antenna at a time at the transmitter. As a result, SM MIMO systems may be thought of as MIMO systems that are converted to single-input multiple-output (SIMO) systems. It is made clear that a part of the incoming sequence of bits is mapped to a transmit antenna index and the antenna corresponding to that index is used for transmission. For an $N_t \times N_r$ MIMO system with SM transmission, the number of bits that can be mapped to an antenna index is given as $n_a = \lfloor \log_2 N_t \rfloor$. Each group of n_a number of bits mapped to an antenna index is thought of as if coming from a set of constellation points known as spatial constellation. This adds to the spectral efficiency of digitally modulated symbols. In general, a digitally modulated symbol can transmit $n_s = \log_2 M$ number of bits per channel use per unit bandwidth. Each group of n_s number of digitally modulated symbols are coming from signal constellation and are physically transmitted. Effectively in an SM MIMO system, by physical transmission of a symbol from signal constellation, $n_a + n_s$ number of bits are transmitted per channel use per unit bandwidth. It is important to note here that the number of spatially modulated bits, $n_a = \lfloor \log_2 N_t \rfloor$, actually puts a restriction on SM MIMO systems that the number of antennas at the transmitter has to be integer power of 2. Otherwise, it may happen that the system never uses $N_t - 2^{\lfloor \log_2 N_t \rfloor}$ antennas for transmission and the extra antennas at transmitter become redundant. This restriction on the number of antennas at transmitter is actually taken care of by the proposal of generalized spatial modulation (GSM) which we will discuss in brief at the end of this chapter [143].

We refer to equation (1.22) for transmitted vector for SM MIMO systems. It can be given as follows when the spatial constellation point is mapped with the kth transmit antenna and the digitally modulated symbol, x_q, is to be transmitted.

$$\mathbf{x} = \begin{bmatrix} 0 & 0 & \cdots & \underset{k^{th}\ posistion}{\underset{\uparrow}{x_q}} & 0 & \cdots & 0 \end{bmatrix}^T \tag{4.1}$$

where $\mathbf{x}$ is the transmitted vector and x_q is the digitally modulated symbol that is to be transmitted from the kth antenna. Note that the transmitted symbol vector has all zeros except at the kth position. In other words, the kth position contains digitally modulated symbol x_q. Thus, the received signal vector for such SM MIMO systems can be given as

$$\mathbf{r} = \mathbf{H}\mathbf{x} + \mathbf{n} \tag{4.2}$$

```
% MATLAB instructions to generate an SM symbol for a MIMO system
%with Nt transmitter antennas and M-PSK modulation
lut = pskmod([0:M-1],M); % Look up table
Eavg = sum(abs(lut).^2)/M; % Average energy of constellation
N_lut = lut/sqrt(Eavg); % Normalized look up table
x=zeros(Nt,1);
x_i = randi([0,1],log2(Nt)+log2(M),1); % SM symbol
x_a = x_i(1:Na);             % bits from spatial constellation
a_i = s_mapping(x_a);        % antenna index
% function s_mapping() is given as an appendix at the end
x_s1 = x_i(Na+1:end);        % bits from signal constellation
x_s = s_mapping(x_s1)-1;     % bits to decimal
x_q = pskmod(x_s,M)/sqrt(Eavg);     % digitally modulated symbol
                                    %with unit energy
x(a_i)=x_q;     % transmitted vector
```

Given that the transmitted symbol vector, $\mathbf{x}$ contains all zeros except at the kth position, the expression for received vector can be simplified as follows.

$$\mathbf{r} = \mathbf{h_k}\mathbf{x} + \mathbf{n} \tag{4.3}$$

where $\mathbf{h_k}$ is the kth column of the channel matrix $\mathbf{H}$. So, the processes of transmission and reception are based on vector operations instead of matrix operations as in spatial multiplexing systems. This has an important role in computational complexity of the detection process which we are going to discuss next.

For SM reception, we consider here optimal detection which is the maximum likelihood (ML) detection rule for joint detection of antenna index (spatially modulated bits) and digitally modulated symbol. The optimal detection rule of SM MIMO systems is already mentioned in equation (1.25) of Chapter 1 and reproduced here.

$$\{\hat{j}, x_{\hat{q}}\} = \arg\min_{j,q} \| \mathbf{r} - \mathbf{h_j} x_q \|^2$$
$$= \arg\min_{j,q} \| \mathbf{h_j} x_q \|^2 - 2Re\{\mathbf{h_j^H} \mathbf{r} x_q^*\} \tag{4.4}$$

```
% MATLAB instructions to detect an SM symbol
    for m=1:M
        z(:,:,m)=H*N_lut(m);
        d(:,:,m)=repmat(r,1,Nt)-z(:,:,m);
        nor(m,:)=sum(abs(d(:,:,m)).^2);
    end
    [mi,ins]=min(nor);
    [mis,ina]=min(mi);
    d_a=ina;      % Antenna index
    d_s=ins(ina)-1; % Digitally modulated symbol
```

In the above expression, analyzing the number of complex multiplications required to detect a symbol, the detection complexity may be given as $2N_rN_t + N_tM + M$ [53]. This detection complexity is considerably less than the detection complexity of equation (3.10). The detection complexity of SM MIMO systems with optimal detection is discussed and compared with sub-optimal detection in [53, 77]. It is claimed that the detection complexity of optimal detection is comparable to that of sub-optimal detection. However, the performance of the optimal detection scheme is much better than that of sub-optimal detection in unconstrained channels.

4.3 Performance analysis of SM MIMO systems

In this section, we will discuss various performance parameters of SM MIMO systems over fading channels. Further, we will analyze various performance metrics over generalized fading channels. The main performance metrics of any wireless communication systems are: outage probability, BER/SER and channel capacity. The performance metrics can be evaluated analytically and through simulations as well. Simulations and analytical evaluations have their own limitations. We will discuss the procedure to evaluate these metrics analytically in the following subsections. In some cases, it may be difficult to arrive at a closed form expression for these performance metrics.

4.3.1 Outage probability

We already discussed in the previous section that the index of antenna transmitting information is also estimated at the receiver. However, this part contributes a little to the signal-to-noise ratio (SNR) level of the received signal for a system in which all the symbols are equally likely. The contribution to SNR due to antenna index is only in the form of the channel coefficients corresponding to that antenna, which are anyway not affecting much because they are independent and follow the same distribution for independent and identically distributed (i.i.d.) channels. A communication system is said to be in outage when the SNR level at the receiver falls below a certain threshold value and correct detection of signal may not be possible at these low SNR levels. This results in a large number of bits/symbols received in error making the system unreliable. In general, outage probability of a wireless communication systems is defined as the probability that the instantaneous SNR is less than a certain value, say, threshold SNR, γ_{th}. Mathematically, it can be represented as

$$P_{out}(\gamma_{th}, \overline{\gamma}) = P(\gamma < \gamma_{th}) \tag{4.5}$$

where P_{out} is the outage probability, γ is the instantaneous SNR at the receiver and $\overline{\gamma}$ is the average received SNR.

In the case of SM MIMO systems, the received signal can be given as

$$\mathbf{r} = \mathbf{h_k x} + \mathbf{n} \tag{4.6}$$

This is similar to the received signal for diversity combining schemes. In the above expression, signal part is characterized by $\mathbf{h_k x}$ and the noise part is $\mathbf{n}$. So, instantaneous ratio of the overall received signal power and the total noise power can be considered. However, this can be further simplified for independent channels in terms of SNR per branch at the receiver. For mutually independent channel coefficients, the total received signal power can be obtained as the sum of powers of received signals at the individual receiver antennas. Thus, it is valid to consider SNR per receiver antenna. Once we obtain the instantaneous SNR at the receiver and its probability density function (PDF), the outage probability can be given as

$$P_{out}(\gamma_{th}, \overline{\gamma}) = \int_0^{\gamma_{th}} p_\gamma(x)dx \tag{4.7}$$

where $p_\gamma(\cdot)$ is the PDF of the received instantaneous SNR. So, outage probability is the cumulative distribution function (CDF) of instantaneous received SNR.

4.3.2 Symbol error rate

When the signal travels through channel, in addition to fading, it also gets affected by the noise. Usually, the noise is considered as a collective effect of the thermal noise generated at various levels in the system. This noise, in most cases, can be modeled as random noise with Gaussian distribution with zero mean, also known as additive white gaussian noise (AWGN) because of its additive nature and equal power spectral density at all the frequencies. Now onward, such channels are referred to as noisy channels. In noisy channels, not all the transmitted symbols can be detected correctly at the receiver. In communications with only signal constellation, i.e., with transmission schemes having only digitally modulated symbols, any amount of added noise shifts the position of constellation point from its original location. This shift depends on the amount of noise that is being added. Because of this, many times it happens that the noise is so large that the constellation point moves into the decision boundary of a neighboring constellation point other than the actually transmitted symbol. In such systems, the received symbol is detected based on the minimum distance of the received noisy constellation point from the actual set of constellation points. It is equivalent to the condition that the constellation plane is divided into regions and each region corresponds to a constellation point. The received noisy symbol is then marked on this constellation plane with regions and the decision is taken in favor of a constellation point if the received symbol falls in its region.

In SM MIMO systems, the detection process is slightly different from other communication systems described earlier. In SM MIMO systems, in addition to the signal constellation, we also have constellation points in the spatial domain which are mapped with the indices of transmitting antennas. So, the distance of received noisy symbol is required to be minimized jointly with respect to the signal constellation and the spatial constellation. The rule for this minimization has been the decision criteria for a received symbol that is spatially modulated in SM MIMO systems. The rule is given as

$$\begin{aligned}\{\hat{j}, x_{\hat{q}}\} &= \arg\min_{j,q} \| \mathbf{r} - \mathbf{h_j} x_q \|^2 \\ &= \arg\min_{j,q} \| \mathbf{h_j} x_q \|^2 - 2Re\{\mathbf{h_j^H} \mathbf{r} x_q^*\}\end{aligned} \tag{4.8}$$

Now, occurrence of error in this case is defined as the event that any of the two, digitally modulated symbol or the spatially modulated data, is detected incorrectly. However, the above expression is for joint detection of antenna index and the digitally modulated symbol. So, in our discussion an SM symbol would be collectively the antenna index (spatial constellation) and the digitally modulated symbol (signal constellation). In this case, probability of error can be obtained by first evaluating pairwise error probability and then averaging the set of pairwise error probabilities for all possible pairs of symbols. To define pairwise error probability, we first define an SM symbol as $x_{SM}(i) = \{x_a(i), x_s(i)\}$, where $x_a(i)$ is the group of bits that are mapped to antenna index of a transmitter antenna and $x_s(i)$ is the group of bits that are transmitted using digital modulation techniques. Now, an SM symbol is said to be received in error if any of $x_a(i)$ or $x_s(i)$ is detected incorrectly at the receiver.

Let us define an event that symbol $x_{SM}(i)$ is transmitted and symbol $x_{SM}(j)$ is detected. The probability of occurrence of such an event is known as pairwise error probability and denoted as $P(x_{SM}(i) \rightarrow x_{SM}(j))$. In other terms, occurrence of error is denoted by

$$\begin{aligned} P(x_{SM}(j, x_q) \rightarrow x_{SM}(\hat{j}, x_{\hat{q}})) &= P\left(\| \mathbf{r} - \mathbf{h}_{\hat{\mathbf{j}}} x_{\hat{q}} \|^2 \leq \| \mathbf{r} - \mathbf{h}_{\mathbf{j}} x_q \|^2\right) \\ P(x_{SM}(j, x_q) \rightarrow x_{SM}(\hat{j}, x_{\hat{q}})) &= P(\| \mathbf{h}_{\hat{\mathbf{j}}} x_{\hat{q}} \|^2 - 2Re\{\mathbf{h}_{\hat{\mathbf{j}}}^{\mathbf{H}} \mathbf{r} \mathbf{x}_{\hat{\mathbf{q}}}^*\} \\ &\leq \| \mathbf{h}_{\mathbf{j}} x_q \|^2 - 2Re\{\mathbf{h}_{\mathbf{j}}^{\mathbf{H}} \mathbf{r} \mathbf{x}_{\mathbf{q}}^*\}) \end{aligned} \tag{4.9}$$

where $\hat{j}$ and $x_{\hat{q}}$ are respectively the estimated antenna index and the detected symbol at receiver when antenna index j was used at transmitter and the symbol x_q was transmitted actually. According to the above expressions, the correct decision is taken at the receiver only in the case that the received vector has the minimum Euclidean distance from the actual transmitted symbols (both spatial and signal constellations) compared to all other possible symbols. Otherwise, the symbols are received in error. For an SM MIMO system with N_t number of transmitter antennas and M-ary modulation scheme, the total number of pairs formed is $(MN_t - 1)!$.

Overall, probability of a symbol received in error can be obtained by applying union bound on the individual pairwise error probability [28, 101]. It can be given as

$$P_e \leq \sum_{j,q} P(x_{SM}(j, x_q) \rightarrow x_{SM}(\hat{j}, x_{\hat{q}})) \tag{4.10}$$

Again, this probability of symbol error is a function of fading channel coefficients. Thus, to find average probability of symbol error or often referred to as SER, it is required to be averaged over the PDF of channel coefficients.

4.3.3 Channel capacity

Channel capacity is another metric that can help to determine the quality of a wireless communication link. Channel capacity gives the maximum rate at which transmission can be done with reliable performance, i.e., sufficiently low probabilities of error. Any rates of transmission higher than the channel capacity results in higher probability of error and the communication link becomes unreliable. The channel capacity is usually defined as follows.

$$C = B \log_2(1 + SNR) \tag{4.11}$$

Lower capacity indirectly means either of two possibilities. 1) The coherence bandwidth of the channel has been reduced or 2) the SNR at receiver has been reduced to a very low value because of fading. Now, in the case of fading channels, the SNR is not constant. SNR in case of fading channels is defined as the instantaneous SNR following some kind of fading statistics. Correspondingly, instantaneous channel capacity is defined for the instantaneous SNR. So, the average channel capacity has to be evaluated by averaging the instantaneous capacity over the PDF of instantaneous SNR.

In case of SM MIMO systems, not all information is transmitted physically. So, the capacity evaluated using the expression above only relates to the bits/information that is physically transmitted. The information of transmit antenna index is carried along with the transmitted symbol without demanding extra channel capacity. Thus, in SM MIMO at some instances, the information is actually transmitted at rates beyond the channel capacity. This gives advantages in overall spectral efficiency of the system. So, SM MIMO systems employ a transmission strategy which enables transmissions at rates beyond the channel capacity.

4.4 Performance of SM systems over Nakagami-m fading channels

In this section, we consider performance of SM MIMO systems over Nakagami-m fading channels. We will discuss the results of simulations for various performance metrics of SM MIMO wireless communication systems. We will begin the discussion with some insights to analytical methods for evaluation of different performance metrics of SM MIMO systems. We begin with the outage probability analysis of SM MIMO systems over Nakagami-m fading channels. As we proceed, we will discuss about SER and channel capacity of SM MIMO systems over Nakagami-m fading channels.

4.4.1 Outage probability

In Section 4.3.1, we discussed that outage probability is the CDF of received instantaneous SNR. To evaluate outage probability, we first need to have statistical information about the received signal and the instantaneous SNR. In SM MIMO systems, the instantaneous SNR depends on the channel coefficients of the channel matrix corresponding to the transmit antenna which is selected for transmission of information. The instantaneous SNR can be obtained from the received signal. The received signal is random in nature mainly because of the noise that is added at various stages of signal processing and the fading that resulted due to mobile environment. The system under consideration has AWGN and Nakagami-m fading. So, the noise is Gaussian distributed. To get instantaneous SNR and its statistics, we need to have statistics of signal part in the received signal vector. The statistics of received signal in SM systems is similar to that of received signals in maximal ratio combining (MRC) techniques.

4.4.2 Symbol error rate

Symbol error rate gives the fraction of information received in error. It is an important metric that defines the quality of wireless communication link. The information about SER can be helpful in deciding the modulation order in the adaptive modulation systems [21,104]. There are adaptive systems for SM MIMO too that use the information about channel conditions and the SER to decide the modulation orders [140]. In fact the SER is mainly dependent upon the channel conditions. If the channels are highly faded, very less signal power is received. This results in increased SER. This SER also depends on the minimum separation between two points on the constellation diagram. We already discussed in Chapter 1 that higher order modulation schemes can improve the transmission rates but as we move to higher order modulation schemes, the minimum distance between two constellation points reduces. This reduced distance between constellation points increases the probability of making error at the receiver while detecting symbols in the noisy environment. In SM MIMO systems, the error in detection at the receiver can be either in the antenna index or in the digitally modulated symbol. However, we have discussed joint detection of antenna index and digitally modulated symbols. So, joint probability of error would be evaluated in this case.

To evaluate probability of error analytically, let us assume that an SM symbol consisting of the antenna index j and the digitally modulated symbol x_q is transmitted. Due to noise and fading, it is detected in error as the transmit antenna index $\hat{j}$ and digitally modulated symbol $x_{\hat{q}}$. In this case, the complete SM symbol is said to have received in error even if either the antenna index or the digitally modulated symbol is received in error. The reason is that at the receiver the information is processed in the form of SM symbols and not the individual units like antenna index and digitally modulated symbols or the individual bits. Again, we considered joint detection of antenna index and digitally modulated symbol while discussing the detection of SM symbols at the receiver. So, we will discuss pairwise error probability for joint detection of SM MIMO symbols over Nakagami-m fading channels. It can be given by

$$P(x_{SM}(j,x_q) \rightarrow x_{SM}(\hat{j},x_{\hat{q}})) = P\left(\| \mathbf{r} - \mathbf{h}_{\hat{\mathbf{j}}} x_{\hat{q}} \|^2 \leq \| \mathbf{r} - \mathbf{h}_{\mathbf{j}} x_q \|^2\right) \tag{4.12}$$

The above expression can be represented in the form of Frobenius norm. From the fact that norm is a positive real number, the above condition on the squared norms holds true even for the Frobenius norms. So, the pairwise error probability can be given as

$$P(x_{SM}(j,x_q) \rightarrow x_{SM}(\hat{j},x_{\hat{q}})) = P\left(\| \mathbf{r} - \mathbf{h}_{\hat{\mathbf{j}}} x_{\hat{q}} \|_F \leq \| \mathbf{r} - \mathbf{h}_{\mathbf{j}} x_q \|_F\right) \tag{4.13}$$

where $\| \cdot \|_F$ represents Frobenius norm of a vector.

Now, substituting the values of received vector from expression of equation (4.3), the expression for pairwise error probability can be simplified to

$$P(x_{SM}(j,x_q) \rightarrow x_{SM}(\hat{j},x_{\hat{q}})) = P\left(\|\mathbf{h_j}x_q+\mathbf{n}-\mathbf{h_{\hat{j}}}x_{\hat{q}}\|_F \leq \|\mathbf{h_j}x_q+\mathbf{n}-\mathbf{h_j}x_q\|_F\right) \tag{4.14}$$

$$P(x_{SM}(j,x_q) \rightarrow x_{SM}(\hat{j},x_{\hat{q}})) = P\left(\| \left(\mathbf{h_j}x_q-\mathbf{h_{\hat{j}}}x_{\hat{q}}\right) - \mathbf{n} \|_F \leq \| \mathbf{n} \|_F\right) \tag{4.15}$$

It can be further simplified by using triangle inequality for vectors. The triangle inequality for vectors relates the sum and differences of the vector norms as follows.

$$\|\left(\mathbf{h_j}x_q-\mathbf{h_{\hat{j}}}x_{\hat{q}}\right)\|_F - \| \mathbf{n} \|_F \leq \| \left(\mathbf{h_j}x_q-\mathbf{h_{\hat{j}}}x_{\hat{q}}\right) + \mathbf{n}\|_F \leq \| \left(\mathbf{h_j}x_q-\mathbf{h_{\hat{j}}}x_{\hat{q}}\right) \|_F + \| \mathbf{n} \|_F \tag{4.16}$$

This says that the pairwise error probability can be represented as

$$P(x_{SM}(j,x_q) \rightarrow x_{SM}(\hat{j},x_{\hat{q}})) \leq P\left(\left\|\left(\mathbf{h_j}x_q-\mathbf{h_{\hat{j}}}x_{\hat{q}}\right)\right\|_F - \| \mathbf{n} \|_F \leq \| \mathbf{n} \|_F\right) \tag{4.17}$$

This reduces the expression for pairwise error probability as

$$P(x_{SM}(j,x_q) \rightarrow x_{SM}(\hat{j},x_{\hat{q}})) \leq P\left(\left\|\frac{\left(\mathbf{h_j}x_q-\mathbf{h_{\hat{j}}}x_{\hat{q}}\right)}{2}\right\|_F \leq \|\mathbf{n}\|_F\right) \tag{4.18}$$

The above expression relates the probability of error with the distance of noise vector from origin and the distance of the difference vector $\frac{\left(\mathbf{h_j}x_q-\mathbf{h_{\hat{j}}}x_{\hat{q}}\right)}{2}$ from the origin. In other words, the occurrence of error depends only on the position of noise vector $\mathbf{n}$ relative to the vector $\left(\mathbf{h_j}x_q-\mathbf{h_{\hat{j}}}x_{\hat{q}}\right)$. If noise vector is nearer to origin, there is no error and if it is nearer to $\left(\mathbf{h_j}x_q-\mathbf{h_{\hat{j}}}x_{\hat{q}}\right)$, there is error in the detection. It is to be noted that $\mathbf{n}$ is a vector of Gaussian distributed white noise samples. In this case, the pairwise error probability can be given as

$$P(x_{SM}(j,x_q) \rightarrow x_{SM}(\hat{j},x_{\hat{q}})) \leq Q\left(\sqrt{\frac{\gamma\left\|\left(\mathbf{h_j}x_q-\mathbf{h_{\hat{j}}}x_{\hat{q}}\right)\right\|_F^2}{2}}\right) \tag{4.19}$$

where $\gamma = \frac{E_s}{N_0}$ is the ratio of energy per SM symbol, E_s to N_0 which is the noise power spectral density and $Q(\cdot)$ is the Q function which gives area under the tail of Gaussian curve and is defined as

$$Q(x) = \frac{1}{\sqrt{2\pi}}\int_x^{\infty} e^{-\frac{\alpha^2}{2}}\,d\alpha \tag{4.20}$$

Note that the expression for pairwise error probability represented by equation (4.19) is dependent on the channel coefficients which are randomly distributed, i.e., the expression represents conditional pairwise error probability and it is conditioned on the channel matrix, $\mathbf{H}$. This conditional pairwise error probability needs to be averaged over the PDF of the random variable which is a function of channel coefficients of jth column and $\hat{j}^{th}$ column of channel matrix. The function may be represented as $k = \left(\left\|\mathbf{h_j}x_q - \mathbf{h_{\hat{j}}}x_{\hat{q}}\right\|\right)$. To evaluate the average pairwise error probability, we need to derive PDF of k using the information that each element of $\mathbf{h_j}$ and $\mathbf{h_{\hat{j}}}$ is independent and identically distributed Nakagami-m random variable. For a particular pair of SM symbols, x_q and $x_{\hat{q}}$ may be considered as complex constants (having real and imaginary parts) that do scaling of the corresponding channel coefficients. For the statistical analysis of k, we represent the difference vector in a simplified way as follows.

$$\mathbf{h_j}x_q - \mathbf{h_{\hat{j}}}x_{\hat{q}} = (\mathbf{h_{jR}} + j\mathbf{h_{jI}})(x_{qR} + jx_{qI}) - \left(\mathbf{h_{\hat{j}R}} + j\mathbf{h_{\hat{j}I}}\right)(x_{\hat{q}R} + jx_{\hat{q}I}) \quad (4.21)$$

where $\mathbf{a_R}$ and $\mathbf{a_I}$ are respectively the vectors of real and imaginary components of each elements of vector $\mathbf{a}$, and b_R and b_I are respectively the real and imaginary components of a complex number b.

The above expression can be further simplified as

$$\begin{aligned}\mathbf{h_j}x_q - \mathbf{h_{\hat{j}}}x_{\hat{q}} &= \{(\mathbf{h_{jR}}x_{qR} - \mathbf{h_{jI}}x_{qI}) + j(\mathbf{h_{jI}}x_{qR} + \mathbf{h_{jR}}x_{qI})\} \\ &\quad - \left\{\left(\mathbf{h_{\hat{j}R}}x_{\hat{q}R} - \mathbf{h_{\hat{j}I}}x_{\hat{q}I}\right) + j\left(\mathbf{h_{\hat{j}I}}x_{\hat{q}R} + \mathbf{h_{\hat{j}R}}x_{\hat{q}I}\right)\right\} \quad (4.22) \\ &= \left\{(\mathbf{h_{jR}}x_{qR} - \mathbf{h_{jI}}x_{qI}) - \left(\mathbf{h_{\hat{j}R}}x_{\hat{q}R} - \mathbf{h_{\hat{j}I}}x_{\hat{q}I}\right)\right\} \\ &\quad + j\left\{(\mathbf{h_{jI}}x_{qR} + \mathbf{h_{jR}}x_{qI}) - \left(\mathbf{h_{\hat{j}I}}x_{\hat{q}R} + \mathbf{h_{\hat{j}R}}x_{\hat{q}I}\right)\right\} \quad (4.23)\end{aligned}$$

Finally, the real and imaginary vector components of the difference vector can be given as

$$\left(\mathbf{h_j}x_q - \mathbf{h_{\hat{j}}}x_{\hat{q}}\right)_R = (\mathbf{h_{jR}}x_{qR} - \mathbf{h_{jI}}x_{qI}) - \left(\mathbf{h_{\hat{j}R}}x_{\hat{q}R} - \mathbf{h_{\hat{j}I}}x_{\hat{q}I}\right) \quad (4.24)$$

$$\left(\mathbf{h_j}x_q - \mathbf{h_{\hat{j}}}x_{\hat{q}}\right)_I = (\mathbf{h_{jI}}x_{qR} + \mathbf{h_{jR}}x_{qI}) - \left(\mathbf{h_{\hat{j}I}}x_{\hat{q}R} + \mathbf{h_{\hat{j}R}}x_{\hat{q}I}\right) \quad (4.25)$$

Given that the channel coefficients are Nakagami-m distributed, using methods of random variable transformation, the PDF of real and imaginary parts of the difference vector, $\left(\mathbf{h_j}x_q - \mathbf{h_{\hat{j}}}x_{\hat{q}}\right)_R$ and $\left(\mathbf{h_j}x_q - \mathbf{h_{\hat{j}}}x_{\hat{q}}\right)_I$, can be obtained. It can eventually be used to finally get the PDF of the norm/squared norm of the difference vector. Let us proceed with the consideration of the real part of the difference vector, $\left(\mathbf{h_j}x_q - \mathbf{h_{\hat{j}}}x_{\hat{q}}\right)_R$, to begin with. It is a difference of two random variable terms. The first term consists of the column of channel matrix corresponding to the transmit antenna used for spatial constellation and the part of actual symbol transmitted, while the later consists of a column of channel matrix corresponding to the transmit antenna erroneously estimated at receiver from spatial constellation and the part of the symbol detected at the receiver.

To describe the method for obtaining the PDF of the difference vector, let us assume that the system uses one dimensional constellation, i.e., only real valued digitally modulated symbols. Despite the real valued signal constellations, the difference vector still would have the non-zero imaginary components on account of reflected multipaths having different phases, i.e., the complex channel coefficients. But, for real valued signal constellations, the real part of the difference vector would reduce to $\left(\mathbf{h_j}x_q - \mathbf{h_{\hat{j}}}x_{\hat{q}}\right)_R = \mathbf{h_{jR}}x_q - \mathbf{h_{\hat{j}R}}x_{\hat{q}}$. Each element of this difference vector is a combination of two scaled random variables with the scaling factors x_q and $x_{\hat{q}}$ respectively. Each element of the difference vector can therefore be considered of the form $aX_1 - bX_2$ with a and b as the scaling factors of variables X_1 and X_2 respectively. The PDF of a scaled random variable sX in general can be given as

$$p_{sX}(x) = \frac{1}{|s|} p_X\left(\frac{x}{s}\right) \tag{4.26}$$

where $p_X(\cdot)$ is the PDF of unscaled random variable and s is the scaling factor.

Note that the squared norm of difference vector, $\left\|\left(\mathbf{h_j}x_q - \mathbf{h_{\hat{j}}}x_{\hat{q}}\right)\right\|_F^2$, can thus be represented as

$$\left\|\left(\mathbf{h_j}x_q - \mathbf{h_{\hat{j}}}x_{\hat{q}}\right)\right\|_F^2 = \sum_{i=1}^{N_r} \left((h_{iR}x_q - h_{\hat{i}R}x_{\hat{q}})^2 + (h_{iI}x_q - h_{\hat{i}I}x_{\hat{q}})^2\right) \tag{4.27}$$

So, the resulting squared norm, $\left\|\left(\mathbf{h_j}x_q - \mathbf{h_{\hat{j}}}x_{\hat{q}}\right)\right\|_F^2$, is a random variable that is the sum of random variables which are obtained as a result of scaling, difference and square operations. The PDF of the resulting squared norm can be obtained using the relations of transformation of random variables and the concepts of functions of random variables. Note that once PDF of either $(h_{iR}x_q - h_{\hat{i}R}x_{\hat{q}})^2$ or $(h_{iI}x_q - h_{\hat{i}I}x_{\hat{q}})^2$ is obtained, the method of moment generating function (MGF) can also be used to obtain the MGF of resultant squared norm. The final MGF can be used to evaluate average pairwise error probability.

Further, we discuss the results obtained from MATLAB® simulation of SM systems over Nakagami-m fading channels. For the simulation, we considered a 4×2 SM MIMO system with quadrature phase shift keying (QPSK) modulation. This results in the transmission of 4 bits/s/Hz, 2 bits/s/Hz using digital modulation and 2 bits/s/Hz using spatial modulation for four transmitter antennas. Results are shown in Figure 4.1 for different values of fading parameter, m. It can be observed that the SER performance degrades for higher values of fading parameter, m. This degradation is on account of reduced variability of the fading coefficients for higher values of m which results in lower values of the squared norm of difference vector $\left(\mathbf{h_j}x_q - \mathbf{h_{\hat{j}}}x_{\hat{q}}\right)$. Also, the higher values of m move the channel environment toward non-fading conditions ($m \to \infty$ models non-fading case). Thus, detection becomes more erroneous for higher values of fading parameter, m, of Nakagami-m fading channels.

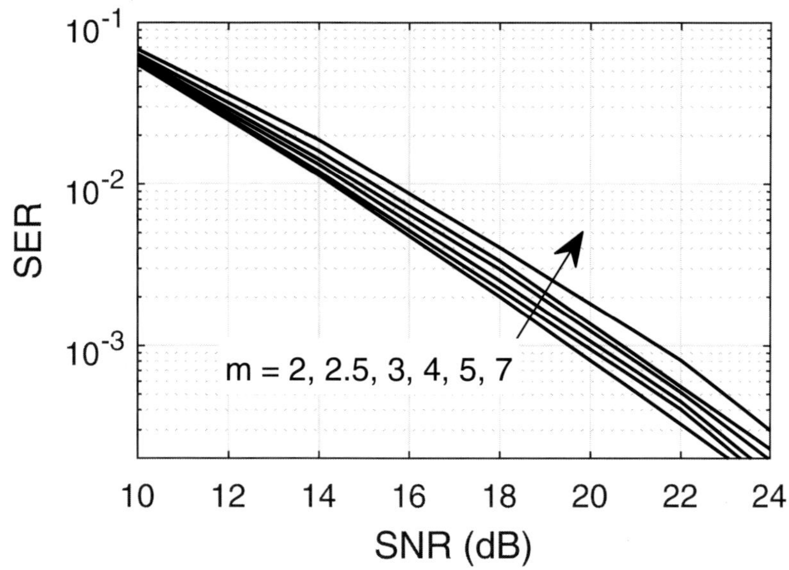

FIGURE 4.1
SER of 4×2 SM MIMO systems over Nakagami-m fading channels for different values of m.

4.5 Error performance of SM systems over $\kappa - \mu$ fading channels

Nakagami-m fading channels consider the received signal to be coming in the form of clusters of multipath components with equal strengths. In $\kappa - \mu$ fading channel model, a dominant component is assumed to be a part of every cluster of multipath components. Thus, $\kappa - \mu$ fading model is more suitable to model the fading environments with possibility of direct line of sight (LOS) path from transmitter to receiver. The analytical evaluation of error performance can be done exactly by following the steps that are discussed in the case of error performance of SM MIMO systems over Nakagami-m fading channel that is discussed in the previous section. It involves transformation of the in-phase and quadrature phase components of the received signal by scaling followed by other basic operations (like difference, squaring and adding) on them. This gives us a new random variable which is the squared norm of the difference vector. It is given as

$$\left\| \left(\mathbf{h}_{\mathrm{j}} x_q \quad \mathbf{h}_{\hat{\mathrm{j}}} x_{\hat{q}} \right) \right\|_F^2 = \sum_{i=1}^{N_r} \left((h_{iR} x_q - h_{\hat{i}R} x_{\hat{q}})^2 + (h_{iI} x_q - h_{\hat{i}I} x_{\hat{q}})^2 \right) \qquad (4.28)$$

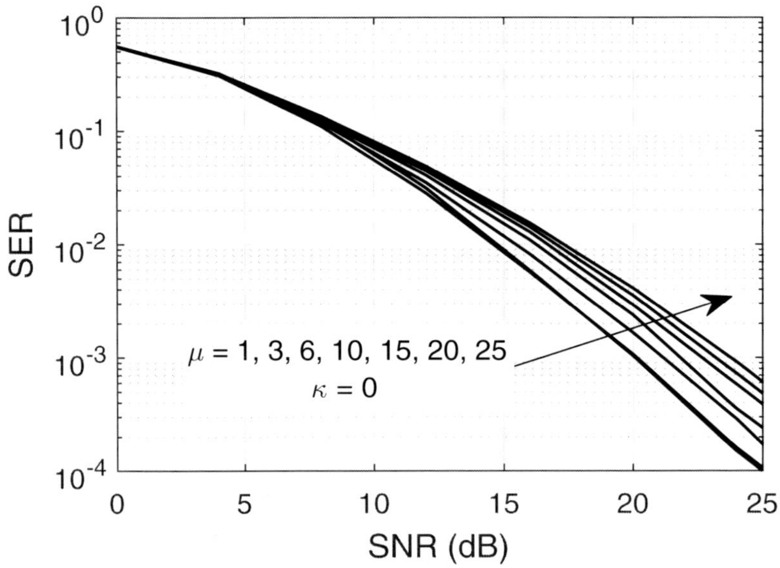

FIGURE 4.2
SER performance of 4×2 SM MIMO systems over $\kappa - \mu$ fading channels with QPSK modulation for $\kappa = 0$.

The PDF of the random variable on the right-hand side of the above expression is required to derive an analytical expression for error performance of SM MIMO systems over $\kappa - \mu$ fading channels. For this, we use the above function of random variables involving the channel coefficients with $\kappa - \mu$ distribution and obtain PDF of $\left\| \left(\mathbf{h_j} x_q - \mathbf{h_{\hat{j}}} x_{\hat{q}} \right) \right\|_F^2$ which will be used to find average pairwise error probability.

In the following discussions of this section, we will discuss the effect of fading parameters, κ and μ, on the SER performance of SM MIMO systems. Eventually the effect of stronger LOS components on SER will also be discussed. MATLAB script file for simulation of SM MIMO systems and MATLAB function to generate $\kappa - \mu$ distributed channel coefficients are given at the end of this book in Appendix B. They can be used to obtain SER results of SM MIMO systems over various fading channels using appropriate values of fading coefficients and suitable modulation schemes for any number of transmitter and receiver antennas.

A 4×2 SM MIMO system with QPSK modulation is simulated for its SER performance over $\kappa - \mu$ fading channels. The SER results are obtained for various values of fading parameter μ and two values of κ, $\kappa = 0$ (no LOS component) and $\kappa = 20$ (strong LOS component). For both these cases, results are shown in Figure 4.2 and Figure 4.3 respectively. SER for different values

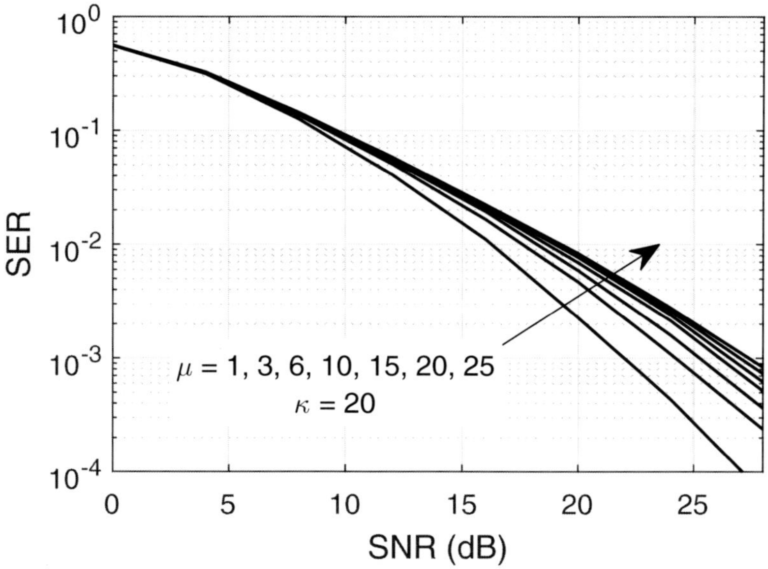

FIGURE 4.3
SER performance of 4×2 SM MIMO systems over $\kappa - \mu$ fading channels with QPSK modulation for $\kappa = 20$.

of μ ranging from 1 to 25 are shown for both the cases. It is observed that the SER performance degrades as the value of μ increases in both the cases (strong LOS and non-LOS). However, an interesting observation is that for $\kappa = 0$, the degradation in SER is low for initial increment in the values of μ while it is opposite for the case of $\kappa = 20$. For example, in Figure 4.2, it is observed that the SER curves for $\mu = 1$ and $\mu = 3$ are almost overlapped and for further increment in the values of μ increase the gap between successive SER curves, while in Figure 4.3, the observation is just opposite. The gap between successive SER curves decreases for higher values of μ. It is maximum between the SER curves for $\mu = 1$ and $\mu = 3$ for $\kappa = 20$. This is because the variability of the power of channel coefficients reduces for higher values of κ and μ. In the strong LOS case, the reduction in variability is already introduced by a large LOS factor, κ; further reduction in variability because of μ reduces for higher values of μ and this is reflected in the SER curves in Figure 4.3. Further, when $\kappa \to \infty$ and/or $\mu \to \infty$, the channel becomes a non-fading AWGN channel for which all the fading coefficients are unity (no variation). The reduced variability of fading coefficients with increasing values of κ and/or μ is shown in Figure 4.4.

Further, the results are plotted to see the degradation in SER performance with increasing values of both μ and κ. In Figure 4.5, the SER results are plotted by varying μ for different values of κ and SNR. The values of μ are

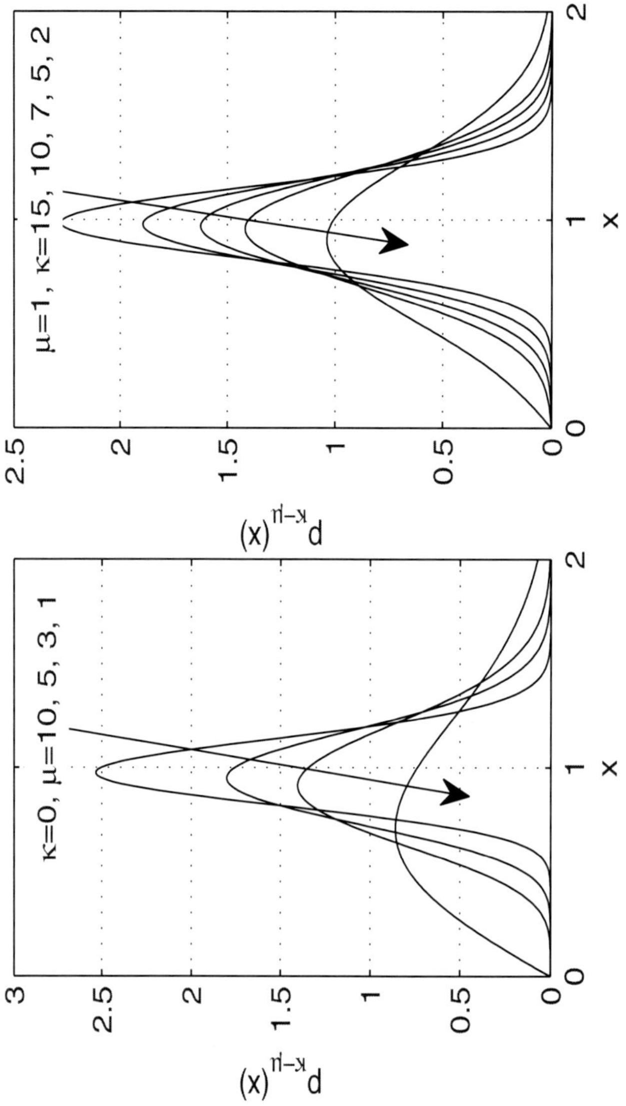

FIGURE 4.4
Figure to show reduction in the variability of fading coefficients with increasing values of κ and/or μ.

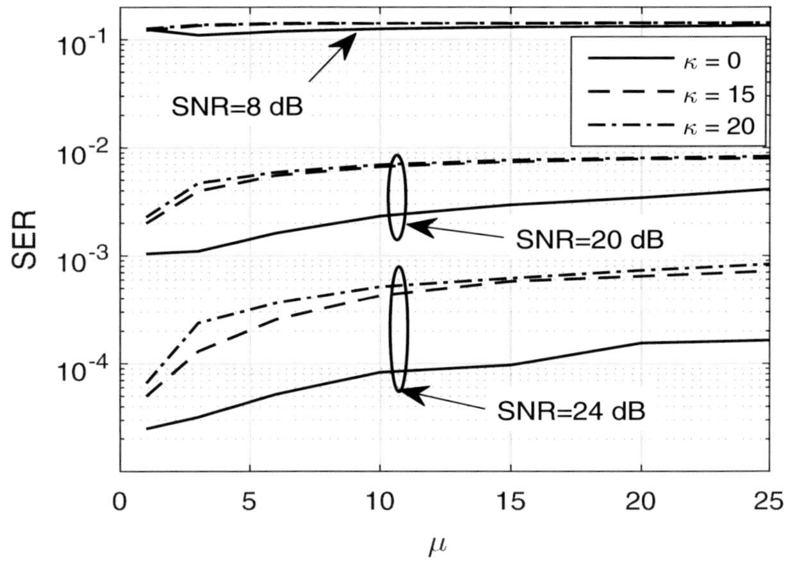

FIGURE 4.5
Variation in SER performance of SM MIMO systems over $\kappa-\mu$ fading channels with varying μ for different values of κ.

considered in the range from 0.5 to 25. The results are plotted for three values of SNR, 8 dB, 20 dB and 24 dB. The only exceptional case observed is for 8 dB SNR and $\kappa = 0$. In this case, SER shows a small improvement for the initial increase in the value of μ and then degrades slightly. This is because of dominance of noise over fading conditions for low SNR regime. Similar behavior has been observed in the case of Nakagami-m fading channels and spatially multiplexed MIMO channels too, while for other cases, a degradation is observed with increasing values of μ. The degradation is more for initial increase in the values of μ and then SER becomes almost constant specially for strong LOS cases (say for $\mu > 15$ for SNR = 20 dB).

4.6 Error performance of SM systems over $\eta - \mu$ fading channels

In the case of Nakagami-m fading channels, we have discussed the method to obtain required PDF for evaluation of average pairwise error probability and the overall error performance of SM MIMO systems. Following the same steps, an analytical expression for error performance of SM MIMO systems

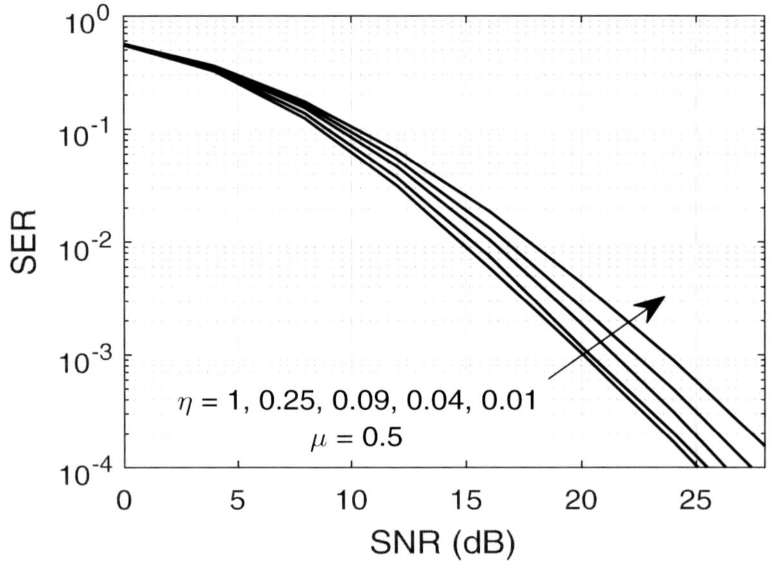

FIGURE 4.6
SER performance of 4×2 SM MIMO systems over $\eta - \mu$ fading channels with QPSK modulation for $\mu = 0.5$.

over $\eta - \mu$ fading channels can be derived. In this section, we will discuss the effect of fading parameters, η and μ, on the SER performance of SM MIMO systems. MATLAB script file for simulation of SM MIMO systems is given at the end of this book in Appendix B. It can be used to obtain SER results of SM MIMO systems over various fading channels using appropriate function to generate fading channel coefficients.

In this section, we will consider 2×2 SM MIMO systems with QPSK modulation scheme for different values of fading parameters, η and μ. We consider different values of η in the range from 0.01 to 1 and three values of μ, $\mu = 0.5$, $\mu = 2.5$ and $\mu = 5$. The results for three cases are shown respectively in Figure 4.6, Figure 4.7 and Figure 4.8. The variation in SER performance for decreasing values of η is found different for $\mu = 0.5$ than that in the cases of $\mu = 2.5$ and $\mu = 5$. $\mu = 0.5$ represents the Hoyt fading channels as a special case of $\eta - \mu$ fading. In this case (Figure 4.7 and Figure 4.8), it is observed that the SER performance degrades with increasing values of fading parameter η. Actually, increasing values of η for $\eta < 1$ represent less severe fading and in general, the SER performance shall be better for less severe fading. But, in SM MIMO systems, the antenna index corresponding to the transmitter antenna is also a part of the information. Estimation of these spatially modulated information bits requires severe fading to detect the information more accurately. Thus, as in the case of spatial multiplexing MIMO systems,

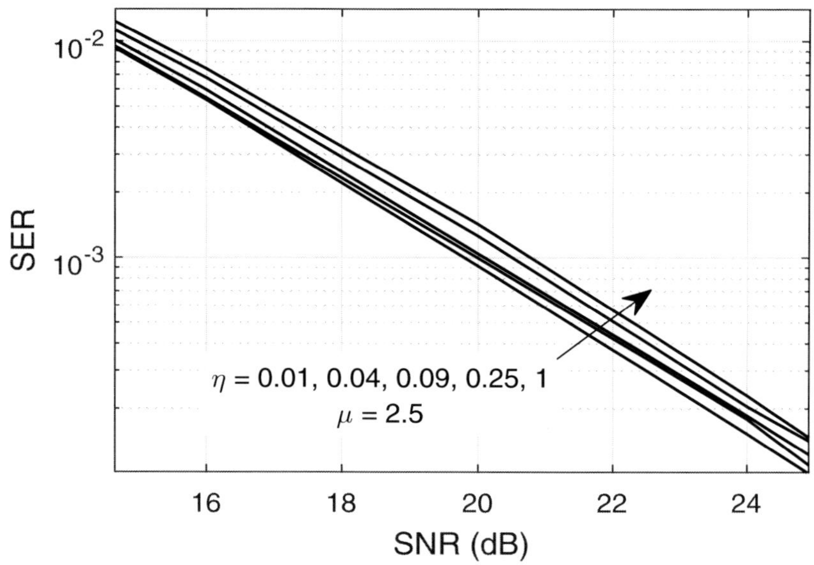

FIGURE 4.7
SER performance of 4×2 SM MIMO systems over $\eta - \mu$ fading channels with QPSK modulation for $\mu = 2.5$

dominance of noise over fading environment plays a role that defines variation of SER with respect to changing values of fading parameters. In the case of $\mu = 0.5$, the fading is severe and detection of a digitally modulated symbol is dominant over the detection of antenna index. Thus, SER performance of SM MIMO systems degrades for lower values of η, while in the case of $\mu = 2.5$ and $\mu = 5$, the situation is just opposite and hence the SER performance improves with lower values of η. In addition to that, if we observe the SER for $\eta = 1$ at some SNR (say 20 dB) in all three figures, it can be observed that the SER performance degrades for increasing values of μ. This is again because of lower variation in the fading coefficients for higher values of μ.

4.7 Error performance of SM systems over $\alpha - \mu$ fading channels

Non-linear fading environments are modeled well by $\alpha - \mu$ distribution. So far in this chapter, we have discussed the performance of SM MIMO systems over fading channels including LOS and non-LOS fading environments. In this section, we will discuss the SER performance of SM MIMO systems over $\alpha - \mu$

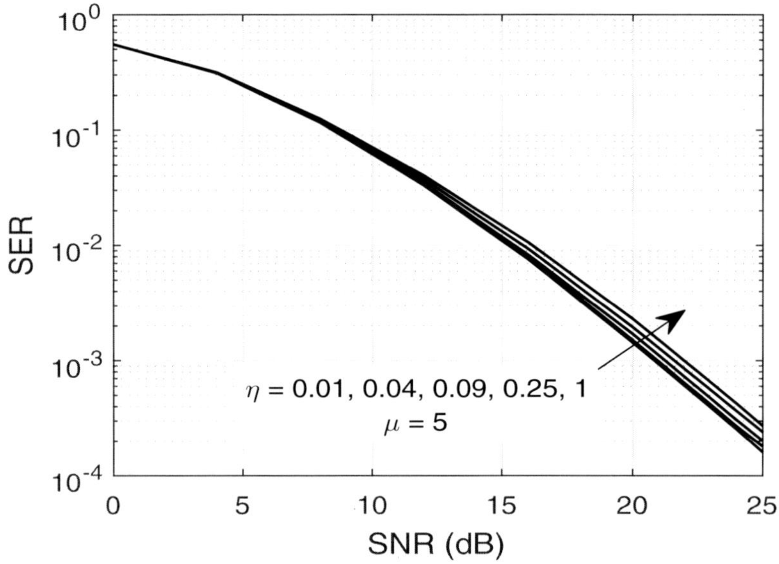

FIGURE 4.8
SER performance of 4×2 SM MIMO systems over $\eta - \mu$ fading channels with QPSK modulation for $\mu = 5$.

fading channels in which the fading parameter, α-accounts for non-linearity of fading environment. So, this section will deal with the discussion on effects of non-linearity in the fading environment on the SER performance of SM MIMO systems.

To discuss SER performance of SM MIMO systems and observe the effect of non-linearities in the fading environment, a 4×2 SM MIMO system is simulated with 8-PSK modulation scheme. The SER results are shown for $\mu = 2.5$ in Figure 4.9 for 12 dB, 18 dB and 22 dB SNRs. An initial improvement is observed in the SER performance with increasing values of α followed by degradation in the SER for a given value of SNR. This indicates that there is an optimum non-linearity for which SER is the minimum. Unlike in the case of spatially multiplexed MIMO systems, the optimum value of α for minimum SER value is almost constant and lies somewhere around $\alpha \simeq 1.5$. This is because both the extreme cases, $\alpha \to 0$ and $\alpha \to \infty$, approach the cases such that the channel is deterministic or a non-fading case.

The effect of μ on the SER performance of SM MIMO systems has been discussed for the case of $\eta - \mu$ and $\kappa - \mu$ fading channels. And the way μ is defined is the same even for $\alpha - \mu$ fading distribution. Thus, a similar observation is seen in Figure 4.10. However, in this case, we have explicitly marked the points of minimum SER for different SNR values. It is observed that the optimum value of μ for minimum SER is $\simeq$1.5. Again, the variation

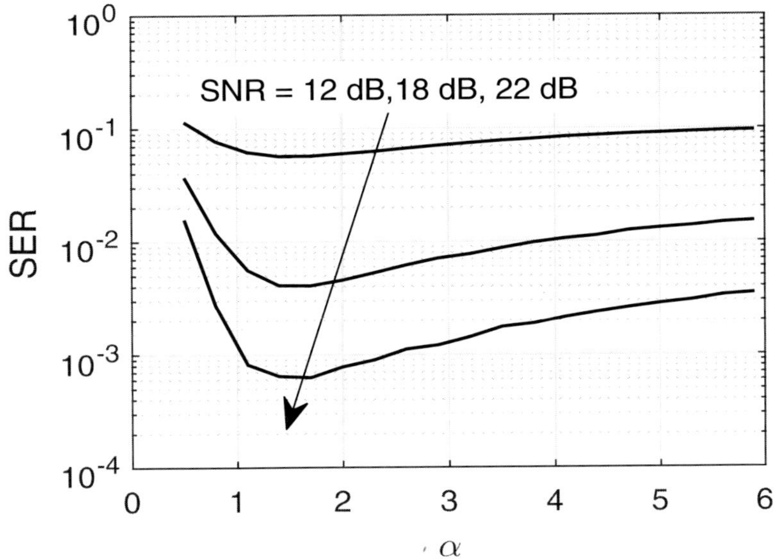

FIGURE 4.9
Variation in SER of SM MIMO systems over $\alpha-\mu$ fading channels with fading parameter, α.

in SER with the values of μ is as expected. Initially, with increasing values of μ, the SER performance improves and it degrades on further increase in the values of μ. Note that for our observations in Figure 4.10, we kept $\alpha = 2$ so that the same results and observations are valid for $\kappa-\mu$, $\eta-\mu$ and Nakagami-m fading channels but with appropriate multiplying factor on the X-axis, i.e., the scaling of μ on account of differences in their physical model.

Further, we will discuss some aspects of generalized spatial modulation which is an advanced form of spatial modulation.

4.8 Generalized spatial modulation

GSM is proposed to overcome the condition that the number of transmitter antennas has to be an integer exponent of 2. In GSM, multiple antennas are used to transmit a digitally modulated symbol. Different combinations of antennas at the transmitter side are mapped to spatial constellation points. So, in GSM, with added complexity of hardware and computations at receiver, more gain in spectral efficiency is obtained. The transmission scheme in GSM is similar to the transmission methodology of SM MIMO systems.

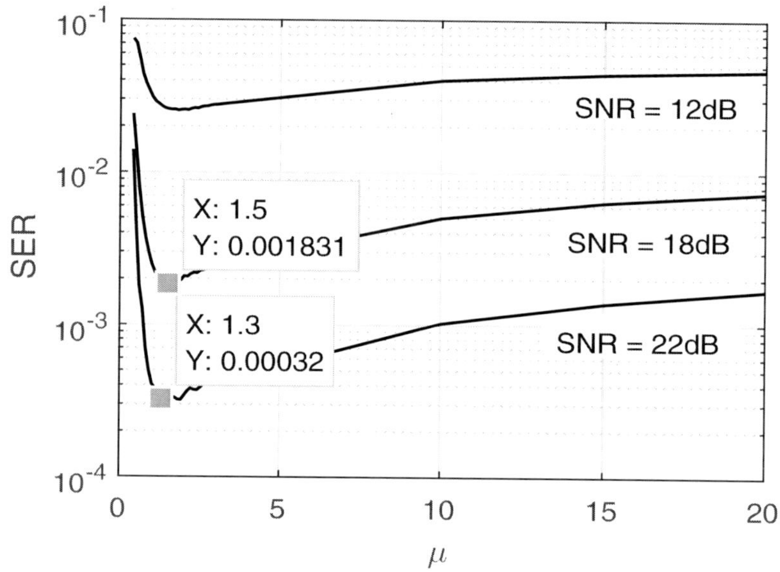

FIGURE 4.10
Variation in SER of SM MIMO systems over $\alpha-\mu$ fading channels with fading parameter, μ.

The difference is about mapping of spatial constellation points. In GSM, each spatial constellation point is mapped to a combination of antennas available at transmitter instead of single antenna in SM. Since multiple antennas are transmitting simultaneously, the number of RF chains required in GSM is more than that required in SM. SM required a single RF chain.

In general, for MIMO systems with N_t antennas at the transmitter and N_{rf} number of RF chains, the total number of possible ways in which antennas can be selected for transmission becomes ${}^{N_t}C_{N_{rf}}$. Let's take an example of the transmission strategy for GSM. Consider a MIMO system with five antennas at the transmitter and assume that the transmitter has two RF chains, i.e., the transmission can be done from two transmitting antennas simultaneously. In this case, the total number of possible combinations of two antennas can be ${}^5C_2 = 10$. But, not all 10 combinations of antennas can be used for transmission. A particular combination of antennas is selected based on the incoming sequence. So, as the required number of antennas at transmitter has to be an integer exponent of 2 in SM, in GSM it is the number of combinations of antennas that has to be an integer exponent of 2. Otherwise, all possible combinations may not be used for transmission. In the case under consideration, we have 10 possible ways of selecting two antennas out of five total antennas. So, only eight of the combinations can be used for transmission. Again, there are algorithms to decide which combinations to use and which to omit. Once the antenna combinations are selected for transmission, there can be two transmission strategies. One is to achieve better diversity and the other is for better spectral efficiency. In the technique with better diversity order, the same symbol from signal constellation is transmitted from all the selected antennas while different symbols from signal constellation are transmitted from different antennas to improve the spectral efficiency.

5

Transmit Antenna Selection

CONTENTS

5.1 Introduction

Transmit antenna selection (TAS) systems have become popular mainly for their reduced complexity of the system hardware and detection complexity. We already discussed in Chapter 1 that in TAS systems, a single antenna is made active for transmission at a given instant. This has advantages like requirement of a single radio frequency (RF) chain, no need of inter antenna synchronization, inherent avoidance of inter antenna interference, etc. The hardware at the transmitter is simplified to a great extent due to the need of a single RF chain which is very bulky hardware consisting of the setup of amplifiers, ADC/DACs, filters, modulators, etc. This reduction in hardware complexity is especially important as far as mobile devices are concerned. Though VLSI technologies have been so advanced that most hardware is in the form of integrated circuits with considerably small sizes and low weights, even then at the RF level the number of RF chains makes the hardware very bulky for the mobile devices. However, the bulky hardware may not be the problem of that big concern at the base stations or the immobile control units and the devices which do not belong to the family of hand-held devices.

TAS with maximal ratio combining at the receiver (TAS/MRC) has been widely studied by the researchers across the globe. Especially, with advancement of full duplex multiple-input multiple-output (MIMO) systems and proposal of large MIMO systems [19], TAS has been accepted as the promising technology for implementation in large MIMO systems to achieve reduced hardware and computational complexity [64]. The computational complexity of TAS systems for detection at the receiver is the same as that for the receiver diversity combining schemes. Apart from the detection complexity, TAS systems have slightly more computational complexity which accounts for the identification of the antenna which is to be selected at the transmitter. However, this added complexity pays in terms of diversity gain. TAS systems are capable of achieving full diversity order. Performance of TAS systems has already been studied over various fading channels including Nakagami-m fading channels, $\kappa - \mu$ fading channels and $\eta - \mu$ fading channels, also taking the cases of multiuser environment and correlated fading channels [14, 15, 24, 25, 49, 56, 74, 79, 95, 96, 126, 129, 138].

In this chapter, we will discuss TAS systems in detail with the explanations on spectral efficiency, diversity order and detection complexity. Further, performance of TAS systems with selection combining (SC) and MRC at receiver over various generalized fading channels will be discussed. In this book, we will mainly present the results and discussion on bit error rate/symbol error rate (BER/SER) of TAS systems. Outage probability and ergodic capacity of TAS systems over generalized fading channels will also be discussed.

5.2 Antenna selection criteria

The motivation behind transmit antenna selection in MIMO systems is to achieve reduction in the hardware complexity of the system and computational complexity of the detector/receiver. We already discussed that this can be achieved by selecting a reduced number of antennas at the transmitter for parallel transmission of bitstreams. This will reduce the hardware requirement. This also simplifies the detection process. Two commonly used metrics for antenna selection at the transmitter are channel capacity and received signal power. Basically, the selection is based on maximization of the selected metric. Maximization of received signal power is directly related to the instantaneous signal-to-noise (SNR). Selection of a subset of total available antennas can be done at both the ends, the transmitter and the receiver or on either side. But, in this chapter, the emphasis is on antenna selection at the transmitter. The concern raised by researchers for antenna selection at the transmitter is the requirement of feedback link either to transmit back the information about the channel coefficients or to transmit back the information about the best antenna/the best set of antennas as per the chosen metric to be maximized. However, to our belief, feedback link is of less concern for full duplex systems where both-way communication happens simultaneously. So, sending the required information for antenna selection is not that difficult.

5.2.1 Received power based antenna selection

For all our discussions, it is assumed that the energy per transmitted symbol remains the same as long as the transmission continues. Also, it is assumed that even if multiple antennas are selected at a time, all the selected transmitter antennas are assigned equal power. Let's assume that N antennas are to be selected at the transmitter in an $N_t \times N_r$ MIMO system. This reduces the original channel matrix of dimension $N_r \times N_t$ to $N_r \times N$. In this case, the antenna selection is done based on the combination that gives maximum received energy per transmitted symbol. Pilot based single antenna selection has been discussed in Chapter 1 (see Section 1.4.2). Here, we will discuss the criteria for general antenna selection including multiple antenna selections. In this case, the received signal becomes

$$\mathbf{r} = \mathbf{H_r x} + \mathbf{n} \tag{5.1}$$

where $\mathbf{H_r}$ is the channel matrix with reduced dimensions and $\mathbf{x}$ is the pilot signal vector transmitted for antenna selection. For each of the transmissions, received energy is calculated and the combination of transmitter antennas for which the maximum energy is received is selected for communication. In this case, the number of combinations of transmitter antennas would be ${}^{N_t}C_N$, as the TAS system is assumed to select only N antennas out of N_t antennas at the transmitter. Energy received per transmission can be given by $E_r = \mathbf{r^H r}$. This received energy is primarily dependent on $\mathbf{H_r^H H_r}$. This can be shown by considering the case of a noise-free condition. In a noise-free case, the received signal energy per transmission can be given by

$$\begin{aligned} E_r &= (\mathbf{H_r x})^H \mathbf{H_r x} \\ &= \mathbf{x^H H_r}^{\mathbf{H}} \mathbf{H_r x} \end{aligned} \tag{5.2}$$

where the pilot signal vector $\mathbf{x}$ is known at the receiver. This process facilitates the channel matrix estimation in addition to identification of the best combination of transmitter antennas.

Let $\mathbb{C}$ be the set of all possible reduced channel matrices. The antenna selection criteria can be represented as

$$I = \underset{\mathbf{H_{r,i}} \in \mathbb{C}}{\arg\max} \left\{ E_{r,i} = \mathbf{x^H H_{r,i}}^{\mathbf{H}} \mathbf{H_{r,i} x} \right\} \tag{5.3}$$

where I represents the combination of transmitter antennas for which the received energy per transmission is the highest among all possible combinations of N antennas out of a total N_t antennas. This antenna selection criteria can be further simplified for single antenna selection at the transmitter. In this case, the number of combinations of transmitter antennas would be ${}^{N_t}C_1 = N_t$ as the TAS system is now assumed to select single antenna out of N_t antennas at the transmitter. The pilot vector $\mathbf{x}$ would be reduced to a scaler pilot to be transmitted from a single antenna at a time. Thus, the antenna selection criteria on expression in equation (5.3) can be given as

$$I = \underset{\mathbf{H_{r,i}} \in \mathbb{C}}{\arg\max} \left\{ E_{r,i} = x^* \mathbf{H_{r,i}}^H \mathbf{H_{r,i}} x \right\} \tag{5.4}$$

Also, the reduced channel matrix $\mathbf{H_r}$ becomes a vector which is a column of the channel matrix corresponding to the selected transmitting antenna. So, the antenna selection criteria get further simplified. Now, it can be given as follows.

$$\begin{aligned} I &= \underset{1 \le i \le N_t}{\arg\max} \left\{ E_{r,i} = |x|^2 \mathbf{h_i}^H \mathbf{h_i} \right\} \\ &= \underset{1 \le i \le N_t}{\arg\max} \left\{ E_{r,i} = \sum_{j=1}^{j=N_r} |h_{j,i}|^2 \right\} \end{aligned} \tag{5.5}$$

where $\mathbf{h_i}$ is the ith column of channel matrix. The above expression is the same as what we discussed in Chapter 1 for antenna selection criteria with MRC at receiver (see expression of equation (1.38)).

5.2.1.1 Norm based transmit antenna selection

The pilot based antenna selection needs to transmit pilot sequence for each instant of antenna selection. In addition, pilot transmission is also required for channel matrix estimation. The information of channel coefficients is nonetheless required at the receiver while detecting the received symbol. So, to reduce the transmission overhead and computations, antenna selection is done using the channel matrix. This method is referred to as a norm based method for TAS [22, 139]. It uses the norm of columns of channel matrix as a parameter for antenna selection. This involves calculation of norms of each column of the channel matrix, then these norms are arranged in ascending order and the antennas that correspond to the highest values of norms are selected for transmission. The technique of norm based antenna selection too maximizes the received signal power. But in this case, the power received from transmission by a single antenna at a time is considered separately while the pilot based selection discussed earlier maximizes the power received when all the antennas to be selected are active. It shall be noted that for noiseless systems, the pilot based antenna selection criteria for single TAS systems are exactly the same as the norm based antenna selection. In general, selecting a single antenna gives maximum benefit in terms of reduction in hardware complexity and reduction in computational complexity. So, norm based antenna selection has become popular among the researchers. But with various transmission strategies in MIMO systems, different algorithms may be used for antenna selections.

5.2.2 Capacity based antenna selection

Apart from received signal energy, the other criterion for antenna selection is the channel capacity. Selecting antennas using maximum received energy does not necessarily guarantee the highest channel capacity. As far as quality of service (QoS) of communication systems is concerned, channel capacity is an important parameter. The MIMO channel capacity when all transmit and receive antennas are in operation is given by

$$C_{MIMO} = \log_2 \det \left(\mathbf{I}_{N_t} + \gamma \mathbf{H^H H} \right) \tag{5.6}$$

where $\mathbf{I_{N_t}}$ is the $N_t \times N_t$ identity matrix and γ is the instantaneous SNR.

But when antenna selection is done either at transmitter or receiver, the matrix size gets reduced accordingly. In this case, we consider antenna selection at the transmitter. While doing transmit antenna selection, we get reduction in the number of columns of the channel matrix. As in the case of received power based antenna selection (Section 5.2.1), we consider N antennas to be selected at the transmitter out of a total of N_t transmitter antennas. So, the number of possible ways in which antennas can be selected is given as ${}^{N_t}C_N$ and the set of all possible combinations of antennas that may be selected is represented as $\mathbb{C}$. So, for the MIMO system under consideration, the antenna selection criteria that maximize the channel capacity after antenna selection can be given as follows.

$$I = \underset{\mathbf{H_{r,i}} \in \mathbb{C}}{\arg\max} \left\{ C_i = \log_2 \det \left(\mathbf{I}_N + \gamma \mathbf{H^H_{r,i} H_{r,i}} \right) \right\} \tag{5.7}$$

where $\mathbf{I_N}$, the $N \times N$ identity matrix for N antennas, are selected at the transmitter. The above expression represents generalized criteria for antenna selection to select any number of antennas at the transmitter, $N < N_t$. However, as discussed in earlier sections/chapters of this book, there are practical challenges with MIMO systems when more than one antenna are selected and transmitting simultaneously. This includes inter antenna interference, requirements of transmit antenna synchronization and complexity of detection process. All these problems mainly arise due to simultaneous transmission and can be avoided if simultaneous transmission is avoided. Thus, it is a better idea to select a single antenna at the transmitter. So, we take a special case that single antenna is to be selected at the transmitter in the antenna selection criteria given by the expression of equation (5.7). Again, in this case, $\mathbb{C}$ would contain only N_t elements and the antenna selection criteria can be simplified to

$$I = \underset{1 \leq i \leq N_t}{\arg\max} \left\{ C_i = \log_2 \left(1 + \gamma \mathbf{h_i^H h_i} \right) \right\} \tag{5.8}$$

where $\mathbf{h_i}$ is the ith column of the channel matrix. Further, in the above expression, it can be noted that the capacity, C_i, contains logarithmic function and $1 + \gamma \mathbf{h_i^H h_i}$ as the argument of that logarithmic function. It can be maximized by only maximizing the value of $\mathbf{h_i^H h_i}$ which is the same as maximizing the squared Frobenius norm of ith column of the channel matrix, $\sum_{j=1}^{N_r} |h_{j,i}|^2$, i.e., the antenna selection criteria reduce to that given in expression of equation (5.5). This leads to an interesting fact that if a single transmit antenna is selected at the transmitter using maximum received signal power, it automatically maximizes the channel capacity. This may not be always true when multiple antennas are selected at the transmitter.

5.3 TAS system model

A brief description of the TAS system model and the transmission scheme is presented in Section 1.4.2 of Chapter 1. The antenna selection criteria for maximum received useful power based TAS are explained by the expressions of equation (1.37) and equation (1.38) for selection combining and MRC at the receiver respectively. The frame structure for the channel estimation and identification of the antenna for maximum useful power based antenna selection at transmitter is given in Figure 1.9. The frame based pilot transmission for TAS systems explained in Chapter 1 does not involve any role of channel matrix in selection of a transmit antenna. However, the antenna selection criteria for selecting a single antenna eventually depends only on the channel coefficients and not on the symbols or sequence of symbols to be transmitted if the system noise is ignored as shown in equation (5.5). In actuality, transmitting a predefined sequence from each antenna one by one and selecting a

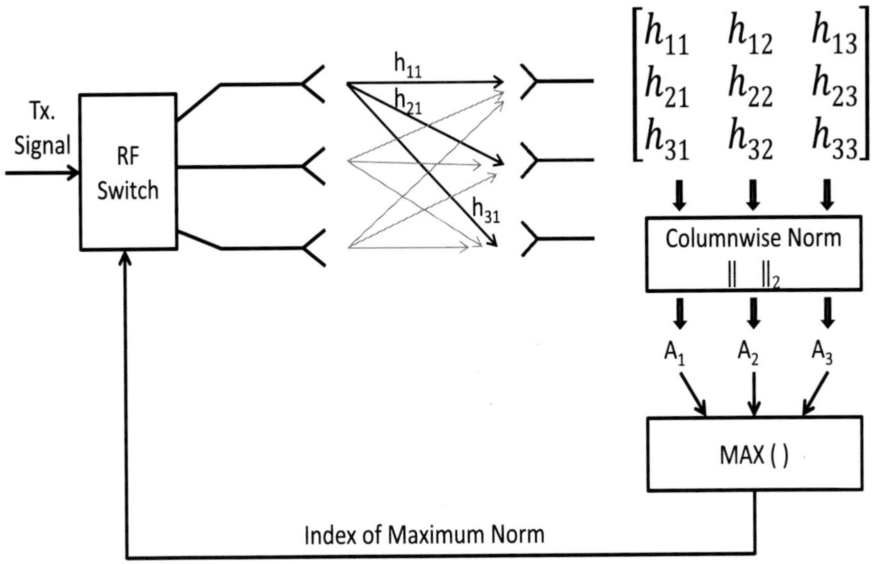

FIGURE 5.1
Norm based TAS system model.

transmit antenna selects the antenna including the effect of noise (additive white Gaussian noise [AWGN] in our case) which necessarily does not maximize the useful signal power all the time. So, most of the times, the decision about the antenna that gives the maximum useful power is based on the norm of the columns of channel matrix, **H**. This method of antenna selection is also referred to as norm based transmit antenna selection by researchers [22, 139]. This method is in no way different from pilot based TAS. A system model for norm based single transmit antenna selection is shown in Figure 5.1. This process begins with the estimation of the channel matrix. The channel matrix is estimated based on pilot signal transmissions. Once the channel matrix is estimated, the norm or squared norm is evaluated for each column. Note that the columnwise norm is calculated because each column of the channel matrix corresponds to a transmitter antenna. All the norms are analyzed and the column corresponding to maximum norm is identified. The information about the antenna corresponding to the column with maximum norm is conveyed to the transmitter via a low rate feedback link from receiver to transmitter. Calculation of the norm of a column of the channel matrix indicates that the diversity combining scheme used at the receiver is MRC. It is because, in MRC, the optimal weights are such that the received signal is the norm times the transmitted symbol after combining. TAS systems with MRC at the receiver are referred to as TAS/MRC systems. In fact, such systems fall under the category of hybrid diversity systems using selection combining at the transmitter and maximal ratio combining at the receiver. If the receiver is designed to use SC, the antennas corresponding to the channel coefficient with the highest magnitude are selected instead of the maximum norm. The column to which the channel coefficient with the highest magnitude belongs gives the transmitter antenna to be selected and the row to which the channel coefficient with the highest magnitude belongs gives the receiver antenna to be selected. TAS systems with SC at the receiver are referred to as TAS/SC systems or joint transmitter and receiver antenna selection systems. Using the information received about the antenna that gives maximum power at the receiver is used at the transmitter and information is transmitted using that particular antenna. Single antenna is used for data transmission at a time. So, the effective channel matrix contains single column and N_r rows, i.e., an $N_r \times 1$ vector.

MRC at the receiver

For the time being, let us assume that MRC is performed at the receiver. The received signal may be given as

$$\mathbf{r} = \mathbf{h}_I x_q + \mathbf{n} \tag{5.9}$$

where $\mathbf{r}$ is the received signal vector of dimension $N_r \times 1$, $\mathbf{h_I}$ is the channel vector corresponding to the antenna selected at the transmitter, x_q is the transmitted symbol which is digitally modulated using suitable M-ary digital modulation scheme and $\mathbf{n}$ is the noise vector of dimension $N_r \times 1$.

The received signal vector is combined using the MRC diversity scheme to get optimum BER/SER performance. The symbol detection rule for MRC with maximum likelihood (ML) detection may be obtained by minimum Euclidean distance between the received symbol and an element in the set of all possible symbols in the M-ary modulation scheme. It can be given by

$$x_{\hat{q}} = \arg\min_{x_{\hat{q}} \in \mathfrak{S}} \{\| \mathbf{r} - \mathbf{h}_I x_{\hat{q}} \|^2\} \tag{5.10}$$

where $\mathfrak{S}$ is the set of all possible symbols at the transmitter. Note that this detection process is the same as the detection in the MRC diversity scheme. Hence, the detection complexity of TAS/MRC MIMO systems is the same as that of the receiver diversity systems, i.e., single-input multiple-output (SIMO) systems. So, by using single TAS, apart from the arrangement of multiple antennas and an RF switch, the hardware complexity at the transmitter is almost equivalent to that of a single antenna system. Moreover, the other challenges of MIMO systems (inter antenna interference, requirement of transmit antenna synchronization, etc.) are also avoided.

Selection combining at the receiver

In SC, a single antenna is selected at the receiver. The receiver selects the antenna that receives the maximum instantaneous power. Again, the selection is done by connecting an antenna to the receiver using an RF switch. This has the advantage of hardware compactness at the receiver. Single TAS and single receiver antenna selection (RAS) convert the MIMO system into a single-input single-output (SISO) system while in operation. In this case, the link corresponding to the transmitter antenna and receiver antenna with the highest magnitude of channel coefficient is selected.

Now, in the next sections of this chapter, we will discuss performance analysis of TAS/MRC systems over various generalized fading channels.

5.4 TAS over Nakagami-m fading channels

The discussion of Nakagami-m distribution and the fading channel models is already presented in Section 2.4.1 of Chapter 2. We will use those discussions and related expressions to evaluate various performance parameters of TAS/MRC MIMO systems in this section. We will mainly focus on the outage probability and bit/symbol error rate. In this section, the fading envelope ($|h_{j,i}|$) of the channel coefficients is assumed to be Nakagami-m distributed and all the links from transmitter to receiver are assumed to be independent and identically distributed (i.i.d.). In all the discussions related to performance analysis of TAS/MRC MIMO systems, we will assume $(N_t, 1; N_r)$ systems, i.e., single transmit antenna selection. Since we consider single transmit antenna selection, the antenna selection criteria are based on maximum instantaneous SNR which is the same as that for maximum channel capacity as discussed in Sections 5.2.1 and 5.2.2. After MRC diversity combining, the resulting instantaneous received SNR is given by $\sum_{n=1}^{N_r} \gamma_n = \overline{\gamma} \sum_{j=1}^{N_r} |h_{j,i}|^2$, where γ_n is the instantaneous SNR of the nth branch. The transmitting antenna that maximizes SNR at the receiver can be determined by using equation (5.5). This involves calculation of squared norms for each column of the channel matrix, i.e., $\gamma_{t,i} = \sum\limits_{j=1}^{N_r} |h_{j,i}|^2$ for $1 \leq i \leq N_t$, where $\gamma_{t,i}$ is squared norm of ith column of the channel matrix which is proportional to the instantaneous received SNR after MRC when the ith antenna is selected at the transmitter.

5.4.1 PDF of received SNR after antenna selection

Probability density function (PDF) of received SNR is required to analytically evaluate the performance of wireless communication systems. We will now discuss the steps involved in obtaining the PDF of received SNR after antenna selection. Each $\gamma_{t,i}$ for $i \in [1, 2, ..., N_t]$ obtained from equation (5.5) is arranged in ascending order such that $\gamma_{t,(1)} \leq \gamma_{t,(2)}, \cdots \leq \gamma_{t,(N)}$, where $\gamma_{t,(l)}$ is the random variable obtained after the arrangement in ascending order and $N = N_t$ for TAS/MRC systems. In the TAS/MRC system, we select the transmitting antenna corresponding to the highest received SNR, i.e., $(\gamma_{t,(N_t)})$. The PDF of received SNR in such a system, assuming all ($|h_{j,i}|$)'s to be i.i.d., can be given by [31]

$$p_{\gamma_{t(N)}}(\gamma) = N\{P_{\gamma_t}(\gamma)\}^{N-1} p_{\gamma_t}(\gamma) \tag{5.11}$$

where $P_{\gamma_t}(\cdot)$ is the cumulative distribution function (CDF) of instantaneous SNR after MRC, $\gamma_{t,i}$, and $p_{\gamma_t}(\cdot)$ is the PDF of instantaneous SNR after MRC, $\gamma_{t,i}$.

We have already discussed that the PDF of SNR is the same as the PDF of the square variate of the envelope. In this case, $\gamma_{t,i} = \sum_{j=1}^{j=N_r} |h_{j,i}|^2$ is the sum of N_r square variates of Nakagami-m distributed random variables. So, referring to the physical model of Nakagami-m fading channels in Section 2.4.1, it can be established that the sum of N_r square variates of i.i.d. Nakagami-m random variables is also Nakagami-m distributed but with parameter m replaced by mN_r. So, the PDF of $\gamma_{t,i}$ can be given by

$$p_{\gamma_{t,i}}(\gamma) = \frac{m^m \gamma^{m-1} e^{-m\frac{\gamma}{\overline{\gamma}}}}{\overline{\gamma}^m \Gamma(m)} \tag{5.12}$$

where $\Gamma(\cdot)$ is the Gamma function defined by $\Gamma(n+1) = \int\limits_0^\infty x^n e^{-x} dx$ and $\overline{\gamma}$ is the average SNR after MRC. Note that $\overline{\gamma}$ is the sum of average SNR of individual links when MRC diversity is used.

The CDF of $\gamma_{t,i}$ can be obtained by integrating the above expression. It can be given by

$$P_{\gamma_{t,i}}(\gamma) = \frac{\gamma_{inc}\left(m, m\frac{\gamma}{\overline{\gamma}}\right)}{\Gamma(m)} \tag{5.13}$$

where $\gamma_{inc}(\cdot,\cdot)$ is the lower incomplete Gamma function defined by $\gamma_{inc}(m,y) = \int\limits_0^y e^{-x} x^{m-1} dx$.

Substituting equation (5.12) and equation (5.13) in equation (5.11), the PDF of received instantaneous SNR, $p_{\gamma_{t(N_t)}}(\cdot)$, for real valued fading parameter, m, can be obtained as

$$p_{\gamma_{t(N_t)}}(\gamma) = N_t \left(\frac{\gamma_{inc}\left(m, m\frac{\gamma}{\overline{\gamma}}\right)}{\Gamma(m)} \right)^{N_t - 1} \frac{m^m \gamma^{m-1} e^{-m\frac{\gamma}{\overline{\gamma}}}}{\overline{\gamma}^m \Gamma(m)} \tag{5.14}$$

The incomplete Gamma function can be represented in the form of a finite sum series for integer values of m as follows.

$$\gamma_{inc}(m,y) = m! - e^{-y} \sum_{k=0}^{m} \frac{m!}{k!} y^k \tag{5.15}$$

So, for integer values of m, the PDF of received SNR can be represented as

$$p_{\gamma_{t(N_t)}}(\gamma) = N_t \left(\frac{m! - e^{-m\frac{\gamma}{\overline{\gamma}}} \sum\limits_{k=0}^{m} \frac{m!}{k!} m \frac{\gamma}{\overline{\gamma}}^{k}}{\Gamma(m)} \right)^{N_t - 1} \frac{m^m \gamma^{m-1} e^{-m\frac{\gamma}{\overline{\gamma}}}}{\overline{\gamma}^m \Gamma(m)} \tag{5.16}$$

The above expression looks more complex than equation (5.14) but it is simpler computationally. Moreover, analytical evaluation of the performance parameters of communication systems involves averaging with respect to PDF of instantaneous SNR which encounters integration of complex functions. Using the later expression for the same makes it difficult to evaluate the integrals of the functions containing powers of incomplete Gamma functions. Hence, equation (5.16) can be helpful in some cases. However, it can only be used for integer values of m. In the next few subsections, we will present analytical expressions for performance parameters of wireless MIMO communication systems with TAS over Nakagami-m fading channels.

5.4.2 Outage probability

Wireless mobile communication is highly dependent on the surroundings. Fading is the most common phenomenon that affects the overall communication process. Due to fading, the received power momentarily degrades to levels that the quality of signal is very poor to get detected correctly. Outage probability is the probability that the received signal strength falls below a certain threshold required for reliable communication. It is also defined in terms of instantaneous channel capacity and the rate of transmission. The probability that the instantaneous channel capacity is below the rate of transmission is known as outage probability [14]. For the analysis of outage probability, we use the definition that outage probability is the probability that the instantaneous SNR at the receiver is less than a certain threshold SNR, γ_{th} [118]. Thus, outage probability can be given by

$$P_{out}(\overline{\gamma}, \gamma_{th}) = P\left(\overline{\gamma}\, \|\mathbf{h}\|_F^2 < \gamma_{th}\right) \tag{5.17}$$

where $\mathbf{h}$ is the column of channel matrix corresponding to the antenna selected at the transmitter. So, the outage probability is the CDF of instantaneous received SNR and can be obtained by integrating the PDF represented by equation (5.14).

$$P_{out}(\overline{\gamma}, \gamma_{th}) = \int\limits_{0}^{\gamma_{th}} \left\{ N_t \left(\frac{\gamma_{inc}\left(m, m\frac{\gamma}{\overline{\gamma}}\right)}{\Gamma(m)} \right)^{N_t - 1} \frac{m^m \gamma^{m-1} e^{-m\frac{\gamma}{\overline{\gamma}}}}{\overline{\gamma}^m \Gamma(m)} \right\} d\gamma \tag{5.18}$$

The integral in the above expression is of the form $I = \int \left\{\int f(x)dx\right\}^M f(x)dx$ as it contains the terms of CDF and PDF of the received SNR before antenna selection. So, the following property of integrations can be used to solve the integral.

$$I = \int \left\{\int f(x)dx\right\}^{\beta} f(x)dx = \frac{1}{\beta+1}\left\{\int f(x)dx\right\}^{\beta+1} \tag{5.19}$$

Thus, the above integral can be solved to get a compact closed form expression for outage probability as

$$P_{out}\left(\overline{\gamma}, \gamma_{th}\right) = \left(1 - \frac{\Gamma\left(m, \frac{m\gamma_{th}}{\overline{\gamma}}\right)}{\Gamma\left(m\right)}\right)^{N_t} \tag{5.20}$$

where $\Gamma(\cdot,\cdot)$ is the upper incomplete Gamma function defined by $\Gamma(m,y) = \int_y^{\infty} e^{-x}x^{m-1}dx$.

For integer values of m, the expression for outage probability can be alternatively represented in the form of finite sum series as

$$P_{out}\left(\overline{\gamma}, \gamma_{th}\right) = \left(1 - e^{-\frac{m\gamma_{th}}{\overline{\gamma}}} \sum_{k=0}^{m-1} \frac{\left(\frac{m\gamma_{th}}{\overline{\gamma}}\right)^k}{k!}\right)^{N_t} \tag{5.21}$$

In equation (5.20) and equation (5.21), the fading parameter m is to be interpreted as mN_r for TAS/MRC systems as the SNR is the sum of square variates of N_r Nakagami-m distributed random variables. Using $m = 1$ in equation (5.21), it gives the outage probability of TAS/MRC systems over Rayleigh fading channels which is a special case of Nakagami-m fading model. It can be given as

$$P_{out}\left(\overline{\gamma}, \gamma_{th}\right) = \left(1 - e^{-\frac{N_r\gamma_{th}}{\overline{\gamma}}} \sum_{k=0}^{N_r-1} \frac{\left(\frac{N_r\gamma_{th}}{\overline{\gamma}}\right)^k}{k!}\right)^{N_t} \tag{5.22}$$

The above expression is computationally efficient. It involves N_r addition and two power operations. It is computationally more efficient than equation (9) of [15], which involves a number of addition and power operations of the order N_tN_r. This difference will become more significant when we consider large MIMO systems [19] with hundreds of antennas at the transmitter and receiver.

Alternatively, the integral of equation (5.18) can be evaluated by using infinite series expansion of incomplete Gamma function as in [14]. But, it gives an expression in terms of sum of infinite series. However, it becomes easy to obtain asymptotic outage probability (outage probability for $\overline{\gamma} \to \infty$) from the infinite series solution. To obtain asymptotic outage probability, we rearrange equation (5.20) and represent in terms of lower incomplete Gamma function as

$$P_{out}\left(\overline{\gamma}, \gamma_{th}\right) = \left(\frac{\gamma_{inc}\left(m, \frac{m\gamma_{th}}{\overline{\gamma}}\right)}{\Gamma\left(m\right)} \right)^{N_t} \tag{5.23}$$

Further, the following set of expressions can be used to bring the lower incomplete Gamma function in the form of sum of infinite series.

$$\gamma_{inc}\left(s, x\right) = \frac{x^s}{s} \Phi\left(s, s+1, -x\right) \tag{5.24}$$

$$\Phi\left(a, b, -x\right) = e^{-x} \Phi\left(b-a, b, x\right) \tag{5.25}$$

$$\Phi\left(a, b, x\right) = \sum_{k=0}^{\infty} \frac{\left(a\right)_k x^k}{\left(b\right)_k k!} \tag{5.26}$$

The above expressions are (9.236.4), (9.212.1) and (9.210.1) of [46]. Now, lower incomplete Gamma function can be represented in the series form as

$$\gamma_{inc}\left(s, x\right) = \frac{e^{-x}}{s} \sum_{k=0}^{\infty} \frac{x^{s+k}}{\left(s+1\right)_k} \tag{5.27}$$

where $(a)_n$ is Pochhammer symbol defined as $(a)_n = \frac{\Gamma(a+n)}{\Gamma(a)}$.

For high values of average SNR, it is valid to truncate the series only to the first term. Thus, asymptotic outage probability can be given by

$$P_{out}^{asympt}\left(\overline{\gamma}, \gamma_{th}\right) = \frac{e^{-\frac{mN_t\gamma_{th}}{\bar{\gamma}}} \left(\frac{m\gamma_{th}}{\bar{\gamma}}\right)^{mN_t}}{\left(m\Gamma\left(m\right)\right)^{N_t}} \tag{5.28}$$

$$= Ce^{-\frac{mN_rN_t\gamma_{th}}{\bar{\gamma}}} \left(\overline{\gamma}\right)^{-mN_rN_t} \tag{5.29}$$

where C is the constant factor containing all the terms independent of the average SNR. Note that while going from equation (5.28) to equation (5.29), we used the fact that the resulting SNR before antenna selection is Nakagami-m distributed with fading parameter mN_r. It can be observed that the outage probability varies as $\left(\overline{\gamma}\right)^{-mN_rN_t}$. This shows that the diversity order of the system is mN_rN_t.

In Figure 5.2, the analytical and simulation results of outage probability are shown for a (3,1;2) TAS/MRC system for a Nakagami-m fading case. The analytical and simulation results are plotted for m = 0.5, 1 and 1.5 for R = 2 bits/s/Hz. It can be found that the results are the same as shown in [14, Figure 1]. It includes the results of outage probability over Rayleigh fading channels which is a special case of Nakagami-m fading channels for $m = 1$.

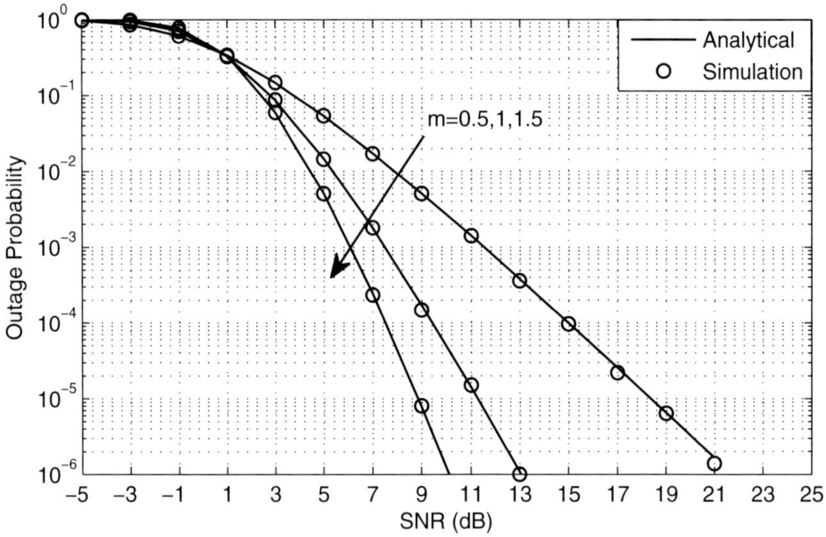

FIGURE 5.2
Outage probability of (3,1;2) TAS/MRC system over Nakagami-m fading channels for $\gamma_{th} = 3$.

5.4.3 Error performance analysis

Error performance is the most important parameter for QoS of wireless communication systems. In this section, we will discuss the method to find expressions for exact SER of TAS/MRC systems for binary phase shift keying (BPSK), binary frequency shift keying (BFSK) and M-ary pulse amplitude modulation (MPAM) schemes and approximate SER of TAS/MRC systems for M-ary quadrature amplitude modulation (MQAM), M-ary phase shift keying (MPSK) and quadrature phase shift keying (QPSK) schemes. BER of a wireless communication system can be calculated by averaging the conditional probability of error (CPE) over the PDF of output SNR. CPE is the error probability in AWGN channels without fading, i.e., when all the channel coefficients are unity. But the random nature of fading coefficients makes CPE too a random variable. So, averaging of CPE is required to get the SER of a wireless communication system. We are not aiming to discuss the ways to get the expression for CPE in this book. The curious readers may follow references [116], [101] or [45] for the same. The exact CPE for various modulation schemes and approximate CPE for MPSK modulation scheme can be given by

$$P_e(\gamma) = aQ\left(\sqrt{b\gamma}\right) - cQ^2\left(\sqrt{b\gamma}\right) \tag{5.30}$$

where $Q(\cdot)$ is the Gaussian Q function which gives the area under the tail of a Gaussian curve and is defined as $Q(\hbar) = \frac{1}{\sqrt{2\pi}} \int_{\hbar}^{\infty} e^{-\left(\frac{u^2}{2}\right)} du$, γ is the instantaneous received SNR and a, b and c are modulation dependent parameters listed in Table A.1.

The SER can be given by

$$\overline{P_e} = \int_{0}^{\infty} P_e(\gamma)\, p_{\gamma_{t(N_t)}}(\gamma)\, d\gamma \tag{5.31}$$

An alternate form of the SER can be obtained by evaluating the above integral by parts as the integrand contains the product of two functions. After doing integration by parts

$$\overline{P_e} = P_e(\gamma)\, P_{\gamma_{t(N)}}(\gamma) - \int_{0}^{\infty} P_e^{'}(\gamma)\, P_{\gamma_{t(N)}}(\gamma)\, d\gamma \tag{5.32}$$

where $P_e^{'}(\cdot)$ is the first derivative of CPE and $P_{\gamma_{t(N)}}(\gamma)$ is the CDF of output SNR. When the first term of the above expression is evaluated at 0 and ∞, the result is zero because $P_e(\gamma) = 0$ at $\gamma \to \infty$ and $P_{\gamma_{t(N)}}(\gamma) = 0$ at $\gamma = 0$. So, eventually, the SER can be obtained as

$$\overline{P_e} = \int_{0}^{\infty} P_e(\gamma)\, p_{\gamma_{t(N)}}(\gamma)\, d\gamma = -\int_{0}^{\infty} P_e^{'}(\gamma)\, P_{\gamma_{t(N)}}(\gamma)\, d\gamma \tag{5.33}$$

So, the same result is obtained either by averaging the conditional probability of error over the PDF of received SNR or by averaging the derivative of conditional probability of error over the CDF of received SNR. By now, we already have the expressions for the PDF of received SNR, equation (5.14) and the CDF of received SNR, equation (5.20).

5.4.3.1 BPSK and MPSK modulation schemes

From equation (5.30) and Table A.1, the CPE for BPSK, MPSK and MPAM can be given in the form $P_e(\gamma) = aQ\left(\sqrt{b\gamma}\right)$ as $c = 0$ and the values of other modulation parameters are $a = 1$, and $b = 2$ for BPSK, $a = 2$, and $b = \sin(\pi/M)$ for MPSK and appropriate values for MPAM modulation schemes, respectively. So, using equation (5.30) and equation (5.23) in (5.33), SER can be evaluated from the following two expressions.

$$\overline{P_e} = \frac{a}{2}\sqrt{\frac{b}{2\pi}} \int_{0}^{\infty} \frac{e^{-\frac{b\gamma}{2}}}{\sqrt{\gamma}} \left(\frac{\gamma_{inc}\left(m, \frac{m\gamma}{\overline{\gamma}}\right)}{\Gamma(m)} \right)^{N_t} d\gamma \tag{5.34}$$

$$\overline{P_e} = \int_0^\infty aQ\left(\sqrt{b\gamma}\right) N_t \left(\frac{\gamma_{inc}\left(m, m\frac{\gamma}{\bar{\gamma}}\right)}{\Gamma(m)} \right)^{N_t-1} \frac{m^m \gamma^{m-1} e^{-m\frac{\gamma}{\bar{\gamma}}}}{\bar{\gamma}^m \Gamma(m)} d\gamma \tag{5.35}$$

Note that Equation (5.34) is used to evaluate the average probability of error using derivative (differentiation) of conditional probability and the CDF of received SNR. The integrand is the product of power of lower incomplete Gamma function and exponential function. Also, equation (5.35) averages CPE with respect to the PDF of received SNR. The integrand contains product of exponential, power of incomplete Gamma function and algebraic functions. So, the case of high SNR or asymptotic SER is considered in this case to simplify the analysis. If one intends to find exact SER, numerical methods for integration or obtaining result in the form of sum of infinite series may be used.

To solve the integrals of equation (5.34) and/or equation (5.35), we need to represent the incomplete gamma function as a sum of infinite series as in equation (5.27). So, the PDF and CDF of received SNR can respectively be given as

$$p_{\gamma_{t(N_t)}}(\gamma) = N_t \left(\frac{e^{-\frac{m\gamma}{\bar{\gamma}}} \sum_{k=0}^{\infty} \frac{\left(\frac{m\gamma}{\bar{\gamma}}\right)^{m+k}}{(m+1)_k}}{m\Gamma(m)} \right)^{N_t-1} \frac{m^m \gamma^{m-1} e^{-m\frac{\gamma}{\bar{\gamma}}}}{\bar{\gamma}^m \Gamma(m)} \tag{5.36}$$

$$P_{\gamma_{t(N_t)}}(\gamma) = \left(\frac{e^{-\frac{m\gamma}{\bar{\gamma}}} \sum_{k=0}^{\infty} \frac{\left(\frac{m\gamma}{\bar{\gamma}}\right)^{m+k}}{(m+1)_k}}{m\Gamma(m)} \right)^{N_t} \tag{5.37}$$

Using equation (5.37) and equation (5.34), the SER can be evaluated as follows.

$$\overline{P_e} = \frac{a}{2}\sqrt{\frac{b}{2\pi}} \int_0^\infty \frac{e^{-\frac{b\gamma}{2}}}{\sqrt{\gamma}} \left(\frac{e^{-\frac{m\gamma}{\bar{\gamma}}} \sum_{k=0}^{\infty} \frac{\left(\frac{m\gamma}{\bar{\gamma}}\right)^{m+k}}{(m+1)_k}}{m\Gamma(m)} \right)^{N_t} d\gamma \tag{5.38}$$

$$= \frac{a}{2(m\Gamma(m))^{N_t}} \sqrt{\frac{b}{2\pi}} \int_0^\infty \frac{e^{-\gamma\left(\frac{b}{2}+\frac{mN_t}{\bar{\gamma}}\right)}}{\sqrt{\gamma}} \left(\sum_{k=0}^{\infty} \frac{\left(\frac{m\gamma}{\bar{\gamma}}\right)^{m+k}}{(m+1)_k} \right)^{N_t} d\gamma \tag{5.39}$$

$$= \frac{a}{2(m\Gamma(m))^{N_t}} \sqrt{\frac{b}{2\pi}} \int_0^\infty \frac{e^{-\gamma\left(\frac{b}{2}+\frac{mN_t}{\bar{\gamma}}\right)}}{\sqrt{\gamma}} \left(\sum_{n_1,n_2,\ldots,n_{N_t}=0}^{\infty} \frac{\left(\frac{m\gamma}{\bar{\gamma}}\right)^{mN_t+n_\Sigma}}{\prod_{i=1}^{N_t} (m+1)_{n_i}} \right) d\gamma \tag{5.40}$$

where $n_\Sigma = \sum\limits_{i=1}^{N_t} n_i$ and $\sum\limits_{n_1,n_2,\ldots,n_{N_t}=0}^{\infty} = \sum\limits_{n_1=0}^{\infty}\sum\limits_{n_2=0}^{\infty}\cdots\sum\limits_{n_{N_t}=0}^{\infty}$. Rearranging the integrand, it can be observed that it only contains an exponential term and an exponent term of the form $\gamma^p e^{-a\gamma}$ with real and positive constants, p and a. This would bring the integral in the form of Gamma functions. However, to simplify the analysis, consider high average SNR regime. The terms with higher power of $\frac{1}{\overline{\gamma}}$ would vanish in this case. To evaluate asymptotic SER, we consider only a single significant term. The significant term for $\overline{\gamma} \to \infty$ corresponds to $n_1 = 0$, $n_2 = 0$, …, $n_{N_t} = 0$. So, asymptotic SER can be given as

$$\overline{P}_{e,asympt} = \frac{am^{(m-1)N_t}}{2(\Gamma(m))^{N_t}}\sqrt{\frac{b}{2\pi}}\left(\frac{1}{\overline{\gamma}}\right)^{mN_t}\int\limits_0^\infty e^{-\gamma\left(\frac{b}{2}+\frac{mN_t}{\overline{\gamma}}\right)}(\gamma)^{mN_t-\frac{1}{2}}d\gamma \quad (5.41)$$

Now, the above integral can be solved using the definition of Gamma function, $\Gamma(\beta) = \int\limits_0^\infty x^{\beta-1}e^{-x}dx$. So, the asymptotic SER for TAS/MRC systems with MPSK and MPAM modulation scheme over Nakagami-m fading channels can be given as

$$\overline{P}_{e,asympt} = \frac{am^{(m-1)N_t}}{2(\Gamma(m))^{N_t}}\sqrt{\frac{b}{2\pi}}\left(\frac{1}{\overline{\gamma}}\right)^{mN_t}\frac{\Gamma\left(mN_t+\frac{1}{2}\right)}{\left(\frac{b}{2}+\frac{mN_t}{\overline{\gamma}}\right)^{mN_t+\frac{1}{2}}} \quad (5.42)$$

The above asymptotic expression is a lower bound to the exact BER/SER and it approaches the exact value at high SNR conditions. The expression of equation (5.42) is valid for any value of fading parameter, number of transmitter antennas and number of receiver antennas. Equation (5.42) gives asymptotic BER of TAS/MRC systems over Nakagami-m fading channels for BPSK and asymptotic SER for MPSK and MPAM modulation schemes applied to TAS/MRC MIMO systems. One may use the appropriate values of a and b as listed in Table A.1 for a specific modulation scheme like BPSK, MPAM and MPSK.

5.4.3.2 MQAM and QPSK modulation schemes

In this subsection, we derive expressions to analyze approximate error performance of TAS/MRC MIMO systems with MQAM and QPSK modulation schemes. We use an approach similar to the moment generating function (MGF) based approach to obtain the expressions for SER. Note that in the case of MQAM and QPSK modulations, the CPE contains two terms as the modulation parameter c bears non-zero value. The first term is in the form of Gaussian Q function and the second term is in the form of squared Gaussian Q function. The CPE as given by equation (5.30) can be approximated by representing the Gaussian Q function as a sum of two exponential functions [16] as

$$Q(\hbar) \approx \frac{1}{12}e^{-\frac{\hbar^2}{2}} + \frac{1}{4}e^{-\frac{2\hbar^2}{3}} \quad (5.43)$$

More details on the approximation of Gaussian Q function can be found in Appendix A.2. Using the above approximation, the derivative of CPE can be approximated as follows.

$$P_e(\gamma) = aQ\left(\sqrt{b\gamma}\right) - cQ^2\left(\sqrt{b\gamma}\right) \tag{5.44}$$

$$P_e^{'}(\gamma) = -\frac{ab}{2\sqrt{2\pi b}}\frac{e^{-\frac{b\gamma}{2}}}{\sqrt{\gamma}} + 2cQ\left(\sqrt{b\gamma}\right)\frac{b}{2\sqrt{2\pi b}}\frac{e^{-\frac{b\gamma}{2}}}{\sqrt{\gamma}} \tag{5.45}$$

$$P_e^{'}(\gamma) \cong -\frac{ab}{2\sqrt{2\pi b}}\frac{e^{-\frac{b\gamma}{2}}}{\sqrt{\gamma}} + 2c\left(\frac{1}{12}e^{-\frac{b\gamma}{2}} + \frac{1}{4}e^{-\frac{2b\gamma}{3}}\right)\frac{b}{2\sqrt{2\pi b}}\frac{e^{-\frac{b\gamma}{2}}}{\sqrt{\gamma}} \tag{5.46}$$

$$P_e^{'}(\gamma) \cong -\frac{ab}{2\sqrt{2\pi b}}\frac{e^{-\frac{b\gamma}{2}}}{\sqrt{\gamma}} + \frac{bc}{12\sqrt{2\pi b}}\frac{e^{-b\gamma}}{\sqrt{\gamma}} + \frac{bc}{4\sqrt{2\pi b}}\frac{e^{-\frac{7b\gamma}{6}}}{\sqrt{\gamma}} \tag{5.47}$$

Substituting $P_e^{'}$ of equation (5.47) in the expression of SER in equation (5.33) and comparing the resulting expression with the definition of Laplace transform, one can find that SER can be represented in the form of Laplace transform of $\frac{P_{\gamma_{t(N)}}(\gamma)}{\sqrt{\gamma}}$.

The Laplace transform of $\frac{P_{\gamma_{t(N)}}(\gamma)}{\sqrt{\gamma}}$ can be given as

$$M_{\gamma_t}(s) = \frac{1}{\left(m\Gamma(m)\right)^{N_t}} \sum_{n_1,n_2,\ldots,n_{N_t}=0}^{\infty} \frac{\left(\frac{m}{\bar{\gamma}}\right)^{mN_t+n_\Sigma}\Gamma\left(mN_t+n_\Sigma+\frac{1}{2}\right)}{\left(\frac{b}{2}+\frac{mN_t}{\bar{\gamma}}\right)^{mN_t+n_\Sigma+\frac{1}{2}}\prod_{i=1}^{N_t}(m+1)_{n_i}} \tag{5.48}$$

Then, using the above equation, the SER can be given by

$$\overline{P}_e \cong \frac{ab}{2\sqrt{2\pi b}}M_{\gamma_t}\left(\frac{b}{2}\right) - \frac{bc}{12\sqrt{2\pi b}}M_{\gamma_t}(b) - \frac{bc}{4\sqrt{2\pi b}}M_{\gamma_t}\left(\frac{7b}{6}\right) \tag{5.49}$$

From (5.48) and (5.49), the asymptotic error performance can be evaluated by truncating the values of all summation variables to be zero, i.e., $(n_1, n_2, ..., n_{N_t} = 0)$.

The simulations were performed for $(3, 1; N_r)$ and $(2, 1; 1)$ TAS/MRC systems over Nakagami-m fading channels with BPSK modulation scheme. The exact, asymptotic and simulation results of BER for different values of SNR are given in Figure 5.3 for various values of fading parameter, m. It is observed that with increase in the value of fading parameter (m), BER performance of TAS/MRC systems shows improvement on account of reduced severity of fading. Moreover, increasing the value of m also increases the diversity order. This can be observed from the higher rate of fall in the BER with SNR for larger value of m, which is directly related to the slope of the BER curve. Also, each added antenna on either transmitter or receiver improves performance and diversity order. It is also important to note that the asymptotic

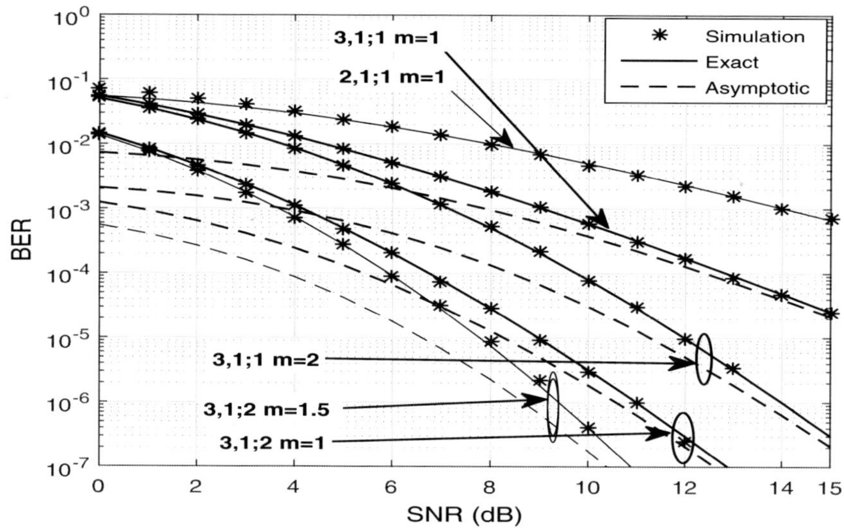

FIGURE 5.3
BER performance of TAS/MRC systems over Nakagami-m fading channels for BPSK modulation.

performance shown is a lower bound on BER on account of truncating the infinite series expression only to single term, i.e., considering only the term with the values of all summation variables 0. The MATLAB script files for simulation of TAS systems over generalized fading channels are available at the end of this book in Appendix B. A generic script file is given at the end. Just by making required changes in the channel coefficient part and modulation scheme, simulations of TAS systems can be performed over any generalized fading scenario and various modulation schemes.

5.5 TAS over $\kappa - \mu$ fading channels

The discussion of $\kappa - \mu$ distribution and the fading channel models is already presented in Section 2.4.3 of Chapter 2. We will use those discussions and related expressions to evaluate various performance parameters of TAS/MRC MIMO systems over $\kappa - \mu$ fading channels in this section. The fading envelopes ($|h_{j,i}|$) of the channel coefficients are assumed to be $\kappa - \mu$ distributed and all the links from transmitter to receiver are assumed to be i.i.d. In all the discussions related to performance analysis of TAS/MRC MIMO systems over $\kappa - \mu$ fading channels, we will assume $(N_t, 1; N_r)$ systems. Since we consider single transmit antenna selection, the antenna selection criteria are based on maximum instantaneous SNR which is the same as that for maximum channel capacity based TAS as discussed in Sections 5.2.1 and 5.2.2. Further discussions on performance analysis of TAS based MIMO systems over $\kappa-\mu$ fading channels are followed from [62]. After MRC diversity combining, the resulting instantaneous received SNR is given by $\sum_{n=1}^{N_r}\gamma_n=\overline{\gamma}\sum_{j=1}^{N_r}|h_{j,i}|^2$, where γ_n is the instantaneous SNR of the nth branch. The transmitting antenna that maximizes SNR at the receiver can be determined by using equation (5.5). This involves calculation of squared norms for each column of the channel matrix, i.e., $\gamma_{t,i}=\sum\limits_{j=1}^{N_r}|h_{j,i}|^2$ for $1\leq i\leq N_t$ where $\gamma_{t,i}$ is the squared norm of ith column of channel matrix which is proportional to the instantaneous received SNR after MRC when i^{th} antenna is selected at the transmitter.

5.5.1 PDF of received SNR after antenna selection

PDF of received SNR is required to analytically analyze the performance of wireless communication systems. We will now discuss the steps involved in obtaining the PDF of received SNR after antenna selection. Each $\gamma_{t,i}$ for $i\in[1,2,...,N_t]$ obtained from equation (5.5) is arranged in ascending order such that $\gamma_{t,(1)}\leq\gamma_{t,(2)},\dots\leq\gamma_{t,(N)}$, where $\gamma_{t,(l)}$ is the random variable obtained after the arrangement in ascending order and $N=N_t$ for TAS/MRC systems. In a TAS/MRC system, we select the transmitting antenna corresponding to the highest received SNR, i.e., $(\gamma_{t,(N_t)})$. The PDF of received SNR in such a system assuming all $(|h_{j,i}|)$'s are to be i.i.d. can be given by equation (5.11).

In our case, $\gamma_{t,i}=\sum\limits_{j=1}^{j=N_r}|h_{j,i}|^2$ is the sum of N_r square variates of i.i.d. $\kappa-\mu$ distributed random variables. So, referring to the physical model of $\kappa-\mu$ fading channels in Section 2.4.3, it can be established that the sum of N_r square variates of i.i.d. $\kappa-\mu$ random variables is also $\kappa-\mu$ distributed but with parameters κ and μ replaced by μN_r. So, the PDF of $\gamma_{t,(N_t)}$ can be given by

$$\begin{aligned}p_{\gamma_{t(N)}}(\gamma)&=N\{P_{\gamma_t}(\gamma)\}^{N-1}p_{\gamma_t}(\gamma)\\&=N\left(1-Q_\mu\left(\sqrt{2\kappa\mu},\sqrt{\frac{2\mu(1+\kappa)\gamma}{\overline{\gamma}}}\right)\right)^{N-1}\\&\quad\times\frac{\mu(1+\kappa)^{\frac{\mu+1}{2}}\gamma^{\frac{\mu-1}{2}}}{\kappa^{\frac{\mu-1}{2}}\overline{\gamma}^{\frac{\mu+1}{2}}e^{\mu\kappa}}e^{-\frac{\mu(1+\kappa)\gamma}{\overline{\gamma}}}I_{\mu-1}\left(2\mu\sqrt{\frac{\kappa(1+\kappa)\gamma}{\overline{\gamma}}}\right)\end{aligned}\quad(5.50)$$

It should be noted that equation (5.50) applies to TAS/MRC as well as joint transmit and receive antenna selection schemes with $N=N_t$ for TAS/MRC and $N=N_tN_r$ for joint transmit and receive antenna selection. For TAS/MRC γ_t is $\kappa-N_r\mu$ distributed as it is the sum of N_r i.i.d. $\kappa-\mu$

square variates [133]. Hence, in all subsequent expressions related to TAS systems over $\kappa-\mu$ fading channels, μ shall be interpreted as $N_r\mu$ for TAS/MRC systems and it remains as it is for joint transmit and receive antenna selection systems.

5.5.2 Outage probability

In this section, a closed form expression for outage probability of MIMO systems with antenna selection is derived for $\kappa-\mu$ fading channels. The closed form expression is evaluated for different fading scenarios as special cases of $\kappa-\mu$ fading. Outage probability is the probability that the instantaneous SNR at the receiver is less than a certain threshold SNR, γ_{th} [118]. Thus, outage probability can be given by

$$P_{out}\left(\overline{\gamma},\gamma_{th}\right)=P\left(\overline{\gamma}\left\|h\right\|_F^2<\gamma_{th}\right) \tag{5.51}$$

So, the outage probability which is the CDF of equation (5.50) can be given as

$$P_{out}\left(\overline{\gamma},\gamma_{th}\right)=\int\limits_0^{\gamma_{th}}\left\{N\left(1-Q_\mu\left(\sqrt{2\kappa\mu},\sqrt{\frac{2\mu\left(1+\kappa\right)x}{\overline{\gamma}}}\right)\right)^{N-1}\right.$$
$$\left.\times\frac{\mu(1+\kappa)^{\frac{\mu+1}{2}}x^{\frac{\mu-1}{2}}}{\kappa^{\frac{\mu-1}{2}}\overline{\gamma}^{\frac{\mu+1}{2}}e^{\mu\kappa}}e^{-\frac{\mu(1+\kappa)x}{\overline{\gamma}}}I_{\mu-1}\left(2\mu\sqrt{\frac{\kappa\left(1+\kappa\right)x}{\overline{\gamma}}}\right)\right\}dx \tag{5.52}$$

Using the relation between PDF and CDF in equation (5.19), the integral of equation (5.52) can be solved to get a compact closed form expression for outage probability as

$$P_{out}\left(\bar{\gamma},\gamma_{th}\right)=\left(1-Q_\mu\left(\sqrt{2\kappa\mu},\sqrt{\frac{2\mu\left(1+\kappa\right)\gamma_{th}}{\overline{\gamma}}}\right)\right)^N \tag{5.53}$$

The above expression of outage probability is in the form of Marcum Q function which can be easily evaluated using the computation tools like MATLAB and MATHEMATICA. Using [116, (4.71)] and $\kappa\to 0$ and the fading parameter values $\mu=m$, the outage probability of TAS/MRC systems over Nakagami-m fading channels can be given by

$$P_{out}\left(\overline{\gamma},\gamma_{th}\right)=\left(1-\frac{\Gamma\left(m,\frac{m\gamma_{th}}{\overline{\gamma}}\right)}{\Gamma\left(m\right)}\right)^N \tag{5.54}$$

where $\Gamma\left(a,b\right)$ is the upper incomplete Gamma function. The above expression is exactly the same as equation (5.20) obtained in the previous section for Nakagami-m fading channels. The above expression for outage probability is computationally simpler than equation (8) of [14] which is a summation of infinite series. The expression of equation (5.54) is valid for integer and non-integer values of Nakagami fading parameter, m. For integer values of m, it can be alternatively represented in the form of finite sum series as

$$P_{out}\left(\overline{\gamma},\gamma_{th}\right)=\left(1-e^{-\frac{m\gamma_{th}}{\overline{\gamma}}}\sum_{k=0}^{m-1}\frac{\left(\frac{m\gamma_{th}}{\overline{\gamma}}\right)^{k}}{k!}\right)^{N} \tag{5.55}$$

In equation (5.54) and equation (5.55), the fading parameter m is to be interpreted as mN_r and m for TAS/MRC systems and joint transmit and receive antenna selection systems respectively. Using $m = 1$ in equation (5.55), the outage probability of TAS/MRC systems over Rayleigh fading channels can be given as

$$P_{out}\left(\overline{\gamma},\gamma_{th}\right)=\left(1-e^{-\frac{N_r\gamma_{th}}{\overline{\gamma}}}\sum_{k=0}^{N_r-1}\frac{\left(\frac{N_r\gamma_{th}}{\overline{\gamma}}\right)^{k}}{k!}\right)^{N} \tag{5.56}$$

The above expression is computationally more efficient than equation (9) of [15] which involves a number of addition and power operations of the order N_tN_r. This difference will become more significant when we consider large MIMO systems [19] with hundreds of antennas at the transmitter and receiver. To bring completeness to the discussion, the comparison of results for outage probability of TAS systems with MRC and SC at the receiver is presented. The simulation results for outage probability of TAS/MRC and joint transmit and receive antenna selection can be observed in Figure 5.4. For this comparison, 2×2 MIMO systems are under consideration. It is observed from the slopes of curves that both TAS/MRC and joint transmit and receive antenna systems give full diversity order but TAS/MRC systems are superior than the joint transmit and receive antenna selection systems. TAS/MRC systems in the presented case give 1.5-2 dB SNR gain over joint transmit and receive antenna selection systems.

5.5.3 Error performance

In this section, we discuss a method to obtain expressions for exact SER of TAS/MRC systems for BPSK, BFSK and MPAM schemes and approximate SER of TAS/MRC systems for MQAM, MPSK and QPSK schemes. BER of a wireless communication system can be calculated by averaging the CPE over the PDF of output SNR. The exact CPE for various modulation schemes and

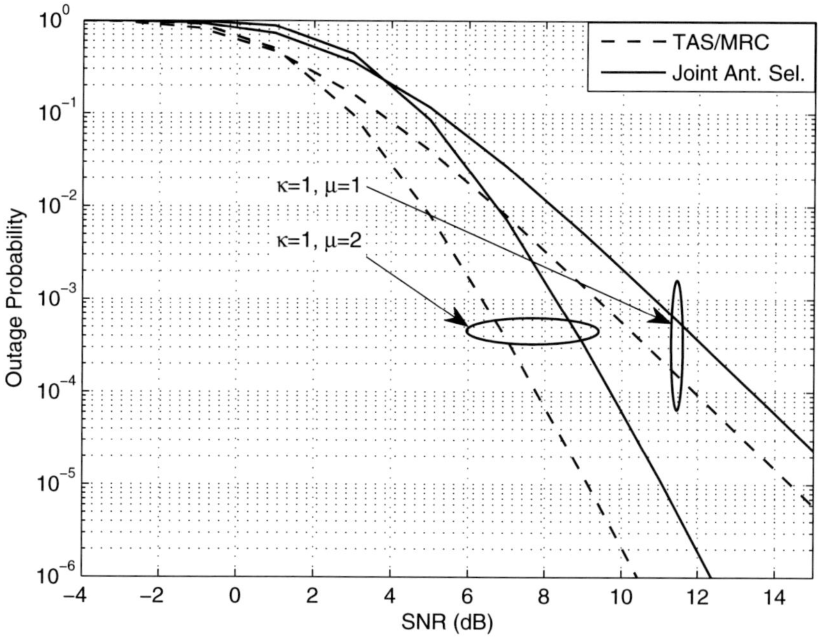

FIGURE 5.4
Outage probability comparison of TAS/MRC and joint transmit and receive antenna selection systems over $\kappa - \mu$ fading channels for $\gamma_{th} = 3$.

approximate CPE for MPSK can be given by

$$P_e(\gamma) = aQ\left(\sqrt{b\gamma}\right) - cQ^2\left(\sqrt{b\gamma}\right) \tag{5.57}$$

where $Q(\cdot)$ is the Gaussian Q function which gives the area under the tail of a Gaussian curve and is defined as $Q(\hbar) = \frac{1}{\sqrt{2\pi}} \int_{\hbar}^{\infty} e^{-\left(\frac{u^2}{2}\right)} du$, γ is SNR and a, b and c are modulation dependent parameters.

Again, using the expression for averaging the conditional probability of error, SER can be obtained as

$$\overline{P_e} = \int_0^{\infty} P_e(\gamma)\, p_{\gamma_{t(N)}}(\gamma)\, d\gamma = -\int_0^{\infty} P_e^{'}(\gamma)\, P_{\gamma_{t(N)}}(\gamma)\, d\gamma \tag{5.58}$$

where $P_e^{'}(\cdot)$ is the first derivative of CPE and $P_{\gamma_{t(N)}}(\gamma)$ is the CDF of output SNR.

5.5.4 BPSK and MPSK modulation schemes

From equation (5.57) and Table A.1, the CPE for BPSK, MPSK and MPAM modulation schemes can be given in the form of Gaussian Q function by $P_e(\gamma) = aQ\left(\sqrt{b\gamma}\right)$ as $c = 0$. The values of other modulation parameters are $a = 1$, and $b = 2$ for BPSK, $a = 2$, and $b = \sin(\pi/M)$ for MPSK and appropriate values for MPAM modulation schemes, respectively. So, using the CDF based approach, SER can be given as

$$\overline{P_e} = \frac{a}{2}\sqrt{\frac{b}{2\pi}}\int_0^\infty \frac{e^{-\frac{b\gamma}{2}}}{\sqrt{\gamma}}\left(1 - Q_\mu\left(\sqrt{2\kappa\mu}, \sqrt{\frac{2\mu(1+\kappa)\gamma}{\overline{\gamma}}}\right)\right)^N d\gamma \tag{5.59}$$

Again, as in the case of Nakagami-m fading channels, a closed form solution for the above integral may not be possible. In this section, we find an infinite series solution. However, numerical evaluation of the integral may also be used to obtain exact values of SER. To obtain series solution, $Q_{\ddot{m}}(\cdot,\cdot)$ can be represented in the form of infinite series for integer as well as non-integer values of $\ddot{m}$ as [116]

$$Q_{\ddot{m}}(\alpha, \beta) = 1 - e^{-\frac{\alpha^2+\beta^2}{2}}\sum_{r=\ddot{m}}^{\infty}\left(\frac{\beta}{\alpha}\right)^r I_r(\alpha\beta) \tag{5.60}$$

and the modified Bessel function in the series representation of $Q_{\ddot{m}}(\cdot,\cdot)$ can also be represented as an infinite series as follows [46, (8.406.1), (8.402)].

$$I_r(\omega) = \sum_{s=0}^{\infty}\frac{1}{s!\Gamma(r+s+1)}\left(\frac{\omega}{2}\right)^{r+2s} \tag{5.61}$$

Using equation (5.60) and equation (5.61) in equation (5.59), the SER can be given by

$$\overline{P_e} = \frac{a}{2}\sqrt{\frac{b}{2\pi}}\sum_{R_N}\sum_{S_N}\frac{(\kappa\mu)^{\sum_{i=1}^{N}s_i}\left(\frac{K'}{N}\right)^{N'_\mu}e^{-N\kappa\mu}}{\prod_{i=1}^{N}s_i!\Gamma(s_i+r_i+\mu+1)}\int_0^\infty\frac{e^{-\left(\frac{b}{2}+K'\right)\gamma}}{\sqrt{\gamma}}\gamma^{N'_\mu}d\gamma \tag{5.62}$$

where $N'_\mu = N\mu + \sum_{i=1}^{N} r_i + \sum_{i=1}^{N} s_i$, $K' = \frac{N\mu(1+\kappa)}{\overline{\gamma}}$, $\sum_{R_N} = \sum_{r_1=0}^{\infty}\sum_{r_2=0}^{\infty}\cdots\sum_{r_N=0}^{\infty}$ and $\sum_{S_N} = \sum_{s_1=0}^{\infty}\sum_{s_2=0}^{\infty}\cdots\sum_{s_N=0}^{\infty}$. Then SER can be evaluated as

$$\overline{P_e} = \frac{a}{2}\sqrt{\frac{b}{2\pi}}\sum_{R_N}\sum_{S_N}\frac{(\kappa\mu)^{\sum_{i=1}^{N}s_i}\left(\frac{K'}{N}\right)^{N'_\mu}e^{-N\kappa\mu}}{\prod_{i=1}^{N}s_i!\Gamma(s_i+r_i+\mu+1)}\frac{\Gamma\left(N'_\mu+\frac{1}{2}\right)}{\left(\frac{b}{2}+K'\right)^{N'_\mu-\frac{1}{2}}} \tag{5.63}$$

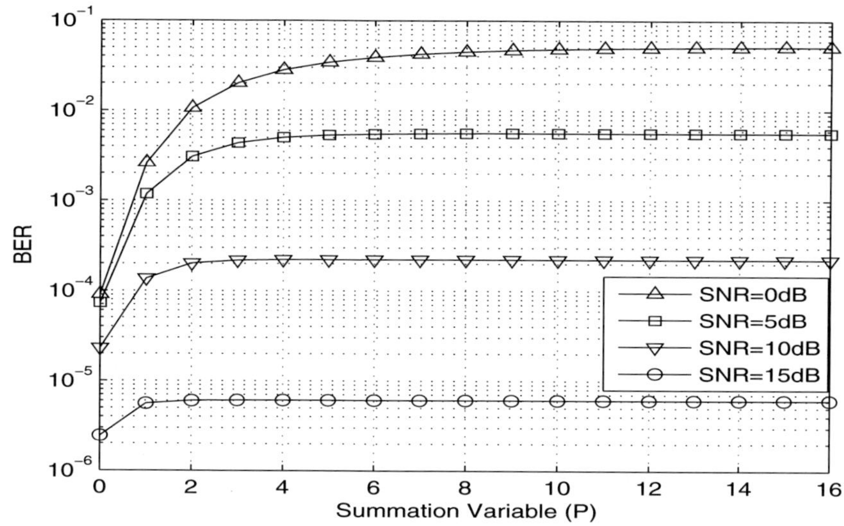

FIGURE 5.5
Plot showing fast convergence of (5.63), BER of (3,1;1) TAS/MRC system for $\kappa = 1.5$ and $\mu = 1$.

Evaluation of the above sum of infinite series may not be possible using computational softwares all the times. So, in such a case, the evaluation may be done by truncation of the summation to some finite terms of each summation variable. To check for the fastness of convergence, each summation is truncated to first P terms, i.e., $\sum\limits_{R_N} \cong \sum\limits_{r_1=0}^{P}\sum\limits_{r_2=0}^{P}\ldots\sum\limits_{r_N=0}^{P}$ and $\sum\limits_{S_N} \cong \sum\limits_{s_1=0}^{P}\sum\limits_{s_2=0}^{P}\ldots\sum\limits_{s_N=0}^{P}$. Figure 5.5 shows the BER for different values of P for the SNR values of 0 dB, 5 dB, 10 dB and 15 dB for the BPSK modulation scheme. It is observed that the BER values converge very fast and $P \geq 10$ gives sufficient accuracy in the numerical values.

The expression of equation (5.63) can be simplified by truncating P to zero for each of the summation variables (r_1, r_2, ..., r_N and s_1, s_2, ..., s_N) to give the asymptotic error probability.

$$\overline{P_e} \geq \frac{a}{2}\sqrt{\frac{b}{2\pi}}\frac{\left(\frac{K'}{N}\right)^{N\mu} e^{-N\kappa\mu}}{\Gamma(\mu+1)}\frac{\Gamma\left(N\mu+\frac{1}{2}\right)}{\left(\frac{b}{2}+K'\right)^{N\mu-\frac{1}{2}}} \tag{5.64}$$

The above asymptotic expression is a lower bound to the exact BER/SER and it approaches the exact value at high SNR conditions. The expressions of equation (5.63) and equation (5.64) are valid for any values of fading parameters, number of transmitter antennas and number of receiver antennas.

Equation (5.63) gives exact BER for BPSK modulation scheme, exact SER for MPAM modulation scheme and approximate SER for MPSK modulation scheme. Equation (5.64) gives asymptotic BER for BPSK and asymptotic SER for MPSK and MPAM modulation schemes applied to TAS/MRC MIMO systems. One may use the appropriate values of a and b as listed in Table A.1 for specific modulation schemes like BPSK and MPSK.

5.5.5 MQAM and QPSK modulation schemes

In this subsection, we discuss a method to obtain expressions to analyze approximate error performance of TAS/MRC MIMO systems with MQAM and QPSK modulation schemes. We use an approach similar to the MGF based approach as used in a previous section for Nakagami-m fading channels to obtain the expressions for SER. The CPE of equation (5.57) can be approximated as a sum of two exponential functions [16]. The derivative of CPE after representing Q function as a sum of exponential functions can be given as

$$P_e^{'} \cong -\frac{ab}{2\sqrt{2\pi b}}\frac{e^{-\frac{b\gamma}{2}}}{\sqrt{\gamma}}+\frac{bc}{12\sqrt{2\pi b}}\frac{e^{-b\gamma}}{\sqrt{\gamma}}+\frac{bc}{4\sqrt{2\pi b}}\frac{e^{-\frac{7b\gamma}{6}}}{\sqrt{\gamma}} \tag{5.65}$$

Substituting $P_e^{'}$ of equation (5.65) in the expression of SER given by equation (5.58) and comparing the resulting expression with the definition of Laplace transform, one can find that SER can be represented in the form of Laplace transform of $\frac{P_{\gamma_{t(N)}}(\gamma)}{\sqrt{\gamma}}$.

The Laplace transform of $\frac{P_{\gamma_{t(N)}}(\gamma)}{\sqrt{\gamma}}$ can be given as

$$M_{\gamma_t}(s)=\sum_{R_N}\sum_{S_N}\frac{(\kappa\mu)^{\sum_{i=1}^{N}s_i}\Gamma\left(N_\mu^{'}+\frac{1}{2}\right)}{\prod_{i=1}^{N}s_i!\Gamma\left(s_i+r_i+\mu+1\right)}\left(\frac{e^{-N\kappa\mu}\sqrt{\overline{\gamma}}}{\sqrt{\mu\left(1+\kappa\right)}}\right)\left(\frac{\mu\left(1+\kappa\right)}{N\mu\left(1+\kappa\right)+s\overline{\gamma}}\right)^{\left(N_\mu^{'}+\frac{1}{2}\right)} \tag{5.66}$$

Then, using the above equation, the SER can be given by

$$\overline{P}_e \cong \frac{ab}{2\sqrt{2\pi b}}M_{\gamma_t}\left(\frac{b}{2}\right)-\frac{bc}{12\sqrt{2\pi b}}M_{\gamma_t}(b)-\frac{bc}{4\sqrt{2\pi b}}M_{\gamma_t}\left(\frac{7b}{6}\right) \tag{5.67}$$

From equation (5.66) and equation (5.67), the asymptotic error performance (lower bound) can be evaluated by truncating the values of all summation variables to be zero, i.e., $(r_1,\ r_2,\ ...,\ r_N=0)$ and $(s_1,\ s_2,\ ...,\ s_N=0)$.

5.5.6 Channel capacity

The maximum data rate for reliable transmission or the channel capacity is one of the measures for performance analysis of wireless communication systems. In this section, we consider ergodic channel capacity as a performance measure and compare ergodic capacities of TAS/MRC and joint transmit and receive antenna selection systems over $\kappa - \mu$ fading channels. The instantaneous capacity of the channel (C_{inst}) can be given by [116]

$$C_{inst} = B\log_2(1+\gamma) \qquad \text{bits/s} \tag{5.68}$$

where B is the channel bandwidth and γ is the instantaneous SNR at the receiver. Note that γ makes the capacity a random variable because its value depends on the randomly varying channel coefficients. So, it has to be averaged over the PDF of received SNR to obtain ergodic capacity per unit bandwidth as

$$\frac{\bar{C}_{erg}}{B} = \int_0^\infty \log_2(1+\gamma) p_{\gamma_{t(N)}}(\gamma)\, d\gamma \tag{5.69}$$

Using equation (5.60) and equation (5.61) in equation (5.50), the PDF of received SNR, $p_{\gamma_{t(N)}}(\gamma)$, can be given as

$$p_{\gamma_{t(N)}}(\gamma) = Ne^{-N\mu\kappa} \sum_{S_N} \sum_{R_{N-1}} \frac{e^{-\left(K'\gamma\right)} \gamma^{\left(N'_\mu - 1\right)}}{\prod_{i=1}^{N-1} \Gamma(s_i + r_i + \mu + 1)} \frac{(\kappa\mu)^{\sum_{i=1}^{N} s_i} \left(\frac{K'}{N}\right)^{N'_\mu}}{\Gamma(s_N + \mu) \prod_{i=1}^{N} s_i!} \tag{5.70}$$

Using equation (5.70) in equation (5.69), the ergodic capacity per unit bandwidth can be evaluated using *Mathematica* as

$$\begin{aligned}\frac{\bar{C}_{erg}}{B} = &\frac{Ne^{-N\mu\kappa}}{\ln(2)} \sum_{S_N} \sum_{R_{N-1}} \frac{(\kappa\mu)^{\sum_{i=1}^{N} s_i}}{\Gamma(s_N + \mu) \prod_{i=1}^{N} s_i!} \frac{\left(\frac{1}{N}\right)^{N'_\mu}}{\prod_{i=1}^{N-1} \Gamma(s_i + r_i + \mu + 1)} \\ &\left(-\frac{\Gamma\left(N'_\mu, K'\right)}{\sin\left(N'_\mu \pi\right)} + (-1)^{N'_\mu} K' \, {}_2F_2\left(\{1,1\}, \left\{2, 2 - N'_\mu\right\}, K'\right) \right. \\ &\left. + \Gamma\left(N'_\mu\right) \left(\frac{\pi}{\sin\left(N'_\mu \pi\right)} - (-1)^{N'_\mu} \left(\ln\left(K'\right) + \psi^{(0)}\left(N'_\mu\right) \right) \right) \right)\end{aligned} \tag{5.71}$$

where $\ln(\cdot)$ is natural logarithm, $\psi^{(i)}(\cdot)$ is the $(i+1)^{th}$ derivative of logarithm of Gamma function and ${}_pF_q(\{a_1, a_2, ...a_p\}, \{b_1, b_2, ..., b_q\}, z)$ is the generalized hypergeometric function.

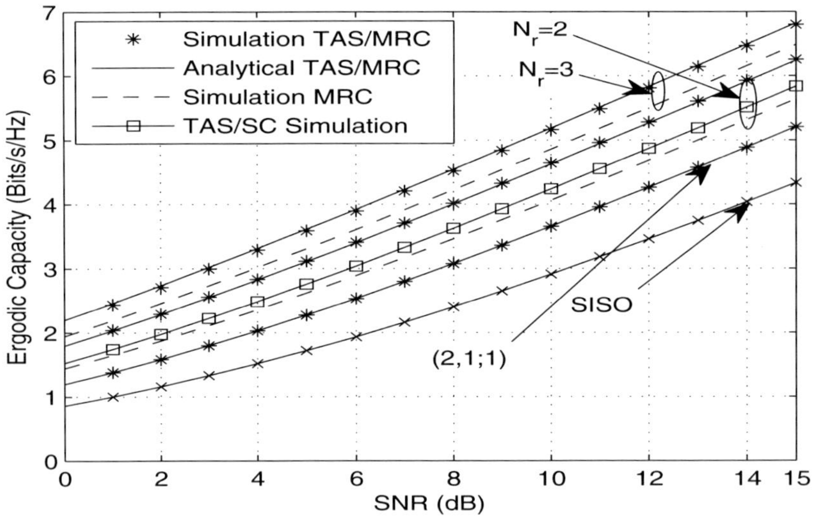

FIGURE 5.6
Ergodic capacity comparison of SISO, MRC, TAS/SC and TAS/MRC systems.

However, for integer values of μ, following the steps of [3], ergodic capacity can be evaluated as

$$\frac{\bar{C}_{erg}}{B} = \frac{Ne^{-N\mu\kappa}}{\ln(2)} \sum_{S_N} \sum_{R_{N-1}} \frac{(\kappa\mu)^{\sum_{i=1}^{N} s_i} \left(\frac{K'}{N}\right)^{N'_\mu}}{\Gamma(s_N+\mu) \prod_{i=1}^{N} s_i!} \frac{\Gamma\left(N'_\mu\right) e^{K'}}{\prod_{i=1}^{N-1} \Gamma(s_i + r_i + \mu + 1)}$$
$$\times \sum_{m=1}^{N'_\mu} \frac{\Gamma\left(m - N'_\mu\right)}{(K')^m} \tag{5.72}$$

Figure 5.6 shows results for ergodic capacity of TAS/MRC and joint transmit and receive antenna selection systems. To see a comparison of channel capacity of various communication systems, the ergodic capacity of the MRC diversity scheme and SISO system is also shown in the figure. The results are shown considering $1 \times N_r$ MRC system, $(2, 1; N_r)$ TAS/MRC system and $(2, 1; N_r, 1)$ joint transmit and receive antenna selection system. It can be observed that the ergodic capacities improve as we move from system to system following the order SISO systems, MRC diversity scheme, TAS/SC systems and TAS/MRC systems. A (2,1;3) TAS/MRC system was simulated to investigate the variation in ergodic capacity for different values of fading parameter μ. The results are plotted in Figure 5.7. The results are obtained for the case, $\kappa = 0$ and μ varying from 1 to 10. It is observed that the ergodic capacity

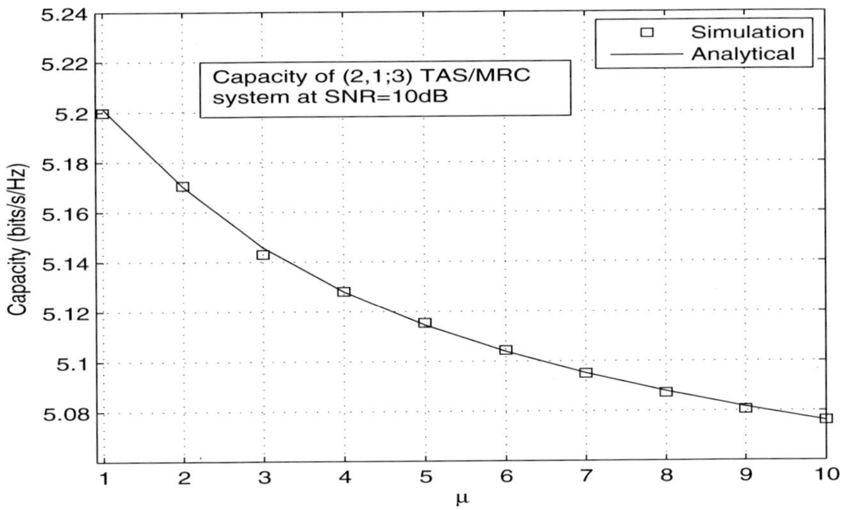

FIGURE 5.7
Capacity variation of $(2,1;3)$ TAS/MRC system with different values of μ.

of TAS/MRC systems degrades when the value of μ increases. From these results, it can be concluded that the capacity degrades when the fading severity decreases. This can be explained by the fact that when $\mu \to \infty$, the channel becomes an AWGN channel and hence the variability of fading coefficients reduces with increasing values of μ. Thus, TAS is no longer useful in AWGN channels (this is common to spatial multiplexing and SM MIMO systems). A similar observation can be made for the capacity with increasing values of the parameter κ, as a non-fading AWGN channel can be modeled as a special case of $\kappa - \mu$ fading model for $\kappa \to \infty$. The results for variation in capacity of TAS systems with κ and μ are given in [62]. Here, we have reproduced them only for increasing values of μ.

Next, we will discuss performance of TAS systems over $\eta - \mu$ fading channels.

5.6 TAS over $\eta - \mu$ fading channels

The discussions on performance analysis of TAS based MIMO systems over $\eta - \mu$ fading channels in this section are followed from [61]. In this section, we assume the fading envelope ($|h_{j,i}|$) to be $\eta - \mu$ distributed. We already discussed about the PDFs of $\eta - \mu$ distributed envelope and square variate in

Section 2.4.2 of Chapter 2. The PDF of $\eta-\mu$ distributed instantaneous SNR can be obtained by scaling of $\eta-\mu$ distributed random variable as in equation (2.25) by a factor of average SNR, $\bar{\gamma}$ [132]. After suitable scaling, the PDF of received instantaneous SNR at each antenna can be given as

$$p_{\gamma_{\eta-\mu}}(\gamma) = \frac{2\sqrt{\pi}(\mu)^{\mu+\frac{1}{2}}h^{\mu}\gamma^{\mu-\frac{1}{2}}}{\Gamma(\mu)\,H^{\mu-\frac{1}{2}}(\bar{\gamma})^{\mu+\frac{1}{2}}}e^{\left(-\frac{2\mu\gamma h}{\bar{\gamma}}\right)}I_{\mu-\frac{1}{2}}\left(\frac{2\mu\gamma H}{\bar{\gamma}}\right) \tag{5.73}$$

where η and $\mu > 0$ are fading parameters, and h and H are functions of the parameter η as defined in Section 2.4.2.

The CDF of received SNR at any single antenna output for $\eta-\mu$ fading channels is given as

$$P_{\gamma_{\eta-\mu}}(x) = \frac{2\sqrt{\pi}(\mu)^{\mu+\frac{1}{2}}h^{\mu}}{\Gamma(\mu)\,H^{\mu-\frac{1}{2}}(\bar{\gamma})^{\mu+\frac{1}{2}}}\int_{0}^{x}\gamma^{\mu-\frac{1}{2}}e^{\left(-\frac{2\mu\gamma h}{\bar{\gamma}}\right)}I_{\mu-\frac{1}{2}}\left(\frac{2\mu\gamma H}{\bar{\gamma}}\right)d\gamma \tag{5.74}$$

The integral in the above expression of CDF for $\eta-\mu$ faded SNR is represented in the form of Yacoub's integral as defined in Chapter 2 by equation (2.28). Also, a closed form expression for Yacoub's integral is reported in [81] but it is not in easily computable form. So, the integral in the above expression can be solved by using series expansion of modified Bessel function and definition of incomplete Gamma function [46, 8.445, 3.351.1], i.e.

$$I_r(\omega) = \sum_{s=0}^{\infty}\frac{1}{s!\Gamma(r+s+1)}\left(\frac{\omega}{2}\right)^{r+2s} \tag{5.75}$$

$$\int_{0}^{u}e^{-\varrho x}x^{\xi}dx = \frac{\gamma_{inc}(\xi+1,u\varrho)}{\varrho^{\xi+1}} \tag{5.76}$$

where $\gamma_{inc}(\cdot,\cdot)$ is lower incomplete Gamma function.

Use of equation (5.75) and equation (5.76) in equation (5.74) and further mathematical simplifications lead to the expression of CDF as follows.

$$P_{\gamma_{\eta-\mu}}(x) = \frac{2^{1-2\mu}\sqrt{\pi}}{h^{\mu}\Gamma(\mu)}\sum_{i=0}^{\infty}\frac{\gamma_{inc}\left(2\mu+2i,\frac{2\mu h x}{\bar{\gamma}}\right)}{i!\Gamma(\mu+i+0.5)}\left(\frac{H}{2h}\right)^{2i} \tag{5.77}$$

5.6.1 PDF and MGF of received SNR after antenna selection

Each $\gamma_{t,ji}$ or $\gamma_{t,i}$ for integer values of $i \in [1, N_t]$ and $j \in [1, N_r]$ obtained from equation (5.5) is arranged in ascending order such that $\gamma_{t,(1)} \leq \gamma_{t,(2)} \leq \ldots \leq \gamma_{t,(N)}$, where $\gamma_{t,(\cdot)}$ is the random variable obtained after the arrangement in ascending order and $N = N_tN_r$ or N_t for TAS/SC or TAS/MRC systems respectively. In the TAS/MRC system, we select the transmitting antenna corresponding to the highest received SNR $\left(\gamma_{t,(N_t)}\right)$ when MRC diversity technique is used at the receiver. In TAS/SC, the transmit and receive antenna pair that maximizes the received SNR $\left(\gamma_{t,(N_t N_r)}\right)$ is selected. The PDF of SNR in such a system can be given by [31].

$$\begin{aligned} p_{\gamma_{t(N)}}(x) &= N\{P_{\gamma_t}(x)\}^{N-1} p_{\gamma_t}(x) \\ &= N\left[\frac{2^{1-2\mu}\sqrt{\pi}}{h^{\mu}\Gamma(\mu)}\sum_{i=0}^{\infty}\frac{\gamma\left(2\mu+2i,\frac{2\mu h x}{\overline{\gamma}}\right)}{i!\Gamma(\mu+i+0.5)}\left(\frac{H}{2h}\right)^{2i}\right]^{N-1} \\ &\quad \times \frac{2\sqrt{\pi}(\mu)^{\mu+\frac{1}{2}}h^{\mu}\gamma^{\mu-\frac{1}{2}}}{\Gamma(\mu)H^{\mu-\frac{1}{2}}(\overline{\gamma})^{\mu+\frac{1}{2}}} e^{\left(-\frac{2\mu\gamma h}{\overline{\gamma}}\right)} I_{\mu-\frac{1}{2}}\left(\frac{2\mu\gamma H}{\overline{\gamma}}\right) \end{aligned} \tag{5.78}$$

The following set of expressions can be used to bring the lower incomplete Gamma function in the form of sum of an infinite series.

$$\gamma_{inc}(s,x) = \frac{x^s}{s}\Phi(s,s+1,-x) \tag{5.79}$$

$$\Phi(a,b,-x) = e^{-x}\Phi(b-a,b,x) \tag{5.80}$$

$$\Phi(a,b,x) = \sum_{k=0}^{\infty}\frac{(a)_k x^k}{(b)_k k!} \tag{5.81}$$

The above expressions are (9.236.4), (9.212.1) and (9.210.1) of [46]. Now, incomplete Gamma function can be represented in the series form as

$$\gamma_{inc}(s,x) = \frac{e^{-x}}{s}\sum_{k=0}^{\infty}\frac{x^{s+k}}{(s+1)_k} \tag{5.82}$$

where $(a)_n$ is the Pochhammer symbol defined as $(a)_n = \frac{\Gamma(a+n)}{\Gamma(a)}$. So, the PDF and MGF of the received instantaneous SNR can be given by equation (5.83) and equation (5.84) as follows.

$$p_{\gamma_{t(N)}}(x) = 2N\ell\sum_{i,j}\frac{2^{j_\Sigma}H^{2i_\Sigma}}{h^{N\mu+2i_\Sigma}\prod_{p=1}^{N}(i_p!\Gamma(\mu'))}\frac{qe^{-\lambda x}x^{r-1}}{\prod_{p=1}^{N-1}\left((\mu+i_p)(2\mu')_{j_p}\right)} \tag{5.83}$$

$$\begin{aligned} M_\gamma(s) &= \int_0^\infty e^{-sx}p_{\gamma_{t(N)}}(x)\,dx \\ &= 2N\ell\sum_{i,j}\frac{2^{j_\Sigma}H^{2i_\Sigma}q\Gamma(r)(\lambda+s)^{-r}}{h^{N\mu+2i_\Sigma}\prod_{p=1}^{N}(i_p!\Gamma(\mu'))\prod_{p=1}^{N-1}\left((\mu+i_p)(2\mu')_{j_p}\right)} \end{aligned} \tag{5.84}$$

where

$$\sum_{i,j} = \sum_i \sum_j = \sum_{i_1=0}^{\infty} \sum_{i_2=0}^{\infty} \cdots \sum_{i_N=0}^{\infty} \sum_{j_1=0}^{\infty} \sum_{j_2=0}^{\infty} \cdots \sum_{j_{N-1}=0}^{\infty}, r = 2N\mu + 2i_\Sigma + j_\Sigma,$$

$$i_\Sigma = \sum_{p=1}^{N} i_p, \quad j_\Sigma = \sum_{p=1}^{N-1} j_p; \quad \mu' = \mu + i_p + \frac{1}{2}, \quad \ell = \left(\frac{\sqrt{\pi}}{\Gamma(\mu)}\right)^N, \quad \lambda = \frac{2N\mu h}{\bar{\gamma}},$$

and $\mathrm{q} = \left(\frac{\mu \mathrm{h}}{\bar{\gamma}}\right)^{\mathrm{r}}$

It should be noted that equation (5.83) applies to TAS/MRC as well as TAS/SC systems. For TAS/MRC γ_t is $\eta - N_r\mu$ distributed as it is the sum of N_r i.i.d. $\eta - \mu$ square variates [136]. So, in equation (5.83), equation (5.84) and all subsequent expressions for performance analysis of TAS systems over $\eta - \mu$ fading channels, μ shall be interpreted as $N_r\mu$ for TAS/MRC systems and N_r becomes the number of antennas selected at the receiver, i.e., $N_r = 1$ for $(N_t, 1; N_r, 1)$ TAS/SC system.

5.6.2 Probability of error

In this section, we derive expression for approximate and exact SER of TAS/MRC systems over $\eta - \mu$ fading channels for various modulation techniques. SER for a wireless communication system can be calculated by averaging the CPE over the PDF of received SNR. CPE for various modulation schemes can be given by

$$P_e(\gamma) = aQ\left(\sqrt{b\gamma}\right) - cQ^2\left(\sqrt{b\gamma}\right) \tag{5.85}$$

In this section, averaging of CPE is done with respect to PDF of the received SNR, unlike in the case of Nakagami-m and $\kappa - \mu$ fading channels where the averaging of the derivative of CPE with respect to CDF of the received SNR was done to obtain expressions for SER for various modulation schemes. Further, to obtain SER, the MGF based approach would be used. The SER can be given by

$$\overline{P_e} = \int_0^{\infty} P_e(\gamma)\, p_{\gamma_{t(N)}}(\gamma)\, d\gamma \tag{5.86}$$

5.6.2.1 Approximate probability of error

To analyze approximate probability of error, the Gaussian Q-function is approximated as a sum of two exponential functions [16].

$$Q(\hbar) \approx \frac{1}{12}e^{-\frac{\hbar^2}{2}} + \frac{1}{4}e^{-\frac{2\hbar^2}{3}} \tag{5.87}$$

Thus, using equation (5.85) and equation (5.87), approximate CPE can be given in the form of sum of exponential functions as

$$P_e(\gamma) \cong \frac{a}{12}e^{-\frac{b\gamma}{2}} + \frac{a}{4}e^{-\frac{2b\gamma}{3}} - \frac{c}{144}e^{-b\gamma} - \frac{c}{16}e^{-\frac{4b\gamma}{3}} - \frac{c}{24}e^{-\frac{7b\gamma}{6}} \tag{5.88}$$

where first two terms and last three terms represent first term and second term of equation (5.85) respectively.

Now, by averaging the above approximate CPE over the PDF of the received SNR, average approximate SER can be calculated as

$$\overline{P}_{e_{appr.}} \cong \sum_i \zeta_i \int_0^\infty e^{-\psi_i \gamma} p_{\gamma_{t(N)}}(\gamma)\, d\gamma \tag{5.89}$$

where $\zeta_1 = \frac{a}{12}$, $\zeta_2 = \frac{a}{4}$, $\zeta_3 = \frac{-c}{144}$, $\zeta_4 = \frac{-c}{16}$, $\zeta_5 = \frac{-c}{24}$, $\psi_1 = \frac{b}{2}$, $\psi_2 = \frac{2b}{3}$, $\psi_3 = b$, $\psi_4 = \frac{4b}{3}$, and $\psi_5 = \frac{7\text{b}}{6}$.

In the above expression, the integral of equation (5.89) fits the definition of MGF of the received SNR $M_\gamma(\psi_i)$. Thus, approximate SER can be given as the weighted sum of MGFs of received SNR. In this case, it can be given as

$$\overline{P}_{e_{appr.}} \cong \sum_i \zeta_i M_\gamma(\psi_i) \tag{5.90}$$

5.6.2.2 Exact probability of error

In this section, we derive expression for exact SER of TAS/MRC systems over $\eta - \mu$ fading channels for various modulation techniques. The exact SER can be evaluated as [116]

$$\overline{P}_e = \int_0^\infty \left[aQ\left(\sqrt{bx}\right) - cQ^2\left(\sqrt{bx}\right)\right] p_{\gamma_{t(N)}}(x)\, dx \tag{5.91}$$

$$= \underbrace{\frac{a}{\pi}\int_0^{\frac{\pi}{2}} M_\gamma\left(\frac{b}{2\sin^2\theta}\right) d\theta}_{I_1} - \underbrace{\frac{c}{\pi}\int_0^{\frac{\pi}{4}} M_\gamma\left(\frac{b}{2\sin^2\theta}\right) d\theta}_{I_2} \tag{5.92}$$

To obtain the above expression of SER in the form of MGF, the definition of Gaussian Q function and square of Gaussian Q function by Craig is used. They are defined by Craig as the following expressions [28].

$$Q(x) = \frac{1}{\pi}\int_0^{\frac{\pi}{2}} e^{-\frac{x^2}{2\sin^2\theta}} d\theta \tag{5.93}$$

$$Q^2(x) = \frac{1}{\pi}\int_0^{\frac{\pi}{4}} e^{-\frac{x^2}{2\sin^2\theta}} d\theta \tag{5.94}$$

Evaluation of I_1

I_1 is the first term of equation (5.92), i.e.

$$I_1 = \frac{2aN}{\pi}\ell\sum_{i,j}\frac{2^{j_\Sigma}H^{2i_\Sigma}q\Gamma(r)\int_0^{\frac{\pi}{2}}\left(\lambda+\frac{b}{2\sin^2\theta}\right)^{-r}d\theta}{h^{N\mu+2i_\Sigma}\prod_{p=1}^{N}(i_p!\Gamma(\mu'))\prod_{p=1}^{N-1}\left((\mu+i_p)(2\mu')_{j_p}\right)} \tag{5.95}$$

The above integral can be evaluated by bringing it in the form of equation (5A.3) of [116] as

$$I_1 = aN\ell\sqrt{\frac{\zeta}{\pi}}\sum_{i,j}\frac{2^{j_\Sigma}H^{2i_\Sigma}q\,\lambda^{-r}\Gamma\left(r+\frac{1}{2}\right)}{h^{N\mu+2i_\Sigma}\prod_{p=1}^{N}(i_p!\Gamma(\mu'))}\frac{{}_2F_1\left(1,r+\frac{1}{2},r+1,\frac{1}{1+\zeta}\right)}{r(1+\zeta)^{r+\frac{1}{2}}\prod_{p=1}^{N-1}\left((\mu+i_p)(2\mu')_{j_p}\right)} \tag{5.96}$$

where $\zeta = \frac{b}{2\lambda}$ and ${}_pF_q(\{a_1,a_2,...a_p\},\{b_1,b_2,...,b_q\},z)$ is the generalized hypergeometric function.

Evaluation of I_2

I_2 is the second term of equation (5.92), i.e.

$$I_2 = \frac{2cN}{\pi}\ell\sum_{i,j}\frac{2^{j_\Sigma}H^{2i_\Sigma}q\Gamma(r)\int_0^{\frac{\pi}{4}}\left(\lambda+\frac{b}{2\sin^2\theta}\right)^{-r}d\theta}{h^{N\mu+2i_\Sigma}\prod_{p=1}^{N}(i_p!\Gamma(\mu'))\prod_{p=1}^{N-1}\left((\mu+i_p)(2\mu')_{j_p}\right)} \tag{5.97}$$

A detailed solution of the above integral is given in Appendix A.3. However, substituting $\sin^2\theta = t/2$ in the above integral, following some mathematical simplifications and using [103, (7.2.4.42)], I_2 can be given as

$$I_2 = \frac{\sqrt{2}cN}{\pi}\ell\sum_{i,j}\frac{2^{j_\Sigma}H^{2i_\Sigma}q\,\lambda^{-r}\Gamma(r)}{h^{N\mu+2i_\Sigma}\prod_{p=1}^{N}(i_p!\Gamma(\mu'))}\frac{F_A^{(1)}\left(r+\frac{1}{2},\frac{1}{2},r,r+\frac{3}{2},\frac{1}{2},-\frac{1}{2\zeta}\right)}{\prod_{p=1}^{N-1}\left((\mu+i_p)(2\mu')_{j_p}\right)(1+2r)(2\zeta)^r} \tag{5.98}$$

where $F_A^{(1)}(a;b_1,b_2;c;x,y)$ is Appell hypergeometric function of two variables and $\zeta = \frac{b}{2\lambda}$.

The infinite sum series of equation (5.96) and equation (5.98) were truncated while evaluating SER using equation (5.92) up to the first P terms such that it gives accurate result up to the 5th place of decimal digit when the value

TABLE 5.1
Number of terms required for accuracy up to 5th place of decimal digit in scientific notation of the result of equation (5.92)

η, μ, N_t, N_r	SNR = 5 dB		SNR = 10 dB	
	P	SER	P	SER
1,0.5,2,2	10	1.98186×10^{-2}	7	9.27794×10^{-4}
0.5,1.25,2,1	18	6.92964×10^{-2}	11	5.49862×10^{-3}
0.8,1.5,3,1	18	4.58657×10^{-2}	11	1.52017×10^{-3}

is represented in scientific notation, i.e., in the form of $a \times 10^b$. The value of P and corresponding SER for some cases is tabulated in Table 5.1. It is observed that we can get accurate results up to five places of decimal by truncating the infinite sums to the first 20 terms, i.e., the first five places of decimal remain unchanged even on increasing the value of P further.

5.6.2.3 Asymptotic probability of error

The asymptotic probability of error is a measure of performance analysis at high SNR conditions. In general, it is used to distinguish the system behavior at high SNR regime with a very easy evaluation method. From equation (5.96) and equation (5.98), it is clear that I_2 corresponds to the term containing squared Gaussian Q-function and it can be neglected at high SNR. The asymptotic error performance can be evaluated by assuming the values of all summation variables to be zero, i.e., $(i_1,\ i_2,\ ...,\ i_N = 0)$ and $(j_1,\ j_2,\ ...,\ j_{N-1} = 0)$ in equation (5.96). The asymptotic probability of error can be given by

$$P_{e_{asympt}} = \frac{a\ell}{2}\sqrt{\frac{\zeta}{\pi}}\frac{\Gamma\left(2N\mu+\frac{1}{2}\right)}{h^{N\mu}\Gamma\left(\mu+\frac{1}{2}\right)^N\mu^N}\frac{{}_2F_1\left(1, 2N\mu+\frac{1}{2}, 2N\mu+1, \frac{1}{1+\zeta}\right)}{\left(\frac{\mu h}{\bar{\gamma}}\right)^{-2N\mu}\lambda^{2N\mu}(1+\zeta)^{2N\mu+\frac{1}{2}}} \tag{5.99}$$

Using equation (9.200) of [46] which is basically the series representation of Appell hypergeometric function of two variables, the above expression can be simplified to obtain asymptotic probability of error as follows.

$$P_{e_{asympt}} = \frac{a}{2\sqrt{\pi}}\left(\frac{\sqrt{\pi}}{\Gamma(\mu)}\right)^N\frac{(2N)^{-2N\mu}\Gamma\left(2N\mu+\frac{1}{2}\right)}{h^{N\mu}\mu^N\Gamma\left(\mu+\frac{1}{2}\right)^N\left(1+\frac{h\bar{\gamma}}{4N\mu h}\right)^{2N\mu}} \tag{5.100}$$

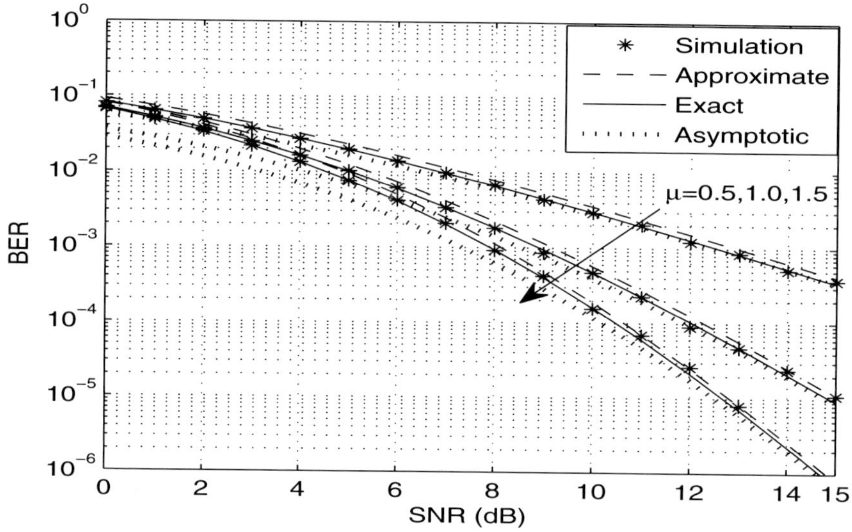

FIGURE 5.8
BER performance of (2,1;1) TAS/MRC system over $\eta-\mu$ fading channels for $\eta = 1$ with BPSK modulation.

Note that the above expression of asymptotic probability of error is in the form of elementary functions and can be easily evaluated using any of the commonly available computational tools.

In Figure 5.8, the results are shown for a (2,1;1) TAS/MRC system for Nakagami-m fading case with $\eta = 1$ and $\mu = m/2$ (a special case of $\eta - \mu$ fading). The analytical (exact, approximate and asymptotic) and simulation results of error performance are plotted for $m = 1, 2$ and 3 for BPSK modulation scheme. As the value of fading parameter m increases, the severity of fading decreases and the BER performance of the system improves. It can be seen from the figure that the asymptotic BER performance is a lower bound and approximate BER is an upper bound to the exact BER performance. The analytical and simulation results can be observed to be in close agreement.

In Figure 5.9, the results are shown for an (N_t,1;1) TAS/MRC system for Rayleigh fading case with $\eta = 1$ and $\mu = 0.5$ (a special case of $\eta - \mu$ fading). The analytical (exact, approximate and asymptotic) and simulation results of error performance are plotted for $N_t = 1$, 2 and 3 for the BPSK modulation scheme.

5.6.3 Channel capacity

In this section, we derive expression for ergodic channel capacity for TAS/MRC systems over $\eta - \mu$ fading channels. The ergodic capacity per unit

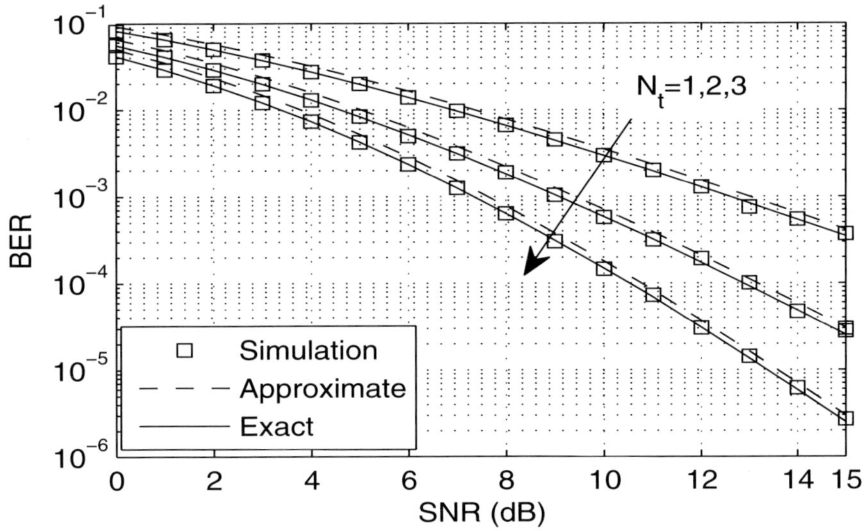

FIGURE 5.9
BER performance of (N_t,1;1) TAS/MRC system over $\eta - \mu$ fading channels for $\eta = 1$ and $\mu = 0.5$ with BPSK modulation.

bandwidth can be given as [116]

$$\frac{\bar{C}_{erg}}{B} = \int_{0}^{\infty} \log_2 (1+\gamma) p_{\gamma_{t(N)}} (\gamma)\, d\gamma \tag{5.101}$$

Using equation (5.83) in the above expression and evaluating it by using the method discussed in [3] for integer values of 2μ, we get the ergodic capacity per unit bandwidth as

$$\frac{\bar{C}_{erg}}{B} = 2N\ell \sum_{i,j} \frac{H^{2i_\Sigma} q\left(2^{j_\Sigma}\right) \Gamma\left(r\right) e^{\,\lambda} \sum\limits_{\alpha=1}^{r} \frac{\Gamma(\alpha - r, \lambda)}{\lambda^\alpha}}{h^{N\mu+2i_\Sigma} \prod\limits_{p=1}^{N} \left(i_p! \Gamma\left(\mu'\right)\right) \prod\limits_{p=1}^{N-1} \left(\left(\mu+i_p\right)\left(2\mu'\right)_{j_p}\right)} \tag{5.102}$$

The variation of ergodic capacity with fading parameters η and μ is also shown to investigate the effect of fading severity on the ergodic capacity. Figure 5.10 shows the variation of ergodic capacity for $0 \le \eta \le 1$ with $\mu = 0.5$ for SNR values of 0 dB, 5 dB and 10 dB with N_t = 2, 3 and 4. It is observed that the ergodic capacity degrades with increasing value of η for $N_t > 2$. This observation is similar to the findings of [96] for Hoyt fading channels. In fact, $\mu = 0.5$ and various values of η models Hoyt fading channels as special case of $\eta - \mu$ fading channel model with the Hoyt fading parameter, $q^2 = \eta$.

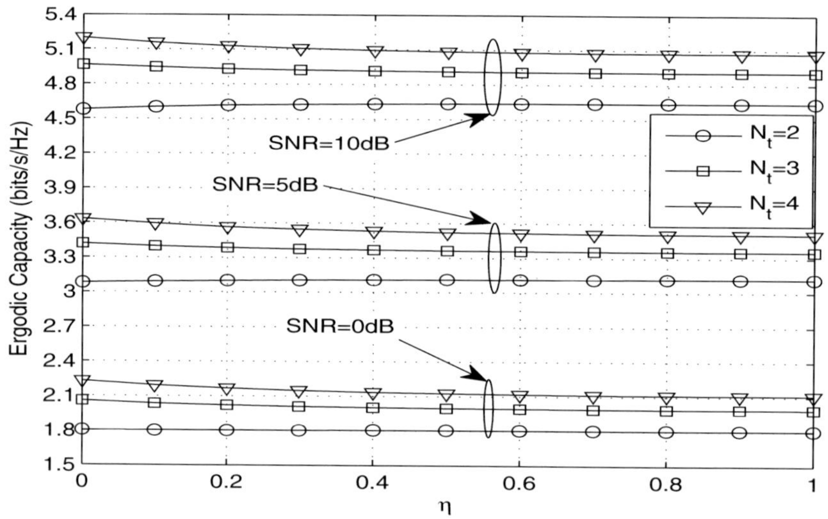

FIGURE 5.10
Capacity of $(N_t, 1; 2)$ TAS/MRC system for different values of η.

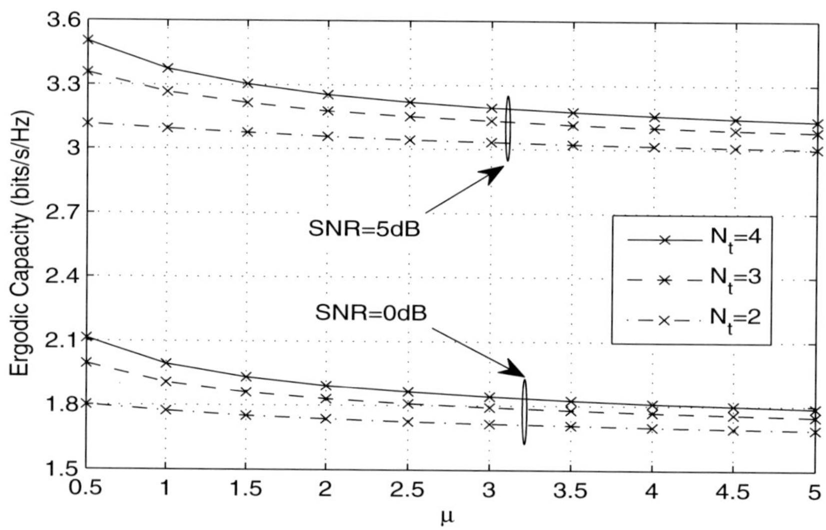

FIGURE 5.11
Capacity of $(N_t, 1; 2)$ TAS/MRC system for different values of μ.

The variation of ergodic capacity for $\eta = 1$ with different values of μ shows a degradation in capacity with increasing values of μ. The simulation results of ergodic capacity are plotted in Figure 5.11 for (N_t,1;2) TAS/MRC system with a different number of transmitting antennas for SNR values of 0 dB and 5 dB. It is observed that the capacity degrades with increasing values of μ irrespective of the number of transmitting antennas.

The observed degradations in ergodic capacity with fading parameters can be explained by the fact that with decreasing severity in fading it is highly likely that the instantaneous SNR will be less varying. In such a case, the PDF of the received SNR after TAS/MRC shifts toward the left indicating a decrease in the average SNR at the receiver. However, this will not degrade the SER performance because this also narrows the PDF of the received SNR which improves the diversity order. The improved diversity order compensates the decrease in the average SNR at the receiver. Hence, an improvement in the SER is observed while the ergodic capacity degrades with decrease in the fading severity.

5.7 TAS over $\alpha - \mu$ fading channels

The analysis of TAS systems over $\alpha - \mu$ fading channels is very much similar to that of Nakagami-m fading channels. Despite the fundamental difference in both these fading models, expressions of PDF and CDF of Nakagami-m and $\alpha - \mu$ random variables are very much similar. This can be verified by comparing the following expressions.

$$p_{\gamma_{Nak-m}}(\gamma) = \frac{m^m \gamma^{m-1}}{\overline{\gamma}^m \Gamma(m)} e^{\frac{-m\gamma}{\overline{\gamma}}}, \qquad \gamma \geq 0 \tag{5.103}$$

$$p_{\gamma_{\alpha-\mu}}(\gamma) = \frac{\alpha \mu^\mu \gamma^{\frac{\alpha\mu}{2}-1}}{2\Gamma(\mu)\overline{\gamma}^{\frac{\alpha\mu}{2}}} e^{-\mu\left(\frac{\gamma}{\overline{\gamma}}\right)^{\frac{\alpha}{2}}} \qquad \mu > 0,\ \alpha > 0 \tag{5.104}$$

$$P_{\gamma_{Nak-m}}(y) = \frac{\gamma\left(m, \frac{my}{\overline{\gamma}}\right)}{\Gamma(m)} \tag{5.105}$$

$$P_{\gamma_{\alpha-\mu}}(y) = \frac{\gamma\left(\mu, \mu\left(\frac{y}{\overline{\gamma}}\right)^{\frac{\alpha}{2}}\right)}{\Gamma(\mu)} \tag{5.106}$$

After antenna selection, the PDF and CDF of the received SNR are functions of the PDF and CDF of the received SNR before antenna selection. So, considering the similarity of PDF and CDF of $\alpha - \mu$ and Nakagami-m fading, the analysis will not be very different. We will use the infinite series expansion of incomplete Gamma function as in the Nakagami-m fading case.

PDF of received SNR after antenna selection can be given as

$$p_{\gamma_{t(N)}}(\gamma) = N\{P_{\gamma_t}(\gamma)\}^{N-1} p_{\gamma_t}(\gamma) \tag{5.107}$$

$$p_{\gamma_{t(N_t)}}(\gamma) = N_t \left(\frac{\gamma_{inc}\left(\mu, \mu\left(\frac{\gamma}{\bar{\gamma}}\right)^{\frac{\alpha}{2}}\right)}{\Gamma(\mu)} \right)^{N_t-1} \frac{\alpha\mu^{\mu}\gamma^{\frac{\alpha\mu}{2}-1} e^{-\mu\left(\frac{\gamma}{\bar{\gamma}}\right)^{\frac{\alpha}{2}}}}{2\bar{\gamma}^{\frac{\alpha\mu}{2}}\Gamma(\mu)} \tag{5.108}$$

The incomplete Gamma function can be represented in the form of a finite sum series for integer values of m as follows.

$$\gamma_{inc}(m, y) = m! - e^{-y} \sum_{k=0}^{m} \frac{m!}{k!} y^k \tag{5.109}$$

So, for integer values of μ, the PDF of received SNR can be represented as

$$p_{\gamma_{t(N_t)}}(\gamma) = N_t \left(\frac{\mu! - e^{-\mu\left(\frac{\gamma}{\bar{\gamma}}\right)^{\frac{\alpha}{2}}} \sum_{k=0}^{\mu} \frac{\mu!}{k!} \mu^k \left(\frac{\gamma}{\bar{\gamma}}\right)^{\frac{k\alpha}{2}}}{\Gamma(\mu)} \right)^{N_t-1} \frac{\alpha\mu^{\mu}\gamma^{\frac{\alpha\mu}{2}-1} e^{-\mu\left(\frac{\gamma}{\bar{\gamma}}\right)^{\frac{\alpha}{2}}}}{\bar{\gamma}^{\frac{\alpha\mu}{2}}\Gamma(\mu)} \tag{5.110}$$

5.7.1 Outage probability

The analysis of outage probability and SER performance of TAS systems over $\alpha - \mu$ fading channels can be easily evaluated in the same way as discussed in Section 5.4. As discussed in earlier sections of this chapter, outage probability is defined as the probability that the received SNR falls below certain threshold SNR and it is the CDF of received SNR. The outage probability of TAS/MRC systems over $\alpha - \mu$ fading channels can be given as

$$P_{out}(\bar{\gamma}, \gamma_{th}) = \left(\frac{\gamma_{inc}\left(\mu, \mu\left(\frac{\gamma_{th}}{\bar{\gamma}}\right)^{\frac{\alpha}{2}}\right)}{\Gamma(\mu)} \right)^{N_t} \tag{5.111}$$

The incomplete gamma function can be expanded as the sum of infinite series to obtain the outage probability for high SNR conditions. For this representation, we use the following set of expressions.

$$\gamma_{inc}(s, x) = \frac{x^s}{s} \Phi(s, s+1, -x) \tag{5.112}$$

$$\Phi(a, b, -x) = e^{-x} \Phi(b-a, b, x) \tag{5.113}$$

$$\Phi\left(a,b,x\right)=\sum_{k=0}^{\infty}\frac{\left(a\right)_k x^k}{\left(b\right)_k k!} \tag{5.114}$$

The above expressions are (9.236.4), (9.212.1) and (9.210.1) of [46]. Now, incomplete Gamma function can be represented in the series form as

$$\gamma_{inc}\left(s,x\right)=\frac{e^{-x}}{s}\sum_{k=0}^{\infty}\frac{x^{s+k}}{\left(s+1\right)_k} \tag{5.115}$$

where $(a)_n$ is the Pochhammer symbol defined as $(a)_n = \frac{\Gamma(a+n)}{\Gamma(a)}$.

Thus, an infinite series expression for outage probability can be given as follows.

$$P_{out}\left(\overline{\gamma},\gamma_{th}\right)=\left(\frac{e^{-\mu\left(\frac{\gamma_{th}}{\overline{\gamma}}\right)^{\frac{\alpha}{2}}}\sum\limits_{n=0}^{\infty}\frac{\left(\mu\left(\frac{\gamma_{th}}{\overline{\gamma}}\right)^{\frac{\alpha}{2}}\right)^{\mu+n}}{(\mu+1)_n}}{\mu\Gamma(\mu)}\right)^{N_t} \tag{5.116}$$

The asymptotic outage probability is defined by the high SNR condition, i.e., $\overline{\gamma}\rightarrow\infty$. In such a case, the summation of the above expression can be truncated to a single term corresponding to $n=0$. So, the asymptotic outage probability can be given by

$$P_{out,asympt}\left(\overline{\gamma},\gamma_{th}\right)=\frac{e^{-\mu N_t\frac{\alpha}{2}\left(\frac{\gamma_{th}}{\overline{\gamma}}\right)}\mu^{\mu N_t}\left(\frac{\gamma_{th}}{\overline{\gamma}}\right)^{\frac{\mu\alpha N_t}{2}}}{\left(\mu\Gamma(\mu)\right)^{N_t}} \tag{5.117}$$

5.7.2 Error performance

Error performance of any wireless communication systems is important to know the amount/fraction of the received symbols being detected in error. For various modulation schemes, probability of error in unfaded channels can be represented in the form of Gaussian Q function when the noise is AWGN in nature. It is also called conditional probability of error because its value is conditioned on the channel coefficients. For various modulation schemes, it can be given as in equation (5.30). It can be averaged over the PDF of received SNR after antenna selection to get the SER.

The PDF of received SNR as in equation (5.118) contains incomplete Gamma function with complicated arguments like $\mu\left(\frac{\gamma_{th}}{\overline{\gamma}}\right)^{\frac{\alpha}{2}}$. A closed form solution to the integral involving such complex arguments of the power of incomplete Gamma function does not exist to the best of our knowledge. So, PDF of received SNR needs to be represented in the form of infinite series summation as in the case of asymptotic outage probability in the previous section (refer equation (5.116)).

So, the PDF of received SNR can be given as

$$p_{\gamma_{t(N_t)}}(\gamma) = N_t \left(\frac{e^{-\left(\mu\left(\frac{\gamma}{\bar{\gamma}}\right)^{\frac{\alpha}{2}}\right)}}{\mu\Gamma(\mu)} \sum_{k=0}^{\infty} \frac{\left(\mu\left(\frac{\gamma}{\bar{\gamma}}\right)^{\frac{\alpha}{2}}\right)^{\mu+k}}{(\mu+1)_k} \right)^{N_t-1} \frac{\alpha\mu^{\mu}\gamma^{\frac{\alpha\mu}{2}-1}e^{-\mu\left(\frac{\gamma}{\bar{\gamma}}\right)^{\frac{\alpha}{2}}}}{2\bar{\gamma}^{\frac{\alpha\mu}{2}}\Gamma(\mu)} \tag{5.118}$$

and the CDF of received SNR for TAS MIMO systems over $\alpha - \mu$ fading channels can be given as

$$P_{\gamma_{t(N_t)}}(\gamma) = \left(\frac{e^{-\left(\mu\left(\frac{\gamma}{\bar{\gamma}}\right)^{\frac{\alpha}{2}}\right)}}{\mu\Gamma(\mu)} \sum_{k=0}^{\infty} \frac{\left(\mu\left(\frac{\gamma}{\bar{\gamma}}\right)^{\frac{\alpha}{2}}\right)^{\mu+k}}{(\mu+1)_k} \right)^{N_t} \tag{5.119}$$

Once CDF and PDF of the received SNR are obtained, any of the methods discussed in previous sections for Nakagami-m, $\kappa - \mu$ or $\eta - \mu$ fading channels can be used to evaluate BER/SER of various digital modulation schemes.

Further, simulation of TAS systems over $\alpha - \mu$ fading channels can be performed for probability of error analysis using the MATLAB script files provided at the end of this book in Appendix B. The only change the users may need to do is in the channel coefficients part. The programs are made generic. Using proper method or function for channel coefficient generation and suitable values of modulation parameters, TAS systems can be simulated for various modulation schemes over generalized fading channels.

6

Space Time Block Coded MIMO Systems

CONTENTS

6.1 Introduction

Space time block coded (STBC) systems began with a purpose of providing transmit diversity for multiple-input multiple-output (MIMO) systems. The foundations of STBC codes were laid down by Alamouti and hence the scheme with two transmit antennas is known as Alamouti STBC scheme [2]. We had brief discussions on basics of transmission and detection of STBC symbols using multiple antennas in Chapter 1. There we considered only the Alamouti transmission scheme which is a special case of STBC codes. Unlike in spatial modulation (SM) and transmit antenna selection (TAS), symbols are transmitted simultaneously in STBC from all antennas at the transmitter. This increases the hardware complexity as simultaneous transmission needs an radio frequency (RF) chain for each transmitting antenna. However, STBC schemes are capable of achieving full diversity orders unlike the SM systems. Though TAS systems can achieve full diversity order, they need feedback from the receiver to the transmitter in order to make the decision of the antenna to be selected for transmission. Thus, each STBC, TAS and SM system has its own merits and may be a good candidate for commercial use. But a system may be chosen as per the requirements for implementation.

STBC codes have been explored widely because of their capability of achieving full diversity order and simplicity of the detection process when they are orthogonal. The Alamouti code for two transmit antennas was generalized for any number of transmit antennas by Tarokh et al. in their publications [122, 123]. However, the main drawback of these designs is that they do not guarantee capacity preservation. In many of the cases, code rate of the design is less than 1, which means such codes will have less spectral efficiency than that of single-input single-output (SISO) systems (serial transmission without coding). To overcome this, it was proposed to use error correcting codes on the incoming bit sequence to combat fading to a certain level and then use STBC for parallel transmission to achieve full diversity order and unity coding gain [125]. These codes are known as space time trellis codes. In addition, SM can be employed for STBC transmission to improve the spectral efficiency of STBC schemes [7]. STBC systems, when used with an SM transmission scheme, is popularly known as space time block coded spatial modulation (STBC SM) systems. However, STBC SM systems introduce complexity in the detection process to achieve the advantage in spectral efficiency. Performance analysis of STBC codes and their integration with SM and TAS systems has also been widely studied by the researchers recently. A comparative study of STBC systems with orthogonal code designs with TAS systems for cognitive radio applications has been reported in [58]. The study has shown that STBC is a promising technique for cognitive radio applications.

Further, this chapter is organized as follows. In the next section, some design criteria for STBC codes will be discussed. This deals with the design of orthogonal STBC for both real and complex constellations. It will be followed by the STBC system model and introduction to performance parameters. Further, the performance of STBC systems over various generalized fading channels will be discussed.

6.2 STBC design criteria

While discussing STBC systems in Chapter 1, it was established that for the STBC codes to perform well, the fading has to be quasi static in nature, i.e., the fading condition of the channel must remain unchanged for the duration of transmission of a complete STBC symbol which comprises multiple time slots. This is required because the detection process involves the set of signals received at different times. The design criteria are nothing but the guidelines to obtain a code that gives optimum diversity gain and coding gain. A good STBC code should ideally give full diversity order and high coding gain. At the beginning, STBC codes were designed to give full diversity. Usually they do not provide any coding gain. In recent years, researchers have come up with STBC code designs for full rate and full diversity [88] for any number of antennas [35, 36].

6.2.1 Rank determinant criteria

Rank determinant criteria help to design an STBC code that gives optimum error performance. In Chapter 1, we considered the Alamouti scheme. The transmission was done using two antennas. The transmission was done in two time slots. In the first time slot, the incoming symbols were transmitted from the transmitter antennas. In the second time slot, these symbols were transmitted with some transformations. The STBC codeword for Alamouti transmission is given by

$$\mathbf{X} = \begin{bmatrix} \mathbf{x_1} & \mathbf{x_2} \end{bmatrix} = \begin{bmatrix} s_1 & -s_2^* \\ s_2 & s_1^* \end{bmatrix} \tag{6.1}$$

The detection process was simplified by some sort of combining of received signals in the two time slots. The two symbols were separated by this process and the detection of both the symbols was made possible separately.

To study the relation of rank determinant criteria with error performance, consider an STBC codeword designed for $N_t \times N_r$ MIMO system. Let the codeword be transmitted over T time slots. So, the STBC codeword is a matrix of dimensions $N_t \times T$ where number of rows gives number of transmit antennas and the number of columns gives the number of time slots over which the STBC codeword is transmitted. An STBC codeword can be given by

$$\mathbf{X}_1 = \begin{bmatrix} s_{11}^1 & s_{12}^1 & \cdots & s_{1T}^1 \\ s_{21}{}^1 & s_{22}^1 & \cdots & s_{2T}^1 \\ \vdots & \vdots & \ddots & \vdots \\ s_{N_t1}^1 & s_{N_t2}^1 & \cdots & s_{N_tT}^1 \end{bmatrix} \tag{6.2}$$

where each element of the codeword matrix, s_{ij}^1, is a digitally modulated symbol for the incoming bit sequence or a transformed version of it as discussed in

Chapter 1. An error is said to have occurred if the codeword $\mathbf{X}_1$ is transmitted and another codeword $\mathbf{X}_2$ is detected at the receiver such that

$$\mathbf{X}_2 = \begin{bmatrix} s_{11}^2 & s_{12}^2 & \cdots & s_{1T}^2 \\ s_{21}{}^2 & s_{22}^2 & \cdots & s_{2T}^2 \\ \vdots & \vdots & \ddots & \vdots \\ s_{N_t1}^2 & s_{N_t2}^2 & \cdots & s_{N_tT}^2 \end{bmatrix} \tag{6.3}$$

Note that there are some differences in the detection process from what we are discussed in Chapter 1.

In this case not necessarily each element of $\mathbf{X}_1$ and $\mathbf{X}_2$ are different. Difference of only some or at least one element is sufficient to consider the detection to be erroneous. And the probability of the event that the transmitted STBC symbol is $\mathbf{X}_1$ and the detected STBC symbol for that transmission is $\mathbf{X}_2$ is called the probability of error for the pair of $\mathbf{X}_1$ and $\mathbf{X}_2$. In general it is known as pairwise error probability (PEP). For the case that an STBC code contains a total of N distinct orthogonal codes, there will be ${}^NC_2$ total possible pairs of the codewords. But given that the codeword $\mathbf{X}_i$ is transmitted, we can have $N-1$ set of probable pairs if the detection is erroneous. Again for the case that $\mathbf{X}_1$ is transmitted, detection of any $\mathbf{X}_j$ other than $\mathbf{X}_1$ results in error. In total, there are $N-1$ such cases, $\mathbf{X}_1 \rightarrow \mathbf{X}_2$, $\mathbf{X}_1 \rightarrow \mathbf{X}_3$, $\cdots$, $\mathbf{X}_1 \rightarrow \mathbf{X}_N$. Using the pairwise error probabilities for each of the pairs, upper bound on the probability of error can be given by union bound as

$$P_e(\mathbf{X}_1) \leq \sum_{i=2}^{N} P(\mathbf{X}_1 \rightarrow \mathbf{X}_i) \tag{6.4}$$

Note that the above expression gives probability that codeword $\mathbf{X}_1$ is received in error. To obtain probability of error for any other codeword, the above expression can be modified accordingly and the overall probability of error can be obtained by averaging the probability of error for each codeword. The maximum likelihood (ML) detection rule used in this case can be given by

$$\hat{\mathbf{X}} = \arg \min_{\mathbf{X} \in \mathbf{X}_{all}} \|\mathbf{R} - \mathbf{H}\mathbf{X}\|_F^2 \tag{6.5}$$

where $\|\cdot\|_F$ is the Frobenius norm of a matrix defined as $\|\mathbf{A}\|_F = \sqrt{tr(\mathbf{A}^H\mathbf{A})}$, $\mathbf{X}_{all}$ is the set of all possible STBC codewords of the dimension $N_t \times T$, $\mathbf{H}$ is the channel matrix of dimension $N_r \times N_t$ and $\mathbf{R} = \mathbf{H}\mathbf{X}_1 + \mathbf{N}$ is the received signal matrix of dimension $N_r \times T$ for the case that $\mathbf{X}_1$ is transmitted.

In this case, pairwise error probability can be given by the following relation.

$$P\left(\mathbf{X}_1 \rightarrow \mathbf{X}_2\right) = P\left(\|\mathbf{R} - \mathbf{H}\mathbf{X}_2\|_F^2 \leq \|\mathbf{R} - \mathbf{H}\mathbf{X}_1\|_F^2\right) \tag{6.6}$$

The above expression can be simplified to obtain the PEP as follows. The detailed simplification is given in [50].

$$P\left(\mathbf{X}_1 \rightarrow \mathbf{X}_2\right) = P\left(\widetilde{N} > \|\left(\mathbf{X_2} - \mathbf{X_1}\right)\mathbf{H}\|_F\right) \tag{6.7}$$

where $\widetilde{N} = tr\left(\mathbf{N}^H\mathbf{H}\left(\mathbf{X_2} - \mathbf{X_1}\right) + \left(\mathbf{X_2} - \mathbf{X_1}\right)^H \mathbf{H}^H\mathbf{N}\right)$ which is a zero mean Gaussian multivariate with variance $2N_0\| \left(\mathbf{X_2} - \mathbf{X_1}\right)\mathbf{H}\|_F$ for a given channel matrix, $\mathbf{H}$.

So, the PEP can be evaluated by using Gaussian Q function as

$$P\left(\mathbf{X}_1 \rightarrow \mathbf{X}_2\right) = Q\left(\sqrt{\frac{\gamma}{2}}\| \left(\mathbf{X_2} - \mathbf{X_1}\right)\mathbf{H}\|_F\right) \tag{6.8}$$

To calculate the PEP using the above expression, Frobenius norm of $\left(\mathbf{X_2} - \mathbf{X_1}\right)\mathbf{H}$ is required. It can be calculated as

$$\| \left(\mathbf{X_2} - \mathbf{X_1}\right)\mathbf{H}\|_F^2 = tr\left(\mathbf{H}\left(\mathbf{X_2} - \mathbf{X_1}\right)\left(\mathbf{X_2} - \mathbf{X_1}\right)^H \mathbf{H}^H\right) \tag{6.9}$$

To calculate it let us define a symbol for the codeword difference matrix $\mathbf{X_n} - \mathbf{X_m}$ in general as $\mathbf{D}^{(m,n)}$. Now the Frobenius norm can be represented as

$$\| \left(\mathbf{X_2} - \mathbf{X_1}\right)\mathbf{H}\|_F^2 = tr\left(\mathbf{H}\left(\mathbf{D}^{(1,2)}\right)\left(\mathbf{D}^{(1,2)}\right)^H \mathbf{H}^H\right) \tag{6.10}$$

Note that the eigenvalues of $\mathbf{Z}^{(m,n)} = \left(\mathbf{D}^{(m,n)}\right)\left(\mathbf{D}^{(m,n)}\right)^H$ will all be real and positive. So, using eigenvalue decomposition (spectral theorem), $\mathbf{Z}^{(m,n)}$ can be given as

$$\mathbf{Z}^{(m,n)} = \mathbf{V}\mathbf{\Lambda}\mathbf{V}^H \tag{6.11}$$

where $\mathbf{V}$ is a unitary matrix and $\mathbf{\Lambda}$ is the eigenvalue matrix of $\mathbf{Z}^{(m,n)}$ which is a diagonal matrix with $\mathbf{Z}^{(m,n)} = diag(\lambda_1, \lambda_2, \cdots)$. So,

$$\| \left(\mathbf{X_2} - \mathbf{X_1}\right)\mathbf{H}\|_F^2 = tr\left(\mathbf{H}\mathbf{V}\mathbf{\Lambda}\mathbf{V}^H\mathbf{H}^H\right) \tag{6.12}$$

Note that all the channel coefficients are mutually independent complex Gaussian random variables and hence if we denote the $(p,q)^{th}$ element of $\mathbf{HV}$ as $\psi_{p,q}$, the $(q,p)^{th}$ element of $\mathbf{V}^H\mathbf{H}^H$ will be the complex conjugate of $\psi_{p,q}$, i.e., $\psi_{p,q}^*$. So, j^{th} diagonal element of $\left(\mathbf{X_2} - \mathbf{X_1}\right)\mathbf{H}\mathbf{H}^H\left(\mathbf{X_2} - \mathbf{X_1}\right)^H$ can thus be given as $\sum_{i=1}^{r} \lambda_i |\psi_{j,i}|^2$. Therefore, the Frobenius norm of $\left(\mathbf{X_2} - \mathbf{X_1}\right)\mathbf{H}$ can be given as

$$\| \left(\mathbf{X_2} - \mathbf{X_1}\right)\mathbf{H}\|_F^2 = \sum_{j=1}^{N_r}\sum_{i=1}^{r} \lambda_i |\psi_{j,i}|^2 \tag{6.13}$$

where r is the rank of $\left(\mathbf{D}^{(1,2)}\right)\left(\mathbf{D}^{(1,2)}\right)^H$ which is the same as the number of non-zero eigenvalues, i.e., there exist only $\lambda_1, \lambda_2, \cdots, \lambda_r$ non-zero eigenvalues and $\lambda_{r+1} = \lambda_{r+2} = \cdots = \lambda_{N_t} = 0$. So, PEP can now be given as

$$P\left(\mathbf{X}_1 \rightarrow \mathbf{X}_2\right) = Q\left(\sqrt{\frac{\gamma}{2}\sum_{j=1}^{N_r}\sum_{i=1}^{r} \lambda_i |\psi_{j,i}|^2}\right) \tag{6.14}$$

Looking at the above expression of PEP, one can comment that the larger the number of non-zero eigenvalues, the lesser will be the PEP because of the larger value of the argument of Gaussian Q function. This larger argument of Q function is on account of the fact that the eigenvalues are non-negative. So, one statement that can be made about STBC codes is that the codes having the codeword difference matrix such that $\left(\mathbf{D}^{(m,n)}\right)\left(\mathbf{D}^{(m,n)}\right)^{H}$ is a full rank matrix for all values of m and n will perform better compared to those not having full rank.

Moreover, in the above expression, we still have random values $|\psi_{j,i}|^2$ dependent on the channel. Note that the values are coming from the product $\mathbf{HV}$. The channel coefficients are assumed to be mutually independent and Rayleigh distributed. Therefore, the independent Gaussian nature of elements of $\mathbf{H}$ makes each of $|\psi_{j,i}|$ to be Rayleigh distributed. So, we take high signal-to-noise ratio (SNR) regime under consideration to approximate Q function as $Q(x) \leq \frac{1}{2}e^{-\frac{x^2}{2}}$. Thus, using this approximation and averaging the PEP expression, we obtain

$$P\left(\mathbf{X}_1 \rightarrow \mathbf{X}_2\right) \leq \prod_{i=1}^{r}\left(1+\frac{\overline{\gamma}\lambda_i}{4}\right)^{-N_r} \tag{6.15}$$

For asymptotic high SNR conditions, one can also be neglected to obtain an upper bound on PEP as

$$P\left(\mathbf{X}_1 \rightarrow \mathbf{X}_2\right) \leq \frac{4^{rN_r}}{\left(\prod_{i=1}^{r}\lambda_i\right)^{M}}\overline{\gamma}^{-rN_r} \tag{6.16}$$

When we consider the overall performance of an STBC system, error probability is dominated by the STBC codeword pair for which the matrix $\mathbf{Z}^{(m,n)}$ has the lowest rank. For a while, it can be assumed that $\mathbf{Z}^{(1,2)}$ has the lowest rank. So, the PEP $P\left(\mathbf{X}_1 \rightarrow \mathbf{X}_2\right)$ becomes the worst case for error performance of the STBC MIMO system. The expression of equation (6.16) can be compared with a standard expression on probability of error, $\left(G_c\gamma\right)^{-G_d}$, for a system with G_c as coding gain and G_d as diversity gain. It can be observed that the diversity gain achieved by STBC systems is $G_d = rN_r$. The full rank of matrix $\mathbf{Z}^{(m,n)}$ can be N_t. So, when an STBC code is designed such that the matrix $\mathbf{Z}^{(m,n)}$ is a full rank matrix for all values of m and n, the system will achieve full diversity. Again, from equation (6.16), it is observed that the coding gain is proportional to the product of non-zero eigenvalues of the matrix $\mathbf{Z}^{(m,n)}$, i.e., $\prod_{i=1}^{r}\lambda_i$ which can be given as $G_c^{(m,n)} = \det\left(\mathbf{Z}^{(m,n)}\right)$.

From the above discussion, the rank determinant criteria for STBC code design can be stated that a well-designed STBC code shall have codeword difference matrix such that $\mathbf{Z}^{(m,n)}$ is a full rank matrix for all values of m and n. This makes an STBC code achieve full diversity order. In addition, for full rank matrices, the code shall have maximum value of det $\left(\mathbf{Z}^{(m,n)}\right)$. The full rank criteria maximize the diversity order and the determinant criteria maximize the coding gain of STBC systems.

6.2.2 Trace criteria

It has already been discussed that the calculation of PEP for the case that an STBC codeword $\mathbf{X}_1$ is transmitted and the codeword $\mathbf{X}_2$ is received in equation (6.8) contains the term $\|(\mathbf{X_2} - \mathbf{X_1})\mathbf{H}\|_F$. This Frobenius norm was evaluated in the previous section using the relation between trace of a matrix and its Frobenius norm. Now, we will follow an alternate way to obtain the required Frobenius norm. It still uses the relation between trace of a matrix and its Frobenius norm. But, now we use the fact that $\|\mathbf{A}\|_F^2 = tr(\mathbf{A}\mathbf{A}^H) = tr(\mathbf{A}^H\mathbf{A})$. It can be given as

$$\|(\mathbf{X_2} - \mathbf{X_1})\mathbf{H}\|_F^2 = tr\left(\left(\mathbf{D}^{(1,2)}\right)^H \mathbf{H}^H\mathbf{H}\left(\mathbf{D}^{(1,2)}\right)\right) \tag{6.17}$$

Note that the only difference between equation (6.9) and equation (6.17) is of the order of hermitian in the trace. For Rayleigh fading coefficients, the expectation of the matrix $\mathbf{H}^H\mathbf{H}$ in the above expression can be given as

$$E\left(\mathbf{H}^H\mathbf{H}\right) = N_r\mathbf{I}_{N_t} \tag{6.18}$$

where $\mathbf{I}_{N_t}$ is the identity matrix of dimensions $N_t \times N_t$. So, the expected value of $\|(\mathbf{X_2} - \mathbf{X_1})\mathbf{H}\|_F^2$ can be obtained as

$$E\left(\|(\mathbf{X_2} - \mathbf{X_1})\mathbf{H}\|_F^2\right) = N_r\ tr\left(\left(\mathbf{D}^{(1,2)}\right)^H\left(\mathbf{D}^{(1,2)}\right)\right) \tag{6.19}$$

$$= N_r\|\mathbf{D}^{(1,2)}\|_F^2 \tag{6.20}$$

Note that the term $\|\mathbf{D}^{(1,2)}\|_F$ can be interpreted as the distance between the codeword matrices $\mathbf{X}_1$ and $\mathbf{X}_2$. In general, the distance between STBC codeword matrices $\mathbf{X}_m$ and $\mathbf{X}_n$ can be obtained in the form of $\|\mathbf{D}^{(m,n)}\|_F$. In the case of SISO systems, the distances between codewords or symbols is defined in terms of Euclidean distance.

Alternatively, for $N_r \to \infty$

$$\frac{\left(\mathbf{H}^H\mathbf{H}\right)}{N_r} = \mathbf{I}_{N_t} \tag{6.21}$$

So, now let us consider the case that there are a large number of antennas available at the receiver. For increasing number of antennas at the receiver, $\mathbf{H}^H\mathbf{H}$ tend to become a diagonal matrix with all the diagonal elements having the same values. And this is the reason massive MIMO systems are not very far from reality. Having a large number of receiver antennas is no longer an unrealistic scenario with developments in large MIMO systems and Terahertz technology [19]. So, for a large number of receiver antennas, the norm relations can be given as

$$\lim_{N_r \to \infty} \frac{\left(\| \left(\mathbf{X_2} - \mathbf{X_1}\right) \mathbf{H}\|_F^2\right)}{N_r} = \|\mathbf{D}^{(1,2)}\|_F^2 \tag{6.22}$$

Thus, PEP can virtually be represented independent of the channel conditions as

$$P\left(\mathbf{X}_1 \to \mathbf{X}_2\right) = Q\left(\sqrt{\frac{N_r \gamma}{2}\|\mathbf{D}^{(1,2)}\|_F^2}\right) \tag{6.23}$$

An upper bound on the PEP can be obtained by approximating the Gaussian Q function as $Q(x) \leq \frac{1}{2}e^{-\frac{x^2}{2}}$.

$$P\left(\mathbf{X}_1 \to \mathbf{X}_2\right) \leq \frac{1}{2}e^{-\frac{N_r \gamma}{4}\|\mathbf{D}^{(1,2)}\|_F^2} \tag{6.24}$$

This shows that the pairwise error probability is related to the distances between STBC codewords and it is dominated by the pair of STBC codewords having minimum distance between them. The distance metric for STBC codewords is defined as $\|\mathbf{D}^{(1,2)}\|_F^2$. It can be observed that an STBC code can be made better in terms of error performance if the minimum distance between two codewords can be maximized. This criterion related to distance is known as trace criteria for STBC design because $tr\left(\mathbf{Z}^{m,n}\right) = \|\mathbf{D}^{(m,n)}\|_F^2$.

6.2.3 Rank determinant criteria verification for Alamouti space time code

We have discussed the Alamouti space time code in Chapter 1. It is mainly designed for two transmit antennas and an STBC codeword transmission in two time slots. So, an Alamouti STBC codeword $\mathbf{X}$ can be given by

$$\mathbf{X} = \left[\begin{array}{cc} \mathbf{x_1} & \mathbf{x_2} \end{array}\right] = \left[\begin{array}{cc} s_1 & -s_2^* \\ s_2 & s_1^* \end{array}\right] \tag{6.25}$$

where $\mathbf{x}_1$ represents the transmitted symbol vector in first time slot, $\mathbf{x}_2$ is the symbol vector transmitted in the second time slot and s_i is a symbol from a constellation for M-ary digital modulation scheme being used.

The rank criteria and the determinant criteria for an STBC code is based on the STBC codeword difference matrix. So, let us define another codeword $\widetilde{\mathbf{X}} \neq \mathbf{X}$ as

$$\widetilde{\mathbf{X}} = \left[\begin{array}{cc} \widetilde{\mathbf{x}}_1 & \widetilde{\mathbf{x}}_2 \end{array}\right] = \left[\begin{array}{cc} \widetilde{s}_1 & -\widetilde{s}_2^* \\ \widetilde{s}_2 & \widetilde{s}_1^* \end{array}\right] \tag{6.26}$$

The codeword difference matrix for these codewords can be given by

$$\mathbf{D}^{(\mathbf{X},\widetilde{\mathbf{X}})} = \left[\begin{array}{cc} \widetilde{s}_1 - s_1 & -\widetilde{s}_2^* + s_2^* \\ \widetilde{s}_2 - s_2 & \widetilde{s}_1^* - s_1^* \end{array}\right] \tag{6.27}$$

Rank of all possible codeword difference matrices is to be evaluated to verify the rank criteria. But, since the two codeword matrices are different, at least one of the elements in the first column of $\mathbf{D}^{(\mathbf{X},\widetilde{\mathbf{X}})}$ is non-zero. Correspondingly, one element in the second column of $\mathbf{D}^{(\mathbf{X},\widetilde{\mathbf{X}})}$ will also be non-zero and the non-zero element in the second column will belong to a different row. Thus, in Alamouti STBC transmission scheme, all the codeword difference matrices are full rank and the rank is 2 which is equal to number of transmit antennas.

Moreover, the codeword difference matrix is such that at least diagonal elements will have non-zero values. This means that determinant of $\mathbf{D}^{(\mathbf{X},\widetilde{\mathbf{X}})}$ will be zero if and only if the codewords are the same, i.e., $\mathbf{X} = \widetilde{\mathbf{X}}$ and the difference matrix has all zero elements. $\det\left(\mathbf{D}^{(\mathbf{X},\widetilde{\mathbf{X}})}\right) \neq 0$ also shows that the difference matrix $\mathbf{D}^{(\mathbf{X},\widetilde{\mathbf{X}})}$ is a full rank matrix. Also, in Chapter 1, it is discussed that Alamouti STBC schemes offer unity code rates, i.e., a symbol is transmitted on every time slot of transmission on average. From the above discussions, it is clear that the Alamouti STBC code is perfectly designed as far as the rank determinant criteria are concerned.

The full rank characteristics and the simple detection process can also be explained by alternate representation of channel matrix. For this purpose, let us refer to the symbols received in the two time slots of Alamouti STBC transmission. Repeating the expressions for the signal received in respective time slots of the transmission as in equation (1.6) and equation (1.7), r_1 and r_2 may be represented as

$$r_1 = h_{11}s_1 + h_{12}s_2 + n_1 \tag{6.28}$$

$$r_2 = -h_{11}s_2^* + h_{12}s_1^* + n_2 \tag{6.29}$$

Taking complex conjugate of the signal received in the second time slot,

$$r_1 = h_{11}s_1 + h_{12}s_2 + n_1 \tag{6.30}$$

$$r_2^* = -s_2h_{11}^* + s_1h_{12}^* + n_2^* \tag{6.31}$$

It leads to an observation that the signals received in both time slots can be represented in matrix form as

$$\begin{bmatrix} r_1 \\ r_2^* \end{bmatrix} = \begin{bmatrix} h_{11} & h_{12} \\ h_{12}^* & -h_{11}^* \end{bmatrix} \begin{bmatrix} s_1 \\ s_2 \end{bmatrix} + \begin{bmatrix} n_1 \\ n_2^* \end{bmatrix} \tag{6.32}$$

In matrix and vector notations, the above two expressions can be represented as single expression as

$$\widetilde{\mathbf{r}} = \mathbf{H}_{eq}\mathbf{s} + \widetilde{\mathbf{n}} \tag{6.33}$$

where the equivalent channel matrix is obtained by rearrangement of the actual channel coefficients h_{11} and h_{21} as

$$\mathbf{H}_{eq} = \begin{bmatrix} h_{11} & h_{12} \\ h_{12}^* & -h_{11}^* \end{bmatrix} \tag{6.34}$$

Note that $\mathbf{H}_{eq}$ in the above expression has an interesting property that leads Alamouti STBC codes to hold the property of separation of the transmitted symbols and the full diversity criterion.

$$\mathbf{H}_{eq}^H\mathbf{H}_{eq} = \begin{bmatrix} {h_{11}}^* & h_{12} \\ {h_{12}}^* & -h_{11} \end{bmatrix} \begin{bmatrix} h_{11} & h_{12} \\ {h_{12}}^* & -{h_{11}}^* \end{bmatrix} \tag{6.35}$$

$$= \begin{bmatrix} |h_{11}|^2 + |h_{12}|^2 & 0 \\ 0 & |h_{11}|^2 + |h_{12}|^2 \end{bmatrix} \tag{6.36}$$

This property of matric $\mathbf{H}_{eq}$ can be useful in detection of STBC codes in general. Multiplying $\widetilde{\mathbf{r}}$ with $\mathbf{H}_{eq}$ separates the transmitted symbols as follows.

$$\mathbf{H}_{eq}^H\widetilde{\mathbf{r}} = \begin{bmatrix} |h_{11}|^2 + |h_{12}|^2 & 0 \\ 0 & |h_{11}|^2 + |h_{12}|^2 \end{bmatrix} \begin{bmatrix} s_1 \\ s_2 \end{bmatrix} + \begin{bmatrix} \widetilde{n_1} \\ \widetilde{n_2^*} \end{bmatrix} \tag{6.37}$$

In the above expression, we can observe that the transmitted symbols s_1 and s_2 are separated as a part of the first row and second row respectively. Also, the multiplying factor that has come into existence due to fading channel coefficients is the same for both symbols and it is $|h_{11}|^2 + |h_{12}|^2$. This means that the signal gets deteriorated due to h_{11} and h_{12} while passing through the channel from transmitter to receiver. But, after doing the operation $\mathbf{H}_{eq}^H\widetilde{\mathbf{r}}$ for detection, a higher value of magnitude of either h_{11} or h_{21} helps in correct detection. This leads to the system offering full diversity order. Also, the same operation leads to separation of the transmitted symbols in an Alamouti STBC codeword. This makes the detection process simple. Now, the separated symbols can be detected using the ML technique for SISO systems, i.e., apart from some initial matrix operations, detection complexity of such STBC MIMO systems is equivalent to that of SISO systems.

In the case of Alamouti STBC, we have just shown that it achieves full diversity. According to rank criteria, only the STBC systems with full rank difference matrix achieve full diversity. Thus, Alamouti STBC design has a full rank codeword difference matrix.

6.3 Generator matrix for STBC design

As in the case of error correcting codes, generator matrix is designed to introduce redundancy such that the resulting codeword system becomes capable of correcting errors as per the design. In the same way, STBC generator matrix introduces redundancy to achieve diversity gain in turn that reduces the error probabilities at greater rates with increasing SNR. Here, let us compare the matrices obtained for equivalent channel coefficients in equation (6.34) and the actual Alamouti STBC codeword.

$$\mathbf{H}_{eq} = \begin{bmatrix} h_{11} & h_{12} \\ h_{12}^* & -h_{11}^* \end{bmatrix} \tag{6.38}$$

$$\mathbf{X} = \begin{bmatrix} s_1 & -s_2^* \\ s_2 & s_1^* \end{bmatrix} \tag{6.39}$$

Note that both $\mathbf{H}_{eq}$ and $\mathbf{X}$ follow the property that $\mathbf{M}^H\mathbf{M}$ would be a diagonal matrix, i.e., they are orthogonal matrices. This allows the symbols transmitted in an STBC codeword to be separately decoded making the detection process computationally efficient. So, any orthogonal matrix of dimension $K \times K$ can be a candidate for STBC design given that it can only have K distinct elements. However, transformed values of these K elements are allowed to be valid entries of the matrix. The transformation may be any operation among negation, complex conjugate or negation of complex conjugate.

6.3.1 Generator matrix for real constellations

The generator matrix of an orthogonal STBC code for real constellation of size $N_t \times N_t$ can have entries only among the elements $x_1, x_2, \cdots, x_{N_t}$ and $-x_1, -x_2, \cdots, -x_{N_t}$. The corresponding generator matrix $\mathbf{G}$ holds the following property.

$$\mathbf{G}\mathbf{G}^T = \left(x_1^2 + x_2^2 + \cdots + x_{N_t}^2\right)\mathbf{I}_{N_t} \tag{6.40}$$

Using this property of generator matrix, the Alamouti STBC design with real constellation can be given by the generator matrix as follows.

$$\mathbf{G}_{Alamouti}^{real} = \begin{bmatrix} x_1 & -x_2 \\ x_2 & x_1 \end{bmatrix} \tag{6.41}$$

It shall be noted that the generator matrix for Alamouti STBC may look different from the one given in [50] as $\begin{bmatrix} x_1 & x_2 \\ x_2 & -x_1 \end{bmatrix}$. But, the design functions in the same way as far as the columns of the matrices are orthogonal to each other. Similarly, some other commonly used generator matrices for real orthogonal STBC designs can be taken as examples. We omit the discussions on designs of such generator matrices as this is not the focus of this book at this moment. Interested readers may follow discussions given in [50, 125]

$$\mathbf{G}_4 = \begin{bmatrix} x_1 & x_2 & x_3 & x_4 \\ x_2 & -x_1 & -x_4 & x_3 \\ x_3 & x_4 & -x_1 & -x_2 \\ x_4 & -x_3 & x_2 & -x_1 \end{bmatrix} \tag{6.42}$$

$$\mathbf{G}_8 = \begin{bmatrix} x_1 & x_2 & x_3 & x_4 & x_5 & x_6 & x_7 & x_8 \\ x_2 & -x_1 & -x_4 & x_3 & -x_6 & x_5 & x_8 & -x_7 \\ x_3 & x_4 & -x_1 & -x_2 & x_7 & x_8 & -x_5 & -x_6 \\ x_4 & -x_3 & x_2 & -x_1 & x_8 & -x_7 & x_6 & -x_5 \\ x_5 & x_6 & -x_7 & -x_8 & -x_1 & -x_2 & x_3 & x_4 \\ x_6 & -x_5 & -x_8 & x_7 & x_2 & -x_1 & -x_4 & x_3 \\ x_7 & -x_8 & x_5 & -x_6 & -x_3 & x_4 & -x_1 & x_2 \\ x_8 & x_7 & x_6 & x_5 & -x_4 & -x_3 & -x_2 & -x_1 \end{bmatrix} \tag{6.43}$$

where $\mathbf{G}_4$ and $\mathbf{G}_8$ are the generator matrices for orthogonal STBC design with four and eight transmitter antennas respectively. The code rate in the above STBC designs is 1. In the discussion of generator matrix until this point, we considered that the number of time slots for an STBC codeword transmission is the same as the number of transmit antennas. But in general this is not a required condition. In general, we can have generalized STBC design such that the generator matrix satisfies the following condition.

$$\mathbf{G}\mathbf{G}^T = c\left(x_1^2 + x_2^2 + \cdots + x_k^2\right)\mathbf{I}_{N_t} \tag{6.44}$$

where c is a constant, $\mathbf{I}_{N_t}$ is an identity matrix of dimension $N_t \times N_t$, and $\mathbf{G}$ is the generator matrix of dimension $N_t \times T$ for an STBC MIMO system with N_t transmit antennas that transmit an STBC codeword over T time slots. The generator matrix has the entries $x_1, x_2, \cdots, x_k$ and $-x_1, -x_2, \cdots, -x_k$. That means k symbols are transmitted over T time slots. Such codes are known to have a code rate of $\frac{k}{T}$. Such generator matrices may be denoted by $\mathbf{G}^{N_t,T,k}$, giving information about number of antennas and code rate of the STBC design. Some generator matrix examples can be given for such generalized STBC designs for real constellation.

$$\mathbf{G}^{3,4,4} = \begin{bmatrix} x_1 & x_2 & x_3 & x_4 \\ x_2 & -x_1 & -x_4 & x_3 \\ x_3 & x_4 & -x_1 & -x_2 \end{bmatrix} \tag{6.45}$$

$$\mathbf{G}^{7,8,8} = \begin{bmatrix} x_1 & x_2 & x_3 & x_4 & x_5 & x_6 & x_7 & x_8 \\ x_2 & -x_1 & -x_4 & x_3 & -x_6 & x_5 & x_8 & -x_7 \\ x_3 & x_4 & -x_1 & -x_2 & x_7 & x_8 & -x_5 & -x_6 \\ x_4 & -x_3 & x_2 & -x_1 & x_8 & -x_7 & x_6 & -x_5 \\ x_5 & x_6 & -x_7 & -x_8 & -x_1 & -x_2 & x_3 & x_4 \\ x_6 & -x_5 & -x_8 & x_7 & x_2 & -x_1 & -x_4 & x_3 \\ x_7 & -x_8 & x_5 & -x_6 & -x_3 & x_4 & -x_1 & x_2 \end{bmatrix} \tag{6.46}$$

Note that the generator matrices to achieve unity code rate $\mathbf{G}^{3,4,4}$ and $\mathbf{G}^{7,8,8}$ can be obtained by removing a row respectively from the generator matrices $\mathbf{G}_4$ and $\mathbf{G}_8$ discussed earlier for the case that the STBC codeword was transmitted for the number of time slots exactly the same as that of the transmitter antennas. Similarly, by just keeping on reducing the number of rows for each reduced number of transmitter antennas, an orthogonal STBC code design with generator matrix $\mathbf{G}^{N_t,T,T}$ of dimension $N_t \times T$ can be obtained accordingly. Next, we will discuss some of the requirements and some properties of generator matrix for complex constellations.

6.3.2 Generator matrix for complex constellations

So far, we discussed generator matrix for orthogonal STBC designs with real constellations. But in practice, complex constellations have been popular for their capability to save bandwidth and power efficiency. Again, for orthogonal STBC design with complex constellation, the required condition on the generator matrix can be given by

$$\mathbf{G}\mathbf{G}^{H} = c\left(x_1^2 + x_2^2 + \cdots + x_k^2\right)\mathbf{I}_{N_t} \tag{6.47}$$

It has already been shown that full rate orthogonal STBC design with complex constellations exists only for $N_t = 2$ [124] and it is Alamouti STBC code. As already mentioned before, nowadays, full rate and full diversity STBC for any number of antennas have been reported in [36]. The generator matrix for Alamouti STBC designs with complex constellations has already been discussed and used in Chapter 1 and Section 6.2.1. It is repeated below to maintain continuity.

$$\mathbf{G}^{complex}_{Alamouti} = \begin{bmatrix} x_1 & -x_2^* \\ x_2 & x_1^* \end{bmatrix} \tag{6.48}$$

Further, orthogonal STBC designs with reduced code rate are possible with complex constellations. Some examples of generator matrices for orthogonal STBC design with complex constellations can be given from [50].

$$\mathbf{G}^{4,8,4} = \begin{bmatrix} x_1 & -x_2 & -x_3 & -x_4 & x_1^* & -x_2^* & -x_3^* & -x_4^* \\ x_2 & x_1 & x_4 & -x_3 & x_2^* & x_1^* & x_4^* & -x_3^* \\ x_3 & -x_4 & x_1 & x_2 & x_3^* & -x_4^* & x_1^* & x_2^* \\ x_4 & x_3 & -x_2 & x_1 & x_4^* & x_3^* & -x_2^* & x_1^* \end{bmatrix} \tag{6.49}$$

Note that the above expression has a close relation between the first four columns and the last four columns. The elements of the first four columns are repeated as the last four columns and the only difference is of conjugate operation. In the other half of the columns all the elements are in complex conjugate form. Also, it can be easily observed that the generator matrix of orthogonal STBC design for real constellations can be easily extended to obtain a generator matrix for orthogonal STBC design with complex constellation. Comparing the above generator matrix of orthogonal STBC design with complex constellation, equation (6.49), with the generator matrix of orthogonal STBC design for real constellations, equation (6.42), it can be observed that the generator matrix for real constellation orthogonal STBC design is the same as the first four columns of the generator matrix for complex constellation orthogonal STBC design.

In the generator matrix given by equation (6.49) and others for real constellation orthogonal STBC designs, all the elements of generator matrix are non-zero and each element is either a symbol being transmitted or has a close relation with some of the symbols to be transmitted. However, it is not always required to have non-zero symbol entry as an element of the generator matrix for STBC designs. Some examples of such a generator matrix with zeros as the elements are given below.

$$\mathbf{G}^{4,4,3} = \begin{bmatrix} x_1 & -x_2^* & x_3^* & 0 \\ x_2 & x_1^* & 0 & x_3^* \\ x_3 & 0 & -x_1^* & -x_2^* \\ 0 & x_3 & x_2 & -x_1 \end{bmatrix} \tag{6.50}$$

$$\mathbf{G}^{8,4,8} = \begin{bmatrix} x_1 & -x_2{}^* & x_3{}^* & 0 & x_4^* & 0 & 0 & 0 \\ x_2 & x_1{}^* & 0 & x_3{}^* & 0 & x_4^* & 0 & 0 \\ x_3 & 0 & -x_1{}^* & -x_2{}^* & 0 & 0 & x_4^* & 0 \\ 0 & x_3 & x_2 & -x_1 & 0 & 0 & 0 & x_4^* \\ x_4 & 0 & 0 & 0 & -x_1^* & -x_2^* & -x_3^* & 0 \\ 0 & x_4 & 0 & 0 & x_2 & -x_1 & 0 & -x_3^* \\ 0 & 0 & x_4 & 0 & -x_3 & 0 & x_1 & -x_2^* \\ 0 & 0 & 0 & x_4 & 0 & -x_3 & x_2 & x_1^* \end{bmatrix} \tag{6.51}$$

Here, using the element 0 as an entry in the generator matrix indicates that the corresponding antenna is sitting idle in the corresponding time slot of the STBC codeword transmission. This actually saves power at the transmitter. For each non-zero entry in the generator matrix, transmitter has to transmit some symbol and the transmit power in addition to the power required for complete processing is needed and for the antennas remaining idle for certain time slots saves power. Instead of proceeding like equation (6.51) for generator matrix with zero elements, if the generator matrix for 8 transmit antennas is designed like in equation (6.49) without any non-zero entries, both would result in code rate of $\frac{1}{2}$. But, in the later case, it would require a block of 16 time slots for transmission of a full STBC codeword while in the case of equation (6.51), 8 time slots are required for transmission of an STBC codeword. However, in the first case 8 symbols are transmitted in 16 time slots and 4 symbols are transmitted in 8 time slots. Thus, code rate is the same in both cases. But, processing the received signals in 8 time slots is always easier than processing the received signals in 16 time slots mainly for the buffer memory and computational demands. So, in the case that 4 symbols are transmitted over 8 time slots, the delay between first symbol arrival at the transmitter and the fourth symbol received at the receiver is small compared to the case that 8 symbols are transmitted over 16 time slots. So, it is better to choose the orthogonal STBC designs with complex constellation having generator matrix with some zero elements than those generator matrices having all the non-zero elements.

We have discussed about orthogonal STBC designs with real and complex constellation signal points. Apart from this, there are other designs for STBC codes which includes quasi orthogonal STBC, diagonally orthogonal STBC, pseudo orthogonal STBC, etc. But in this book our main focus is on performance analysis of orthogonal STBC MIMO systems over generalized fading channels and hence detailed discussion only on orthogonal STBC designs is presented. In the next few sections, beginning with STBC detection, we will discuss performance analysis of STBC MIMO systems over various generalized fading channels. The methods to derive expressions for error probability will be discussed.

6.4 STBC detection

Consider an STBC MIMO system that has N_t antennas at the transmitter and N_r antennas at the receiver. An orthogonal STBC design is used to transmit a codeword in T time slots. In this section, we consider complex constellation signal points so that the discussion remains more generalized. The same detection schemes can be used for STBC systems with real constellation signal points too. In general the relation between transmitted signals and the received signals can be given as

$$\mathbf{R} = \mathbf{HX} + \mathbf{N} \tag{6.52}$$

where $\mathbf{R}$ is the matrix with dimension $N_r \times T$ representing received signals at N_r receiving antennas in T time slots, $\mathbf{H}$ is the channel matrix of dimension $N_r \times N_t$, $\mathbf{X}$ is the STBC codeword matrix of dimension $N_t \times T$ and $\mathbf{N}$ is the noise matrix with noise sample elements at each receiver antenna in all the time slots.

Note that the structure of STBC codewords is such that there are a total of N_tT instances of transmission but there can be maximum T symbols and others are added for redundancy because code rate is lesser than 1 and the aim of using STBC is to achieve full diversity order. The transmitted STBC codeword symbol can be detected at the receiver using the ML detection algorithm. It does extensive search over all possible transmitted STBC codeword matrices and detects the STBC codeword matrix which is most likely to have been transmitted using the following minimization criteria.

$$\hat{\mathbf{X}} = \begin{matrix} \text{arg} \\ \text{min} \\ \mathbf{X} \in \chi \end{matrix} \; \|\mathbf{R} - \mathbf{HX}\|^2 \tag{6.53}$$

where χ is the set of all possible STBC codeword matrix alphabets. These criteria do extensive search over all possible STBC codewords. The complexity grows exponentially with the number of codewords in this case. Also, by ML detection this way, the symbols are not detected separately which is a speciality of orthogonal STBC MIMO systems. The methods are available for each design so that the symbols can be separated from the received signals. In the next subsections, we will discuss these methods for single receiver antennas and multiple receiver antennas.

6.4.1 Detection for single receiver antenna

In this subsection, we discuss the method of separate detection of symbols for STBC designs using $\mathbf{G}^{4,8,4}$ in equation (6.49) for real designs. To explain the detection process, initially we discuss it for the case that the receiver has a single antenna. For a while, it is assumed that the system is noiseless. In this case, the received signal matrix can be given by

$$\mathbf{R} = \begin{bmatrix} h_{11} & h_{12} & h_{13} & h_{14} \end{bmatrix} \begin{bmatrix} x_1 & -x_2 & -x_3 & -x_4 & x_1^* & -x_2^* & -x_3^* & -x_4^* \\ x_2 & x_1 & x_4 & -x_3 & x_2^* & x_1^* & x_4^* & -x_3^* \\ x_3 & -x_4 & x_1 & x_2 & x_3^* & -x_4^* & x_1^* & x_2^* \\ x_4 & x_3 & -x_2 & x_1 & x_4^* & x_3^* & -x_2^* & x_1^* \end{bmatrix} \tag{6.54}$$

In this case, the received signal in each time slot can be given as

$$r_1 = h_{11}x_1 + h_{12}x_2 + h_{13}x_3 + h_{14}x_4 \tag{6.55}$$
$$r_2 = -h_{11}x_2 + h_{12}x_1 - h_{13}x_4 + h_{14}x_3 \tag{6.56}$$
$$r_3 = -h_{11}x_3 + h_{12}x_4 + h_{13}x_1 - h_{14}x_2 \tag{6.57}$$
$$r_4 = -h_{11}x_4 - h_{12}x_3 + h_{13}x_2 + h_{14}x_1 \tag{6.58}$$
$$r_5 = h_{11}x_1^* + h_{12}x_2^* + h_{13}x_3^* + h_{14}x_4^* \tag{6.59}$$
$$r_6 = -h_{11}x_2^* + h_{12}x_1^* - h_{13}x_4^* + h_{14}x_3^* \tag{6.60}$$
$$r_7 = -h_{11}x_3^* + h_{12}x_4^* + h_{13}x_1^* - h_{14}x_2^* \tag{6.61}$$
$$r_8 = -h_{11}x_4^* - h_{12}x_3^* + h_{13}x_2^* + h_{14}x_1^* \tag{6.62}$$

To represent the received signals in the form of equation (6.33) as discussed earlier, take complex conjugate of r_5, r_6, r_7 and r_8. The following relations are obtained.

$$r_1 = h_{11}x_1 + h_{12}x_2 + h_{13}x_3 + h_{14}x_4 \tag{6.63}$$
$$r_2 = -h_{11}x_2 + h_{12}x_1 - h_{13}x_4 + h_{14}x_3 \tag{6.64}$$
$$r_3 = -h_{11}x_3 + h_{12}x_4 + h_{13}x_1 - h_{14}x_2 \tag{6.65}$$
$$r_4 = -h_{11}x_4 - h_{12}x_3 + h_{13}x_2 + h_{14}x_1 \tag{6.66}$$
$$r_5^* = h_{11}^*x_1 + h_{12}^*x_2 + h_{13}^*x_3 + h_{14}^*x_4 \tag{6.67}$$
$$r_6^* = -h_{11}^*x_2 + h_{12}^*x_1 - h_{13}^*x_4 + h_{14}^*x_3 \tag{6.68}$$
$$r_7^* = -h_{11}^*x_3 + h_{12}^*x_4 + h_{13}^*x_1 - h_{14}^*x_2 \tag{6.69}$$
$$r_8^* = -h_{11}^*x_4 - h_{12}^*x_3 + h_{13}^*x_2 + h_{14}^*x_1 \tag{6.70}$$

Taking complex conjugate of r_5, r_6, r_7 and r_8 brings the received signals of all the times slots into a form that can be represented in matrix form. The above set of equations can be represented in matrix form as

$$\begin{bmatrix} r_1 \\ r_2 \\ r_3 \\ r_4 \\ r_5^* \\ r_6^* \\ r_7^* \\ r_8^* \end{bmatrix} = \begin{bmatrix} h_{11} & h_{12} & h_{13} & h_{14} \\ h_{12} & -h_{11} & h_{14} & -h_{13} \\ h_{13} & -h_{14} & -h_{11} & h_{12} \\ h_{14} & h_{13} & -h_{12} & -h_{11} \\ h_{11}^* & h_{12}^* & h_{13}^* & h_{14}^* \\ h_{12}^* & -h_{11}^* & h_{14}^* & -h_{13}^* \\ h_{13}^* & -h_{14}^* & -h_{11}^* & h_{12}^* \\ h_{14}^* & h_{13}^* & -h_{12}^* & -h_{11}^* \end{bmatrix} \begin{bmatrix} x_1 \\ x_2 \\ x_3 \\ x_4 \end{bmatrix} \tag{6.71}$$

The above system of received signals or their conjugates in the respective time slots can be interpreted as a spatial multiplexing MIMO system in which the same signal vector, $\begin{bmatrix} x_1 \\ x_2 \\ x_3 \\ x_4 \end{bmatrix}$, is transmitted from the multiple transmitter antennas and the transmitted symbols can be thought about as they have passed through an equivalent MIMO channel which can be given by the following matrix.

$$\mathbf{H}_{eq} = \begin{bmatrix} h_{11} & h_{12} & h_{13} & h_{14} \\ h_{12} & -h_{11} & h_{14} & -h_{13} \\ h_{13} & -h_{14} & -h_{11} & h_{12} \\ h_{14} & h_{13} & -h_{12} & -h_{11} \\ h_{11}^* & h_{12}^* & h_{13}^* & h_{14}^* \\ h_{12}^* & -h_{11}^* & h_{14}^* & -h_{13}^* \\ h_{13}^* & -h_{14}^* & -h_{11}^* & h_{12}^* \\ h_{14}^* & h_{13}^* & -h_{12}^* & -h_{11}^* \end{bmatrix} \tag{6.72}$$

In any communication system, estimation of channel coefficients is necessary for coherent detection. So, without loss of generality, it is assumed that the channel coefficients are available at the receiver. This makes it easier to obtain $\mathbf{H}_{eq}$ from the coefficients h_{11}, h_{12}, h_{13} and h_{14}. In noisy environments, the matrix form representation can be given as

$$\widetilde{\mathbf{r}} = \mathbf{H}_{eq}\mathbf{x} + \widetilde{\mathbf{n}} \tag{6.73}$$

where each element of $\widetilde{\mathbf{r}}$ represents r_i or r_i^* depending on the value of i and each element of $\widetilde{\mathbf{n}}$ represents effective additive white Gaussian noise (AWGN) in the received signal of the ith time slot.

To detect the symbols, $\widetilde{\mathbf{r}}$ is pre-multiplied by the Hermitian of effective channel matrix, $\mathbf{H}_{eq}^H$. This results in the following relation in which the symbols are separated by virtue of the orthogonal nature of the STBC code design. It can be given as

$$\mathbf{H}_{eq}^H\widetilde{\mathbf{r}} = \begin{bmatrix} h_{11}^* & h_{12}^* & h_{13}^* & h_{14}^* & h_{11} & h_{12} & h_{13} & h_{14} \\ h_{12}^* & -h_{11}^* & -h_{14}^* & +h_{13}^* & h_{12} & -h_{11} & -h_{14} & h_{13} \\ h_{13}^* & h_{14}^* & -h_{11}^* & -h_{12}^* & h_{13} & h_{14} & -h_{11} & -h_{12} \\ h_{14}^* & -h_{13}^* & h_{12}^* & -h_{11}^* & h_{14} & -h_{13} & h_{12} & -h_{11} \end{bmatrix} \mathbf{H}_{eq} \begin{bmatrix} x_1 \\ x_2 \\ x_3 \\ x_4 \end{bmatrix} + \mathbf{H}_{eq}^H\widetilde{\mathbf{n}} \tag{6.74}$$

This results in the expression given as

$$\mathbf{H}_{eq}^H\widetilde{\mathbf{r}} = 2\begin{bmatrix} \|\mathbf{h}\|_F^2 & 0 & 0 & 0 \\ 0 & \|\mathbf{h}\|_F^2 & 0 & 0 \\ 0 & 0 & \|\mathbf{h}\|_F^2 & 0 \\ 0 & 0 & 0 & \|\mathbf{h}\|_F^2 \end{bmatrix} \begin{bmatrix} x_1 \\ x_2 \\ x_3 \\ x_4 \end{bmatrix} + \mathbf{H}_{eq}^H\widetilde{\mathbf{n}} \tag{6.75}$$

$$\begin{bmatrix} \widetilde{r}_1 \\ \widetilde{r}_2 \\ \widetilde{r}_3 \\ \widetilde{r}_4 \end{bmatrix} = 2\begin{bmatrix} \|\mathbf{h}\|_F^2 x_1 \\ \|\mathbf{h}\|_F^2 x_2 \\ \|\mathbf{h}\|_F^2 x_3 \\ \|\mathbf{h}\|_F^2 x_4 \end{bmatrix} + \mathbf{H}_{eq}^H\widetilde{\mathbf{n}} \tag{6.76}$$

where $\|\mathbf{h}\|_F$ is the Frobenius norm of channel matrix which is a row vector in this case for $N_r = 1$ given by $\mathbf{h} = \begin{bmatrix} h_{11} & h_{12} & h_{13} & h_{14} \end{bmatrix}$.

From the above expression, equation (6.76), it is clearly understood that the ML detection rule for x_i can be given as

$$\hat{x}_i = \underset{x \in \Phi}{\arg\min} \left\| \widetilde{r}_i - \|\mathbf{h}\|_F^2 x \right\|^2 \tag{6.77}$$

where Φ is the set of all possible constellation signal points from an M-ary digital modulation scheme.

To summarize the detection process for orthogonal STBC designs,

- The signals received in different time slots are collected and stored in the memory.
- The vector is generated by taking complex conjugate of the signals received in the later half of the time slots.
- The vector obtained is pre-multiplied by the Hermitian of equivalent channel matrix, $\mathbf{H}_{eq}$.
- The result obtained is used in the ML rule of equation (6.77) to detect each of the symbols one by one.

Next, we will consider the same case to discuss STBC detection for multiple receiver antennas.

6.4.2 Detection for multiple receiver antenna

The detection process for orthogonal designs for STBC MIMO systems with multiple receiver antennas is just a simple extension of the process discussed for STBC MIMO systems with single receiver antennas. We consider the same system and the same generator matrix in the discussions of this section. For an $N_t \times N_r$ MIMO system, the channel matrix is of the dimension $N_r \times N_t$. For such a system, the channel matrix can be given as

$$\mathbf{H} = \begin{bmatrix} h_{11} & h_{12} & \cdots & h_{1N_t} \\ h_{21} & h_{22} & \cdots & h_{2N_t} \\ \vdots & \vdots & \ddots & \vdots \\ h_{N_r1} & h_{N_r2} & \cdots & h_{N_rN_t} \end{bmatrix} = \begin{bmatrix} \mathbf{h}_1 \\ \mathbf{h}_2 \\ \vdots \\ \mathbf{h}_{N_r} \end{bmatrix} \tag{6.78}$$

where $\mathbf{h}_j$ represents the jth row of channel matrix, $\mathbf{H}$. Each row corresponds to a receiver antenna and the elements of a row represent the channel coefficients between each transmitter antenna to that receiver antenna. We can take a generalized case of N_r antennas at the receiver, but to make it understood better we will discuss the detection process for $N_r = 2$ and an easy extension of generalizing it for N_r receiver antennas is left as an exercise to the readers.

For the case of orthogonal STBC design given by $\mathbf{G}^{4,8,4}$ as in equation (6.49), $N_t = 4$ and we are discussing the case that $N_r = 2$. Thus, the channel matrix can be given as

$$\mathbf{H} = \begin{bmatrix} h_{11} & h_{12} & h_{13} & h_{14} \\ h_{21} & h_{22} & h_{23} & h_{24} \end{bmatrix} = \begin{bmatrix} \mathbf{h}_1 \\ \mathbf{h}_2 \end{bmatrix} \tag{6.79}$$

Again, consider the system with no noise for a while. The received signal matrix relation can be given as

$$\mathbf{R} = \begin{bmatrix} h_{11} & h_{12} & h_{13} & h_{14} \\ h_{21} & h_{22} & h_{23} & h_{24} \end{bmatrix} \begin{bmatrix} x_1 & -x_2 & -x_3 & -x_4 & x_1^* & -x_2^* & -x_3^* & -x_4^* \\ x_2 & x_1 & x_4 & -x_3 & x_2^* & x_1^* & x_4^* & -x_3^* \\ x_3 & -x_4 & x_1 & x_2 & x_3^* & -x_4^* & x_1^* & x_2^* \\ x_4 & x_3 & -x_2 & x_1 & x_4^* & x_3^* & -x_2^* & x_1^* \end{bmatrix} \tag{6.80}$$

In general, $\mathbf{R}$ is the received signal matrix of dimensions $N_r \times T$. In this case the received signal matrix, $\mathbf{R}$, will have the dimensions 2×8. Let the received signals at antenna 1 be denoted by r_{1i} in the ith time slot and the received signals at antenna 2 be denoted by r_{2i} in the ith time slot. Note that thc signals received at antenna 1, r_{11} to r_{18} in respective time slots, can be exactly given by r_1 to r_8 in the expressions of equation (6.55) to equation (6.62) for the case of single receiver antennas. Similarly, the signals received at antenna 2, r_{21} to r_{28} in respective time slots, can be easily obtained by the same approach.

Further, the set of signals received at the receiver antenna 1 and the set of signals received at the receiver antenna 2 are separately processed as in the case of single receiver STBC MIMO systems. To begin with, the set of received signals is represented as

$$\widetilde{\mathbf{r}}_1 = \mathbf{H}_{eq1}\mathbf{x} + \widetilde{\mathbf{n}}_1 \tag{6.81}$$

and

$$\widetilde{\mathbf{r}}_2 = \mathbf{H}_{eq2}\mathbf{x} + \widetilde{\mathbf{n}}_2 \tag{6.82}$$

where $\widetilde{\mathbf{r}}_1 = \begin{bmatrix} r_{11} & r_{12} & r_{13} & r_{14} & r_{15}^* & r_{16}^* & r_{17}^* & r_{18}^* \end{bmatrix}^T$, $\widetilde{\mathbf{r}}_2 = \begin{bmatrix} r_{21} & r_{22} & r_{23} & r_{24} & r_{25}^* & r_{26}^* & r_{27}^* & r_{28}^* \end{bmatrix}^T$, $\widetilde{\mathbf{n}}_1$ is the effective noise at receiver antenna 1, $\widetilde{\mathbf{n}}_2$ is the effective noise at receiver antenna 2, and $\mathbf{H}_{eq1}$ and $\mathbf{H}_{eq2}$ are the equivalent channel matrices derived respectively from first and second row of $\mathbf{H}$ corresponding to the receiver antennas.

The next step is pre-multiplication of $\widetilde{\mathbf{r}}_1$ and $\widetilde{\mathbf{r}}_2$ respectively with $\mathbf{H}_{eq1}$ and $\mathbf{H}_{eq2}$. This does separation of the transmitted symbols during 8 time slots of a corresponding STBC codeword. Thus, after this processing, the signals received at antenna 1 with separated symbols can be given by

$$\mathbf{H}_{eq1}^H\widetilde{\mathbf{r}}_1 = 2\begin{bmatrix} \|\mathbf{h}_1\|_F^2 & 0 & 0 & 0 \\ 0 & \|\mathbf{h}_1\|_F^2 & 0 & 0 \\ 0 & 0 & \|\mathbf{h}_1\|_F^2 & 0 \\ 0 & 0 & 0 & \|\mathbf{h}_1\|_F^2 \end{bmatrix}\begin{bmatrix} x_1 \\ x_2 \\ x_3 \\ x_4 \end{bmatrix} + \mathbf{H}_{eq}^H\widetilde{\mathbf{n}} \tag{6.83}$$

$$\begin{bmatrix} \widetilde{r}_{11} \\ \widetilde{r}_{12} \\ \widetilde{r}_{13} \\ \widetilde{r}_{14} \end{bmatrix} = 2\begin{bmatrix} \|\mathbf{h}_1\|_F^2 x_1 \\ \|\mathbf{h}_1\|_F^2 x_2 \\ \|\mathbf{h}_1\|_F^2 x_3 \\ \|\mathbf{h}_1\|_F^2 x_4 \end{bmatrix} + \mathbf{H}_{eq1}^H\widetilde{\mathbf{n}}_1 \tag{6.84}$$

Similarly, after processing the signals received at antenna 2 with separated symbols can be given by

$$\mathbf{H}_{eq2}^H\widetilde{\mathbf{r}}_2 = 2\begin{bmatrix} \|\mathbf{h}_2\|_F^2 & 0 & 0 & 0 \\ 0 & \|\mathbf{h}_2\|_F^2 & 0 & 0 \\ 0 & 0 & \|\mathbf{h}_2\|_F^2 & 0 \\ 0 & 0 & 0 & \|\mathbf{h}_2\|_F^2 \end{bmatrix}\begin{bmatrix} x_1 \\ x_2 \\ x_3 \\ x_4 \end{bmatrix} + \mathbf{H}_{eq}^H\widetilde{\mathbf{n}} \tag{6.85}$$

$$\begin{bmatrix} \widetilde{r}_{21} \\ \widetilde{r}_{22} \\ \widetilde{r}_{23} \\ \widetilde{r}_{24} \end{bmatrix} = 2\begin{bmatrix} \|\mathbf{h}_2\|_F^2 x_1 \\ \|\mathbf{h}_2\|_F^2 x_2 \\ \|\mathbf{h}_2\|_F^2 x_3 \\ \|\mathbf{h}_2\|_F^2 x_4 \end{bmatrix} + \mathbf{H}_{eq2}^H\widetilde{\mathbf{n}}_2 \tag{6.86}$$

where $\mathbf{h}_1$ and $\mathbf{h}_2$ are respectively the first and second row of channel matrix which can be obtained from equation (6.79).

Now, as observed from equation (6.84) and equation (6.86), further processing for detection of symbols can be done by adding the two expressions, equation (6.84) and equation (6.86), and then using the ML detection rule.

This results in

$$\begin{bmatrix} \widetilde{r}_{11}+\widetilde{r}_{21} \\ \widetilde{r}_{12}+\widetilde{r}_{22} \\ \widetilde{r}_{13}+\widetilde{r}_{23} \\ \widetilde{r}_{14}+\widetilde{r}_{24} \end{bmatrix} = 2\begin{bmatrix} \left(\|\mathbf{h}_1\|_F^2+\|\mathbf{h}_2\|_F^2\right)x_1 \\ \left(\|\mathbf{h}_1\|_F^2+\|\mathbf{h}_2\|_F^2\right)x_2 \\ \left(\|\mathbf{h}_1\|_F^2+\|\mathbf{h}_2\|_F^2\right)x_3 \\ \left(\|\mathbf{h}_1\|_F^2+\|\mathbf{h}_2\|_F^2\right)x_4 \end{bmatrix} + \mathbf{H}_{eq1}^H\widetilde{\mathbf{n}}_1 + \mathbf{H}_{eq2}^H\widetilde{\mathbf{n}}_2 \qquad (6.87)$$

Overall, let us interpret the signal obtained after all these processing and mathematical manipulations on the received signals at all the receiver antennas in all the time slots corresponding to transmission of a complete STBC codeword as effective received signal. This enables us to represent the expression of equation (6.87) as

$$\begin{bmatrix} \widetilde{r}_1 \\ \widetilde{r}_2 \\ \widetilde{r}_3 \\ \widetilde{r}_4 \end{bmatrix} = \begin{bmatrix} \widetilde{r}_{11}+\widetilde{r}_{21} \\ \widetilde{r}_{12}+\widetilde{r}_{22} \\ \widetilde{r}_{13}+\widetilde{r}_{23} \\ \widetilde{r}_{14}+\widetilde{r}_{24} \end{bmatrix} = 2\begin{bmatrix} \left(\|\mathbf{H}\|_F^2\right)x_1 \\ \left(\|\mathbf{H}\|_F^2\right)x_2 \\ \left(\|\mathbf{H}\|_F^2\right)x_3 \\ \left(\|\mathbf{H}\|_F^2\right)x_4 \end{bmatrix} + \begin{bmatrix} n_{ef1} \\ n_{ef2} \\ n_{ef3} \\ n_{ef4} \end{bmatrix} \qquad (6.88)$$

where $\|\mathbf{H}\|_F$ is the Frobenius norm of channel matrix, $\mathbf{H}$, and n_{efi} is the effective noise in the effective received signal corresponding to the ith symbol.

From the above expression, equation (6.88), it is clearly understood that the ML detection rule for x_i can be given as

$$\hat{x}_i = \underset{x\in\Phi}{\arg\min} \left\|\widetilde{r}_i - 2\|\mathbf{H}\|_F^2 x\right\|^2 \qquad (6.89)$$

where Φ is the set of all possible constellation signal points from an M-ary digital modulation scheme.

So far our discussion was about the STBC detection for two receiver antennas. This detection process can be easily extended to generalized scenario of N_r receiver antennas. The only thing one needs to keep in mind for N_r antennas is that there will be N_r received signals in every time slot. These signals have to be processed in the same way as that discussed for $N_r = 2$ in this section previously. Finally, in a generalized scenario for an $N_t \times N_r$ STBC MIMO system transmitting k symbols with M-ary digital modulation scheme in T time slots, the effective received signal can be given as

$$\begin{bmatrix} \widetilde{r}_1 \\ \widetilde{r}_2 \\ \vdots \\ \widetilde{r}_k \end{bmatrix} = C\begin{bmatrix} \left(\|\mathbf{H}\|_F^2\right)x_1 \\ \left(\|\mathbf{H}\|_F^2\right)x_2 \\ \vdots \\ \left(\|\mathbf{H}\|_F^2\right)x_k \end{bmatrix} + \begin{bmatrix} n_{ef1} \\ n_{ef2} \\ \vdots \\ n_{efk} \end{bmatrix} \qquad (6.90)$$

where C is the constant whose value varies depending on the values of k, T and N_t. From the above expression, equation (6.90), and the expression of ML detection, equation (6.89), an ML detection rule for STBC system with N_r

receiver antennas can be given as

$$\hat{x}_i = \underset{x \in \Phi}{\arg\min} \left\| \widetilde{r}_i - C\|\mathbf{H}\|_F^2 x \right\|^2 = \underset{x \in \Phi}{\arg\min} \left\| \frac{1}{C}\widetilde{r}_i - \|\mathbf{H}\|_F^2 x \right\|^2 \tag{6.91}$$

where Φ is the set of all possible constellation signal points from an M-ary digital modulation scheme. The detection process for orthogonal STBC designs with multiple antennas at the receiver is summarized below.

- The signals received in different time slots at different receiver antennas are collected and stored in the memory.
- The vectors are generated by taking complex conjugate of the signals received in the later half of the time slots at all the receiver antennas.
- The vector obtained for each receiver antenna is pre-multiplied by the Hermitian of equivalent channel matrix corresponding with that receiver antenna, $\mathbf{H}_{eqi}$.
- The result obtained is used in the ML rule of equation (6.91) to detect each of the symbols one by one.

6.5 STBC systems over Nakagami-m fading channels

The discussion of Nakagami-m distribution and the fading channel models is already presented in Section 2.4.1 of Chapter 2. We will use those discussions and related expressions to evaluate various performance parameters of STBC MIMO systems in this section. We will mainly focus on the outage probability and bit/symbol error rate. In this section, the fading envelope ($|h_{j,i}|$) of the channel coefficients is assumed to be Nakagami-m distributed and all the links from transmitter to receiver are assumed to be independent and identically distributed (i.i.d.). In all the discussions related to performance analysis of STBC MIMO systems, we will assume $N_t \times N_r$ MIMO systems, i.e., N_t antennas at the transmitter side and N_r antennas at the receiver side. It is assumed that an STBC codeword is transmitted over T time slots. So, a codeword for complete STBC transmission can be given by a matrix with dimensions $N_t \times T$. It is already discussed that the STBC codewords can be separately processed at different antennas and then the processed signals can be added together to have an expression similar to maximal ratio combining

(MRC). After MRC, various detection techniques may be used to detect the received symbols in the STBC codeword. In general, ML is the optimum detection technique for receivers in communication systems. The symbols can be detected based on the minimum Euclidean distance. Also, from discussions in the previous section and Chapter 1 (Section 1.3.2 and expression of equation (1.21)), the received signal can be given by

$$\widetilde{r}_i = \|\mathbf{H}\|_F^2 x_i + \widetilde{n}_i \tag{6.92}$$

where $\widetilde{r}_i$ is the received signal after performing the required steps of STBC detection processing and performing MRC corresponding to the ith transmitted symbol. Note that the steps discussed in the previous section involve pre-multiplication of the processed received signals with $\mathbf{H}_{eff}$. This step inherently involves the operation that is the same as performing MRC. So, a separate step of MRC is not required in STBC detection. In fact, this property of STBC designs makes them able to achieve full diversity orders.

6.5.1 PDF of SNR at receiver

Note that the effective received signal in equation (6.92) is in the same form of the received signal after MRC in the receiver diversity techniques. In other words, it has the signal part given by $\|\mathbf{H}\|_F^2 x_i$ which is the sum of squared magnitudes of the channel coefficients multiplied by the transmitted symbol. $\|\mathbf{H}\|_F^2 x_i$ represents sum of $N_t N_r$ square variants of Nakagami-m distributed random variables. Referring to the physical model of Nakagami-m fading channels, the sum of $N_t N_r$ square variate of Nakagami-m distributed random variables is also a square variate of Nakagami-m distributed random variable with the fading parameter value mN_tN_r. Thus, from equation (2.11), the probability density function (PDF) of SNR for an STBC MIMO system over Nakagami-m fading channels can be given by

$$p_\gamma(\gamma) = \frac{(mN_tN_r)^{mN_tN_r}\gamma^{mN_tN_r-1}}{\overline{\gamma}^{mN_tN_r}\Gamma(mN_tN_r)} e^{\frac{-mN_tN_r\gamma}{\overline{\gamma}}}, \qquad \gamma \geq 0 \tag{6.93}$$

Note that $\overline{\gamma}$ is the average SNR and is used as scaling factor of $\|\mathbf{H}\|_F^2$ in many books to indicate average SNR at the receiver. The PDF of SNR is further useful in evaluation of various performance parameters of STBC MIMO systems. Using the fact that received instantaneous SNR follows Nakagami-m distribution with the m replaced by mN_tN_r, the further analysis follows from the analysis of SISO systems or receiver MRC diversity techniques which are thoroughly investigated by the researchers. In the next part of this section, we will evaluate outage probability and symbol error rate (SER) (exact and approximate) of STBC MIMO systems.

6.5.2 Outage probability

Outage probability is already defined in the earlier chapters of this book. Here, we will use the same definition to evaluate the outage probability of STBC MIMO systems. Given the average SNR at the receiver is $\overline{\gamma}$, the outage probability is defined as the probability that the instantaneous receiver SNR is below a predefined threshold, γ_{th}. Thus, outage probability can be given by

$$P_{out}\left(\overline{\gamma}, \gamma_{th}\right) = P\left(\overline{\gamma}\left\|\mathbf{H}\right\|_F^2 < \gamma_{th}\right) \tag{6.94}$$

It can be evaluated by substituting equation (6.93) in the above expression as

$$P_{out}\left(\overline{\gamma}, \gamma_{th}\right) = \int_0^{\gamma_{th}} \frac{\left(mN_tN_r\right)^{mN_tN_r} \gamma^{mN_tN_r-1}}{\overline{\gamma}^{mN_tN_r}\Gamma\left(mN_tN_r\right)} e^{\frac{-mN_tN_r\gamma}{\overline{\gamma}}} d\gamma \tag{6.95}$$

As discussed earlier and also can be observed from above, outage probability is the cumulative distribution function (CDF) of the received instantaneous SNR. Thus, from the relation of CDF of Nakagami-m distributed SNR in (2.15), the outage probability of STBC MIMO systems over Nakagami-m fading channels can be given as

$$P_{out}\left(\overline{\gamma}, \gamma_{th}\right) = \frac{\gamma_{inc}\left(mN_tN_r, \frac{mN_tN_r\gamma_{th}}{\overline{\gamma}}\right)}{\Gamma\left(mN_tN_r\right)} \tag{6.96}$$

6.5.3 Error performance analysis

In this section, we will discuss a method to obtain an expression for error performance analysis of STBC MIMO systems over Nakagami-m fading channels. Error performance analysis is primarily about the probability of error in the detected symbols at the receiver. In this book, we consider orthogonal STBC designs. In this case, the instantaneous SNR at the receiver can be given as

$$\gamma = \overline{\gamma}\|\mathbf{H}\|_F^2 \tag{6.97}$$

It has already been discussed that γ is a Nakagami-m distributed random variable with m replaced by mN_tN_r and its PDF can be given as in expression of equation (6.93). In general, to evaluate error performance of any wireless communication system, the channel is assumed to be constant for a while and the error performance is evaluated for those channel conditions. This probability of error is known as conditional error probability. Conditional probability of error depends on the instantaneous received SNR which is a random variable on account of the randomly varying channel coefficients. Conditional probability of error for various modulation schemes can be given by

$$P_e\left(\gamma\right) = aQ\left(\sqrt{b\gamma}\right) - cQ^2\left(\sqrt{b\gamma}\right) \tag{6.98}$$

where $Q(\cdot)$ is the Gaussian Q function which gives the area under the tail of a Gaussian curve and is defined as $Q(\hbar) = \frac{1}{\sqrt{2\pi}} \int\limits_{\hbar}^{\infty} e^{-\left(\frac{u^2}{2}\right)} du$, γ is the instantaneous received SNR and a, b and c are modulation dependent parameters listed in Table A.1. The average probability of error can be obtained by averaging the conditional error probability over the PDF of received instantaneous SNR. For averaging the conditional error probability, we will use the moment generating function (MGF)-based approach in this chapter. It can be given as

$$\overline{P}_e = \underbrace{\frac{a}{\pi} \int\limits_0^{\frac{\pi}{2}} M_\gamma \left(\frac{b}{2\sin^2\theta} \right) d\theta}_{I_1} - \underbrace{\frac{c}{\pi} \int\limits_0^{\frac{\pi}{4}} M_\gamma \left(\frac{b}{2\sin^2\theta} \right) d\theta}_{I_2} \tag{6.99}$$

where $M_\gamma(\cdot)$ is the MGF of received instantaneous SNR. MGF of Nakagami-m distributed instantaneous SNR is given as in equation (2.18).

$$M_\gamma(s) = \int_0^\infty p_\gamma(\gamma) e^{-s\gamma} d\gamma \tag{6.100}$$

$$= \left(\frac{m}{m + \overline{\gamma} s} \right)^m \tag{6.101}$$

For the case of $N_t \times N_r$ MIMO STBC systems, m is replaced by mN_tN_r. Thus, MGF can be given as

$$M_\gamma(s) = \left(\frac{mN_tN_r}{mN_tN_r + \overline{\gamma} s} \right)^{mN_tN_r} \tag{6.102}$$

The probability of error can be obtained by evaluating I_1 and I_2. First, we evaluate I_1 which may be obtained by substituting equation (6.102) in equation (6.99) as

$$I_1 = \frac{a}{\pi} \int\limits_0^{\frac{\pi}{2}} \left(\frac{mN_tN_r}{mN_tN_r + \frac{b\overline{\gamma}}{2\sin^2\theta}} \right)^{mN_tN_r} d\theta \tag{6.103}$$

Similarly, I_2 can be given as

$$I_2 = \frac{c}{\pi} \int\limits_0^{\frac{\pi}{4}} \left(\frac{mN_tN_r}{mN_tN_r + \frac{b\overline{\gamma}}{2\sin^2\theta}} \right)^{mN_tN_r} d\theta \tag{6.104}$$

The integrals I_1 and I_2 can be solved using various substitutions as in [115].

Solution for I_1

To solve I_1, let us change the variable by substitution $\cos^2\theta = t$. This leads to

$$\sin^2\theta = 1 - \cos^2\theta = 1 - t$$

$$-2\cos\theta \sin\theta d\theta = dt \tag{6.105}$$

Thus, the limits of integral would change from 0 to 1 and the integrating variable $d\theta$ changes to

$$d\theta = \frac{dt}{-2\sqrt{t}\sqrt{1-t}} \tag{6.106}$$

Therefore, I_1 can now be given as

$$I_1 = \frac{a}{\pi}\int_0^1 \left(\frac{mN_tN_r}{mN_tN_r + \frac{b\overline{\gamma}}{(1-t)}}\right)^{mN_tN_r} \frac{dt}{2\sqrt{t}\sqrt{1-t}} \tag{6.107}$$

After rearrangements and mathematical manipulations, this integral can be represented in the standard form as

$$I_1 = \frac{a}{2\pi}\left(\frac{2mN_tN_r}{2mN_tN_r + b\overline{\gamma}}\right)^{mN_tN_r} \int_0^1 \frac{(1-t)^{mN_tN_r-\frac{1}{2}}}{\sqrt{t}} \left(1 - \frac{t}{1+\frac{b\overline{\gamma}}{2mN_tN_r}}\right)^{-mN_tN_r} dt \tag{6.108}$$

The above expression represents the integral in the standard form of Gauss hypergeometric function defined as [37]

$${}_2F_1(a,b;c;x) = \frac{\Gamma(c)}{\Gamma(c-a)\Gamma(a)}\int_0^1 t^{b-1}(1-t)^{c-b-1}(1-tx)^{-a}\,dt \tag{6.109}$$

Comparing the above definition of Gauss hypergeometric function with the expression of I_1, the parameters can be obtained as $a = mN_tN_r$, $b = \frac{1}{2}$, $c = mN_tN_r + 1$ and $x = 1 - \frac{1}{1+\frac{b\overline{\gamma}}{2mN_tN_r}}$ can be given as

$$\begin{aligned} I_1 = {} & \frac{a}{2\pi}\left(\frac{2mN_tN_r}{2mN_tN_r + b\overline{\gamma}}\right)^{mN_tN_r} \frac{\Gamma(mN_tN_r)\Gamma\left(\frac{1}{2}\right)}{\Gamma(mN_tN_r+1)} \\ & \times {}_2F_1\left(mN_tN_r, \frac{1}{2}; mN_tN_r+1; 1 - \frac{1}{1+\frac{b\overline{\gamma}}{2mN_tN_r}}\right) \end{aligned} \tag{6.110}$$

Solution for I_2

Similar to the case of I_1, to solve I_2, let us change the variable by substituting $\sin^2\theta = \frac{t}{2}$. Further manipulations of this substitution leads to

$$2\sin\theta\cos\theta d\theta = \frac{dt}{2} \tag{6.111}$$

$$d\theta = \frac{dt}{4\sqrt{\frac{t}{2}}\sqrt{1-\frac{t}{2}}} \tag{6.112}$$

Also, the limits of integral would change from 0 to 1. Therefore, I_2 can now be given as

$$I_2 = \frac{c}{\pi} \int_0^1 \left(\frac{mN_tN_r}{mN_tN_r + \frac{b\overline{\gamma}}{t}} \right)^{mN_tN_r} \frac{dt}{4\sqrt{\frac{t}{2}}\sqrt{1-\frac{t}{2}}} \tag{6.113}$$

$$= \frac{c}{\pi} \int_0^1 \left(1 + \frac{b\overline{\gamma}}{mN_tN_rt} \right)^{-mN_tN_r} \frac{dt}{4\sqrt{\frac{t}{2}}\sqrt{1-\frac{t}{2}}} \tag{6.114}$$

$$= \frac{c}{2\sqrt{2}\pi} \int_0^1 t^{mN_tN_r-\frac{1}{2}} \left(t + \frac{b\overline{\gamma}}{mN_tN_r} \right)^{-mN_tN_r} \frac{dt}{\sqrt{1-\frac{t}{2}}} \tag{6.115}$$

The above integral may be compared with the definition of Appell hypergeometric function of two variables which is defined as [103]

$$F_A^{(1)}\left(\alpha;\beta,\beta';\gamma;x;y\right) = \frac{\Gamma\left(\gamma\right)}{\Gamma\left(\alpha\right)\Gamma\left(\gamma-\alpha\right)} \int_0^1 t^{\alpha-1}\left(1-t\right)^{\gamma-\alpha-1}\left(1-tx\right)^{-\beta}\left(1-ty\right)^{-\beta'}dt \tag{6.116}$$

Thus, comparing the integral I_2 with the above definition of Appell hypergeometric function various parameters can be obtained as $\alpha = mN_tN_r + \frac{1}{2}$, $\beta = \frac{1}{2}$, $\beta' = mN_tN_r$, $\gamma = mN_tN_r + \frac{3}{2}$, $x = \frac{1}{2}$ and $y = -\frac{mN_tN_r}{b\overline{\gamma}}$. Thus, I_2 can be evaluated as

$$\begin{aligned} I_2 =& \frac{c}{2\sqrt{2}\pi} \left(\frac{b\overline{\gamma}}{mN_tN_r} \right)^{mN_tN_r} \frac{\Gamma\left(mN_tN_r + \frac{1}{2}\right)}{\Gamma\left(mN_tN_r + \frac{3}{2}\right)} \\ & \times F_A^{(1)}\left(mN_tN_r + \frac{1}{2}; \frac{1}{2}, mN_tN_r; mN_tN_r + \frac{3}{2}; \frac{1}{2}; -\frac{mN_tN_r}{b\overline{\gamma}} \right) \end{aligned} \tag{6.117}$$

Note that using I_1 and I_2, the error probability of STBC MIMO systems can be evaluated analytically via the closed form expressions of equations (6.110) and (6.117). But these solutions are in the form of Gauss hypergeometric function and Appell hypergeometric function. These functions need to be evaluated numerically in the software computing tools. Some of the hypergeometric functions are readily implemented in tools such as MATLAB, *Mathematica*, Maple, etc. However, the internal evaluations of these functions involve numerical methods of finite integrals. Thus, even though these expressions are closed form expressions, the computation involves evaluation using suitable numerical methods. Instead, with some approximations, a closed form expression for approximate evaluation of error probability can be a good option to use. We look into the approximations that may be considered.

6.5.3.1 Approximate error performance

In the previous chapter, we introduced the following two approximations for Gaussian Q function for positive values of argument.

$$Q(\hbar) \approx \frac{1}{2}e^{-\frac{\hbar^2}{2}} \tag{6.118}$$

$$Q(\hbar) \approx \frac{1}{12}e^{-\frac{\hbar^2}{2}} + \frac{1}{4}e^{-\frac{2\hbar^2}{3}} \tag{6.119}$$

The bound given by equation (6.118) is known as Chernoff bound, named after the proposer of this bound, and the approximation given by equation (6.119) is known as Chiani approximation, named after the researcher who proposed this approximation. Note that these approximations are popularly used in the analysis of communication systems because of the positive values of SNR in all cases. Both these approximations give upper bound on the exact values of Gaussian Q function. The later expression gives an accurate approximation for Gaussian Q function as discussed in Appendix A.2. Note that both these approximations represent Gaussian Q function in terms of exponential functions. This enables the representation of average SER as a sum of MGF terms of the SNR. In addition to Gaussian Q function, the expression of conditional error probability involves the squared Gaussian Q function. Using the above approximations, the squared Gaussian Q functions may be approximated as

$$Q^2(\hbar) \leq \frac{1}{4}e^{-\hbar^2} \tag{6.120}$$

$$Q^2(\hbar) \leq \frac{1}{144}e^{-\hbar^2} + \frac{1}{16}e^{-\frac{4\hbar^2}{3}} + \frac{1}{24}e^{-\frac{7\hbar^2}{6}} \tag{6.121}$$

These approximations can be used in finding the average error probability of any communication system by averaging the approximated Gaussian Q function over the PDF of received SNR. Now, using equation (6.118) and equation (6.120), the approximated Gaussian Q function is in the form of exponential functions. Hence,

$$P_e(\gamma) = aQ\left(\sqrt{b\gamma}\right) - cQ^2\left(\sqrt{b\gamma}\right) \leq \frac{a}{2}e^{-\frac{b\gamma}{2}} - \frac{c}{4}e^{-b\gamma} \tag{6.122}$$

Similarly, using equations (6.119) and (6.121), the approximated error probability can be given as

$$\begin{aligned} P_e(\gamma) &= aQ\left(\sqrt{b\gamma}\right) - cQ^2\left(\sqrt{b\gamma}\right) \\ &\leq \frac{a}{12}e^{-\frac{b\gamma}{2}} + \frac{a}{4}e^{-\frac{2b\gamma}{3}} - \frac{c}{144}e^{-b\gamma} - \frac{c}{16}e^{-\frac{4b\gamma}{3}} - \frac{c}{24}e^{-\frac{7b\gamma}{6}} \end{aligned} \tag{6.123}$$

Taking average of equation (6.122) and equation (6.123), an approximation for average error probability may be obtained. Here, our aim is to make readers

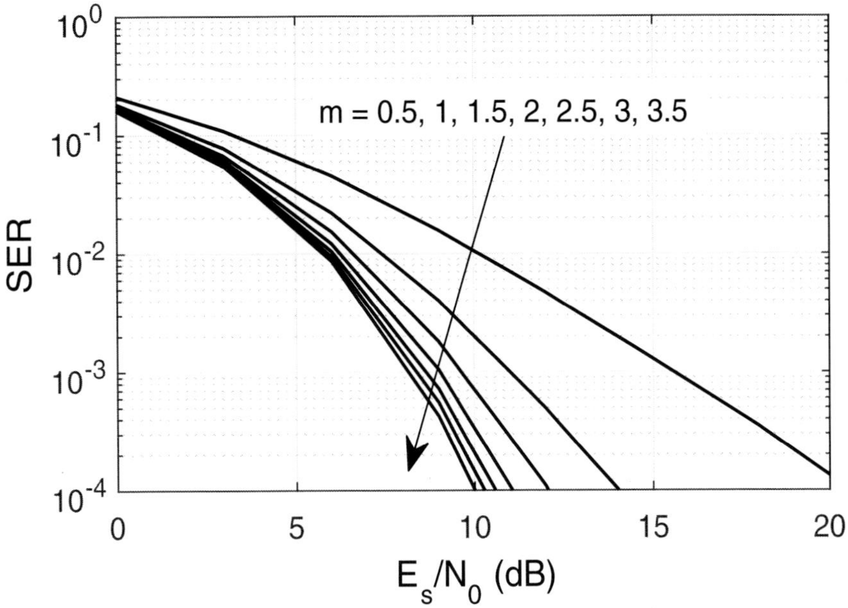

FIGURE 6.1
SER of 2×2 Alamouti STBC MIMO systems over Nakagami-m fading channels

familiar with the approach to proceed for the approximations. So, we will proceed further with averaging of equation (6.122) even though equation (6.122) is a looser approximation as compared to equation (6.123). Therefore, average approximate probability of error can be given by

$$\overline{P}_e \leq \int_0^\infty \left(\frac{a}{2}e^{-\frac{b\gamma}{2}} - \frac{c}{4}e^{-b\gamma}\right) p_\gamma(\gamma)d\gamma \tag{6.124}$$

Comparing the above expression with the definition of MGF of a random variable, an upper bound on the average probability of error can be given by

$$\overline{P}_e \leq \frac{a}{2}M_\gamma\left(\frac{b}{2}\right) - \frac{c}{4}M_\gamma(b) \tag{6.125}$$

where $M_\gamma(\cdot)$ is the MGF of received SNR, γ. Note that the MGF of Nakagami-m distributed SNR has already been discussed in Chapter 2. It is in very simple form involving elementary mathematical operations. Thus, looking at the simplicity of evaluations, one can choose approximate value or exact value of error probability depending on the required accuracy level.

A 2×2 Alamouti STBC MIMO system was simulated using the MATLAB script given in Appendix B at the last of this book for SER performance.

For this simulation, quadrature phase shift keying (QPSK) modulation was considered. In this simulation, we used E_s as the average energy of a digitally modulated symbol per transmitter antenna per time slot. N_0 is one sided noise power spectral density. In this chapter, the ratio E_s/N_0 is used as a parameter to indicate signal strength with respect to that of the noise. SER of STBC MIMO systems over Nakagami-m fading channels with different values of fading parameter is shown in Figure 6.1. An improvement in the diversity order is observed with the increasing values of m. Note that in this case, the SNR at the receiver is not the same as the ratio, E_s/N_0. This is because of the steps involved in the detection after reception of the signal.

6.6 STBC systems over $\kappa - \mu$ fading channels

The discussion of $\kappa - \mu$ distribution and the fading channel models was presented in Section 2.4.3 of Chapter 2. There we discussed about the physical model, PDF and CDF of $\kappa - \mu$ distributed envelope and SNR and the techniques to generate samples that follow $\kappa - \mu$ distribution. We will use those discussions and related expressions to evaluate various performance parameters of STBC MIMO systems in this section. In this section, the fading envelope ($|h_{j,i}|$) of the channel coefficients is assumed to be $\kappa - \mu$ distributed and all the links from transmitter to receiver are assumed to be i.i.d. In all the discussions related to performance analysis of STBC MIMO systems, we will assume $N_t \times N_r$ MIMO systems. It is assumed that an STBC codeword is transmitted over T time slots. So, a codeword for complete STBC transmission can be given by a matrix with dimensions $N_t \times T$. It was already discussed that the STBC codewords can be separately processed at different antennas and then the processed signals can be added together to obtain an effect which is similar to MRC. From discussions in previous sections of this chapter and Chapter 1 (Section 1.3.2 and expression of equation (1.21)), the instantaneous SNR of the received signal can be given by

$$\gamma = \overline{\gamma}||\mathbf{H}||_F^2 \tag{6.126}$$

where each element of $\mathbf{H}$ is $\kappa - \mu$ distributed.

6.6.1 PDF of SNR at receiver

Note that the effective received signal in (6.92) is in the same form of the received signal after MRC in the receiver diversity techniques. In other words, it has the signal part given by $||\mathbf{H}||_F^2 x_i$ which is proportional to the sum of squared magnitudes of the channel coefficients. $||\mathbf{H}||_F^2 x_i$ represents sum of N_tN_r square variants of $\kappa - \mu$ distributed random variables. Referring to the physical model of $\kappa - \mu$ fading channels, the sum of N_tN_r square variate of $\kappa-\mu$ distributed random variables is also a square variate of $\kappa-\mu$ distributed random variable with the fading parameters κ and $\mu N_t N_r$. Thus, from equation (2.36), the PDF of SNR for an STBC MIMO system over $\kappa-\mu$ fading channels can be given by

$$p_\gamma(\gamma)=\frac{\mu(1+\kappa)^{\frac{\mu+1}{2}}}{\kappa^{\frac{\mu-1}{2}}e^{\mu\kappa}\overline{\gamma}^{\frac{\mu+1}{2}}}\gamma^{\frac{\mu-1}{2}}e^{-\frac{\mu(1+\kappa)\gamma}{\overline{\gamma}}}I_{\mu-1}\left(2\mu\sqrt{\frac{\kappa(1+\kappa)\gamma}{\overline{\gamma}}}\right) \tag{6.127}$$

where $\mu=\frac{1}{V(\gamma)}\frac{1+2\kappa}{(1+\kappa)^2}=\mu N_t N_r$. Note that in the above expression, we do not replace μ by $\mu N_t N_r$ to keep the representation simpler. Actually, in all future discussions related to STBC MIMO systems over $\kappa-\mu$ fading channels μ shall be interpreted as $\mu N_t N_r$.

Note that $\overline{\gamma}$ is the average SNR and is used as a scaling factor of $\|\mathbf{H}\|_F^2$ in many books to indicate average SNR at the receiver. The PDF of SNR is further used in evaluation of various performance parameters of any wireless communication system, STBC MIMO systems in this chapter. Using the fact that received instantaneous SNR follows $\kappa-\mu$ distribution, further analysis follows from the analysis of SISO systems or receiver MRC diversity techniques which are thoroughly investigated by the researchers over various fading channels including $\kappa-\mu$ fading channels. In the next part of this section, we will evaluate outage probability and SER of STBC MIMO systems over $\kappa-\mu$ fading channels.

6.6.2 Outage probability

Outage probability is already defined in the earlier chapters of this book. Here, we will use the same definition to evaluate the outage probability of STBC MIMO systems. Outage probability gives the proportion of time for which a communication link is out of order. Given the average SNR at the receiver is $\overline{\gamma}$, the outage probability is defined as the probability that the instantaneous receiver SNR is below a predefined threshold, γ_{th}. Thus, outage probability can be given by

$$P_{out}(\overline{\gamma},\gamma_{th})=P\left(\overline{\gamma}\,\|\mathbf{H}\|_F^2<\gamma_{th}\right) \tag{6.128}$$

It can be evaluated by substituting equation (6.127) in the above expression as

$$P_{out}(\overline{\gamma},\gamma_{th})=\int_0^{\gamma_{th}}\frac{\mu(1+\kappa)^{\frac{\mu+1}{2}}}{\kappa^{\frac{\mu-1}{2}}e^{\mu\kappa}\overline{\gamma}^{\frac{\mu+1}{2}}}\gamma^{\frac{\mu-1}{2}}e^{-\frac{\mu(1+\kappa)\gamma}{\overline{\gamma}}}I_{\mu-1}\left(2\mu\sqrt{\frac{\kappa(1+\kappa)\gamma}{\overline{\gamma}}}\right)d\gamma \tag{6.129}$$

As discussed earlier and also can be observed from above, outage probability is the CDF of the received instantaneous SNR. Thus, from the relation of CDF of $\kappa-\mu$ distributed instantaneous SNR in equation (2.38) and Appendix A.6, the outage probability of STBC MIMO systems over $\kappa - \mu$ fading channels can be given as

$$P_{out}\left(\overline{\gamma}, \gamma_{th}\right) = 1 - Q_{\mu}\left(\sqrt{2\kappa\mu}, \sqrt{\frac{2\mu(1+\kappa)\gamma_{th}}{\overline{\gamma}}}\right) \tag{6.130}$$

where $Q_{\mu}(\alpha, \beta)$ is generalized Marcum Q-function [46]. It is defined as

$$Q_{\mu}(\alpha, \beta) = \int_{\beta}^{\infty} t\left(\frac{t}{\alpha}\right)^{\mu-1} e^{-\frac{t^2+\alpha^2}{2}} I_{\mu-1}\left(\alpha t\right) dt \tag{6.131}$$

6.6.3 Error performance analysis

In this section, we will discuss a method to obtain an expression for error performance analysis of STBC MIMO systems over $\kappa - \mu$ fading channels. Error performance analysis is primarily about the probability of error in the detected symbols at the receiver. Outage probability is directly related to error performance of the system but there is no exact relation between these two performance parameters. For orthogonal STBC designs the instantaneous SNR at the receiver can be given as Frobenius norm of the channel times average SNR as in equation (6.126).

It was already discussed that γ is a $\kappa - \mu$ distributed random variable with fading parameters κ and μ replaced by $\mu N_t N_r$ and its PDF can be given as in the expression of equation (6.127). In general, to evaluate error performance of any wireless communication system, the channel is assumed to be constant for a while and the error performance is evaluated for those channel conditions. This probability of error is known as conditional error probability. Conditional probability of error depends on the instantaneous received SNR which is a random variable on account of the randomly varying channel coefficients. Conditional probability of error for various modulation schemes can be given by

$$P_e\left(\gamma\right) = aQ\left(\sqrt{b\gamma}\right) - cQ^2\left(\sqrt{b\gamma}\right) \tag{6.132}$$

where $Q(\cdot)$ is the Gaussian Q function which gives the area under the tail of a Gaussian curve and is defined as $Q(\hbar) = \frac{1}{\sqrt{2\pi}} \int\limits_{\hbar}^{\infty} e^{-\left(\frac{u^2}{2}\right)} du$, γ is the instantaneous received SNR and a, b and c are modulation dependent parameters listed in Table A.1. The average probability of error can be obtained by averaging the conditional error probability over the PDF of received instantaneous SNR. For averaging the conditional error probability, we will use the MGF-based approach in this chapter. It can be given as

$$\overline{P}_e = \underbrace{\frac{a}{\pi}\int_0^{\frac{\pi}{2}} M_\gamma\left(\frac{b}{2\sin^2\theta}\right)d\theta}_{I_1} - \underbrace{\frac{c}{\pi}\int_0^{\frac{\pi}{4}} M_\gamma\left(\frac{b}{2\sin^2\theta}\right)d\theta}_{I_2} \tag{6.133}$$

where $M_\gamma(\cdot)$ is the MGF of received instantaneous SNR. MGF of $\kappa-\mu$ distributed instantaneous SNR is given as in equation (2.18).

$$M_\gamma(s) = \int_0^\infty p_\gamma(\gamma)e^{-s\gamma}d\gamma \tag{6.134}$$

$$= \left(\frac{\mu(1+\kappa)}{\mu(1+\kappa)+s\overline{\gamma}}\right)^\mu e^{\left(\frac{\mu^2\kappa(1+\kappa)}{\mu(1+\kappa)+s\overline{\gamma}}\right)-\kappa\mu} \tag{6.135}$$

For the case of $N_t \times N_r$ STBC MIMO systems, the fading parameter values are κ and $\mu N_t N_r$. However, as discussed earlier, to keep the expressions simpler, the parameters are κ and μ and interpreted as $\mu N_t N_r$ in the above expression and all expressions henceforth.

The probability of error can be obtained by evaluating I_1 and I_2. First, we evaluate I_1 which may be obtained by substituting equation (6.135) in equation (6.133) as

$$I_1 = \frac{a}{\pi}\int_0^{\frac{\pi}{2}} \left(\frac{\mu(1+\kappa)}{\mu(1+\kappa)+\frac{b\overline{\gamma}}{2\sin^2\theta}}\right)^\mu e^{\left(\frac{\mu^2\kappa(1+\kappa)}{\mu(1+\kappa)+\frac{b\overline{\gamma}}{2\sin^2\theta}}\right)-\kappa\mu} d\theta \tag{6.136}$$

Similarly, I_2 can be given as

$$I_2 = \frac{c}{\pi}\int_0^{\frac{\pi}{4}} \left(\frac{\mu(1+\kappa)}{\mu(1+\kappa)+\frac{b\overline{\gamma}}{2\sin^2\theta}}\right)^\mu e^{\left(\frac{\mu^2\kappa(1+\kappa)}{\mu(1+\kappa)+\frac{b\overline{\gamma}}{2\sin^2\theta}}\right)-\kappa\mu} d\theta \tag{6.137}$$

Integrals I_1 and I_2 are definite integrals with finite limits. It is possible to obtain a numerical solution for them. Alternatively, closed form solutions in terms of hypergeometric functions can be obtained by various substitutions as we did in the case of analysis of STBC MIMO systems over Nakagami-m fading channels.

Solution for I_1

To solve I_1, let us change the variable of integration by taking substitution as

$$t = \frac{\mu(1+\kappa)}{\mu(1+\kappa)+\frac{b\overline{\gamma}}{2\sin^2\theta}} \tag{6.138}$$

The aim of this substitution is to make the exponent term simple and in the form of e^{pt}. Thus, after mathematical simplifications, various terms required in the integral can be represented as

$$\sin^2\theta = \frac{b\bar{\gamma}t}{2\mu(1+\kappa)(1-t)} \tag{6.139}$$

$$2\cos\theta\sin\theta d\theta = \frac{b\bar{\gamma}dt}{2\mu(1+\kappa)(1-t)^2} \tag{6.140}$$

$$\sin\theta = \sqrt{\frac{b\bar{\gamma}t}{2\mu(1+\kappa)(1-t)}} \tag{6.141}$$

$$\cos\theta = \sqrt{1-\frac{b\bar{\gamma}t}{2\mu(1+\kappa)(1-t)}} \tag{6.142}$$

Further, the substitution converts the lower and upper limits of the integral from 0 to 0 and from $\frac{\pi}{2}$ to $\frac{\mu(1+\kappa)}{\mu(1+\kappa)+\frac{b\bar{\gamma}}{2}}$ respectively. Now, using all the above terms for the substitution, the integral can be given as

$$I_1 = \frac{a}{\pi}\int_0^{\frac{\mu(1+\kappa)}{\mu(1+\kappa)+\frac{b\bar{\gamma}}{2}}} \frac{b\bar{\gamma}(\mu\kappa t)^{\mu}e^{t}(1-t)^{-1}}{2\sqrt{b\bar{\gamma}t}\sqrt{2\mu(1+\kappa)(1-t)-b\bar{\gamma}t}}dt \tag{6.143}$$

For further simplification of this integral, it can be brought in the form of confluent hypergeometric function with the substitution

$$y = \frac{2\mu(1+\kappa)+b\bar{\gamma}}{2\mu(1+\kappa)}t \tag{6.144}$$

The above substitution converts the upper limit of the integral to unity without changing the lower limit of the integration. This makes it easy to represent this integral into standard form of confluent hypergeometric function of two variables. Confluent hypergeometric function is defined as

$$\Phi_1(a,b,c,x,y) = \frac{\Gamma(c)}{\Gamma(a)\Gamma(c-a)}\int_0^1 t^{a-1}(1-t)^{c-a-1}(1-xt)^{-b}e^{yt}dt \tag{6.145}$$

Further simplifications after substitution of equation (6.144) into equation (6.143) bring the integral in the form that can be given as

$$I_1 = \frac{a}{\pi}\int_0^1 \frac{b\bar{\gamma}y^{\mu-\frac{1}{2}}e^{\frac{2\mu(1+\kappa)}{2\mu(1+\kappa)+b\bar{\gamma}}y}\left(1-\frac{2\mu(1+\kappa)}{2\mu(1+\kappa)+b\bar{\gamma}}y\right)^{-1}}{2\sqrt{b\bar{\gamma}}\sqrt{2\mu(1+\kappa)(1-y)}}\frac{2\mu(1+\kappa)}{2\mu(1+\kappa)+b\bar{\gamma}}dy \tag{6.146}$$

$$= \frac{a}{\pi}\frac{\sqrt{b\bar{\gamma}}(\mu\kappa)^{\mu}}{2}\left(\frac{2\mu(1+\kappa)}{2\mu(1+\kappa)+b\bar{\gamma}}\right)^{\mu+\frac{1}{2}}\int_0^1\frac{y^{\mu-\frac{1}{2}}e^{\frac{2\mu(1+\kappa)}{2\mu(1+\kappa)+b\bar{\gamma}}y}(1-y)^{-\frac{1}{2}}}{\sqrt{2\mu(1+\kappa)}\left(1-\frac{2\mu(1+\kappa)}{2\mu(1+\kappa)+b\bar{\gamma}}y\right)}dy \tag{6.147}$$

The above expression can be compared with the definition of confluent hypergeometric function given in equation (6.145) to obtain the arguments of the function as

$$a = \mu + \frac{1}{2} \tag{6.148}$$

$$b = 1 \tag{6.149}$$

$$c = \mu + 1 \tag{6.150}$$

$$x = \frac{2\mu(1+\kappa)}{2\mu(1+\kappa) + b\bar{\gamma}} \tag{6.151}$$

$$y = \frac{2\mu(1+\kappa)}{2\mu(1+\kappa) + b\bar{\gamma}} \tag{6.152}$$

Thus, I_1 can be finally evaluated in the form of confluent hypergeometric function as

$$I_1 = \frac{a}{\pi}\frac{\sqrt{b\bar{\gamma}}(\mu\kappa)^{\mu}}{2\sqrt{2\mu(1+\kappa)}}\left(\frac{2\mu(1+\kappa)}{2\mu(1+\kappa)+b\bar{\gamma}}\right)^{\mu+\frac{1}{2}}\frac{\Gamma\left(\mu+\frac{1}{2}\right)\Gamma\left(\frac{1}{2}\right)}{\Gamma(\mu+1)}$$
$$\Phi_1\left(\mu+\frac{1}{2}, 1, \mu+1, \frac{2\mu(1+\kappa)}{2\mu(1+\kappa)+b\bar{\gamma}}, \frac{2\mu(1+\kappa)}{2\mu(1+\kappa)+b\bar{\gamma}}\right) \tag{6.153}$$

Next, we will discuss the solution of I_2.

Solution for I_2

One can proceed for solution of I_2 following the steps used for solution of I_1 earlier. We take the same substitution as that for I_1 in equation (6.138). Thus, the same expressions will be used in the integral as discussed in equations (6.139)-(6.142). However, the upper limit of the integral will now be $\frac{\mu(1+\kappa)}{\mu(1+\kappa)+b\bar{\gamma}}$. Now, it can be represented as

$$I_2 = \frac{c}{\pi}\int_0^{\frac{\mu(1+\kappa)}{\mu(1+\kappa)+b\bar{\gamma}}} \frac{b\bar{\gamma}(\mu\kappa t)^{\mu}e^{t}(1-t)^{-1}}{2\sqrt{b\bar{\gamma}t}\sqrt{2\mu(1+\kappa)(1-t)-b\bar{\gamma}t}}dt \tag{6.154}$$

For further simplification of this integral, it can be brought in the form of confluent Lauricella's hypergeometric function of three variables with the substitution

$$y = \frac{\mu(1+\kappa)+b\bar{\gamma}}{\mu(1+\kappa)}t \tag{6.155}$$

The above substitution converts the upper limit of the integral to unity without changing the lower limit of the integration. This makes it easy to represent this integral into standard form of confluent Lauricella's hypergeometric function of three variables. It is defined as

$$\Phi_1^{(3)}(a,b_1,b_2,c,x,y,z) = \frac{\Gamma(c)}{\Gamma(a)\Gamma(c-a)} \times \int_0^1 t^{a-1}(1-t)^{c-a-1}(1-xt)^{-b_1}(1-yt)^{-b_2}e^{zt}dt \tag{6.156}$$

Further simplifications after substitution of equation (6.155) into equation (6.154) brings the integral in the form that can be given as

$$I_2 = \frac{c}{\pi}\frac{\sqrt{b\bar{\gamma}}(\mu\kappa)^{\mu}}{2}\left(\frac{\mu(1+\kappa)}{\mu(1+\kappa)+b\bar{\gamma}}\right)^{\mu+\frac{1}{2}} \times \int_0^1 \frac{y^{\mu-\frac{1}{2}}e^{\frac{\mu(1+\kappa)}{\mu(1+\kappa)+b\bar{\gamma}}y}\left(1-\frac{2\mu(1+\kappa)+b\bar{\gamma}}{2\mu(1+\kappa)+2b\bar{\gamma}}y\right)^{-\frac{1}{2}}}{\sqrt{2\mu(1+\kappa)}\left(1-\frac{\mu(1+\kappa)}{\mu(1+\kappa)+b\bar{\gamma}}y\right)}dy \tag{6.157}$$

The above expression can be compared with the definition of confluent Lauricella's hypergeometric function given in equation (6.156) to obtain the arguments of the function as

$$a = \mu + \frac{1}{2} \tag{6.158}$$

$$b_1 = 1 \tag{6.159}$$

$$b_2 = \frac{1}{2} \tag{6.160}$$

$$c = \mu + \frac{3}{2} \tag{6.161}$$

$$x = \frac{\mu(1+\kappa)}{\mu(1+\kappa)+b\bar{\gamma}} \tag{6.162}$$

$$y = \frac{2\mu(1+\kappa)}{2\mu(1+\kappa)+2b\bar{\gamma}} \tag{6.163}$$

$$z = \frac{\mu(1+\kappa)}{\mu(1+\kappa)+b\bar{\gamma}} \tag{6.164}$$

Thus, I_2 can be finally evaluated in the form of confluent Lauricella's hypergeometric function as

$$I_2 = \frac{c}{\pi}\frac{\sqrt{b\bar{\gamma}}(\mu\kappa)^{\mu}}{2\sqrt{2\mu(1+\kappa)}}\left(\frac{\mu(1+\kappa)}{\mu(1+\kappa)+b\bar{\gamma}}\right)^{\mu+\frac{1}{2}}\frac{\Gamma\left(\mu+\frac{1}{2}\right)\Gamma\left(\frac{1}{2}\right)}{\Gamma(\mu+1)} \Phi_1^{(3)}\left(\mu+\frac{1}{2},1,\frac{1}{2},\frac{3}{2},\frac{\mu(1+\kappa)}{\mu(1+\kappa)+b\bar{\gamma}},\frac{2\mu(1+\kappa)}{2\mu(1+\kappa)+2b\bar{\gamma}},\frac{\mu(1+\kappa)}{\mu(1+\kappa)+b\bar{\gamma}}\right) \tag{6.165}$$

Note that I_1 and I_2 are respectively in the form of confluent hypergeometric function and confluent Lauricella's function. These functions are not commonly available in the mathematical computation softwares. Thus, numerical evaluation methods for finite integrals may be used. Alternatively, these functions can be numerically evaluated using their series representation. The confluent hypergeometric function in series form can be given as

$$\Phi_1(a,b,c,x,y)=\sum_{m=0}^{\infty}\sum_{n=0}^{\infty}\frac{(a)_{m+n}(b)_m x^m y^n}{(c)_{m+n}m!n!} \tag{6.166}$$

The condition for convergence of this function is $|x|<1$, which is satisfied in our case for all values of average SNR.

The confluent Lauricella's hypergeometric function in series form can be given as

$$\Phi_1^{(3)}(a,b,c,x,y,z)=\sum_{m=0}^{\infty}\sum_{n=0}^{\infty}\sum_{p=0}^{\infty}\frac{(a)_{m+n+p}(b_1)_m(b_2)_n x^m y^n z^p}{(c)_{m+n+p}m!n!p!} \tag{6.167}$$

The conditions for convergence of this function are $|x|<1$ and $|y|<1$, which are satisfied in our case for all values of average SNR. These functions can only be numerically evaluated or the sum of infinite series is required to evaluate these functions. Thus, even though these expressions look to be in closed form, the computation involves evaluation using suitable numerical methods. Instead, with some approximations, a closed form expression for approximate evaluation of error probability can be a good option to use. We look into the approximations that may be considered.

The approximations of Gaussian Q functions have been discussed in the previous section for the case of error performance of STBC MIMO systems over Nakagami-m fading channels. The analysis can be simplified by representing the Gaussian Q functions as a sum of exponential functions. Thus, the expression of average error probability can be obtained in the form of sum of MGF.

Using the similar simulation set-up discussed for SER performance over Nakagami-m fading channels in the previous section, we simulated 2×2 Alamouti STBC MIMO system over $\kappa-\mu$ fading channels. Note that the effect of number of clusters in the received signal has already been discussed by varying the value of m in the case of Nakagami-m fading channels. Now, in this section, we have shown SER results for various strengths of the line of sight (LOS) component in the received signal by simulating the system with different values of κ. The results are shown in Figure 6.2. As it is clear from the earlier discussions, the improvement in SER with a stronger LOS component is observed as expected. This is not the case for SM MIMO systems and spatially multiplexed MIMO systems. This shows that STBC systems can perform well even in the non-fading or LOS conditions.

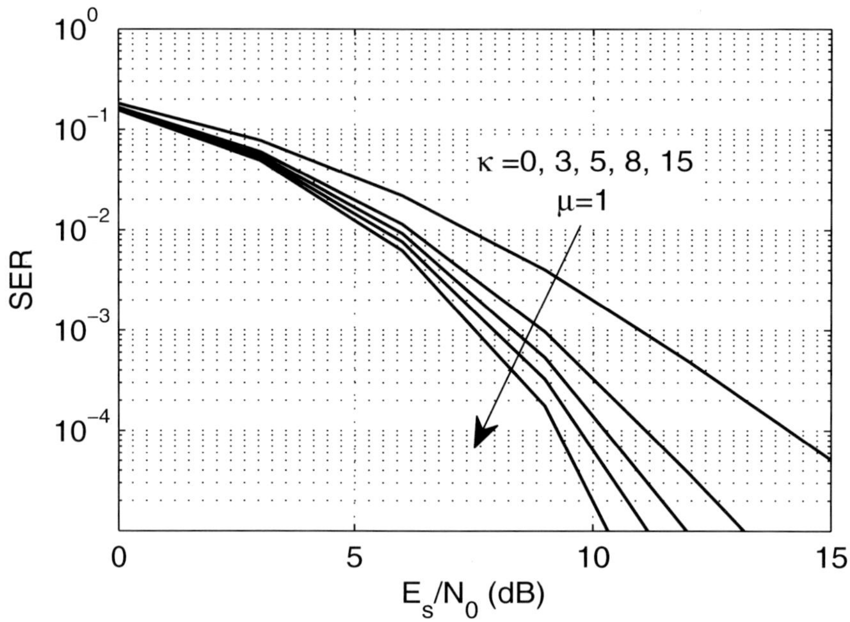

FIGURE 6.2
SER of 2 × 2 Alamouti STBC MIMO systems over $\kappa - \mu$ fading channels

6.7 STBC systems over $\eta - \mu$ fading channels

The discussion of $\eta - \mu$ distribution and the fading channel models is already presented in Section 2.4.2 of Chapter 2. There we discussed about the physical model, PDF and CDF of $\eta - \mu$ distributed envelope and SNR and the techniques to generate samples that follow $\eta - \mu$ distribution. We will use those discussions and related expressions to evaluate various performance parameters of STBC MIMO systems in this section. In this section, the fading envelope ($|h_{j,i}|$) of the channel coefficients is assumed to be $\eta - \mu$ distributed and all the links from transmitter to receiver are assumed to be i.i.d. In all the discussions related to performance analysis of STBC MIMO systems, we will assume $N_t \times N_r$ MIMO systems. It is assumed that an STBC codeword is transmitted over T time slots. So, a codeword for complete STBC transmission can be given by a matrix with dimensions $N_t \times T$. It is already discussed that the STBC codewords can be separately processed at different antennas and then the processed signals can be added together to achieve a similar effect as that of MRC. From discussions in previous sections of this chapter and Chapter 1 (Section 1.3.2 and expression of equation (1.21)), the instantaneous

SNR of the received signal can be given by

$$\gamma = \overline{\gamma}\|\mathbf{H}\|_F^2 \tag{6.168}$$

where each element of $\mathbf{H}$ is $\eta - \mu$ distributed.

6.7.1 PDF of SNR at receiver

Note that the effective received signal in equation (6.92) is in the same form of the received signal after MRC in the receiver diversity techniques. In other words, it has the signal part given by $\|\mathbf{H}\|_F^2 x_i$ which is proportional to the sum of squared magnitudes of the channel coefficients. $\|\mathbf{H}\|_F^2 x_i$ represents sum of $N_t N_r$ square variants of $\eta - \mu$ distributed random variables. Referring to the physical model of $\eta - \mu$ fading channels, the sum of $N_t N_r$ square variate of $\eta - \mu$ distributed random variables is also a square variate of $\eta - \mu$ distributed random variable with the fading parameters η and $\mu N_t N_r$. Thus, from equation (2.25), the PDF of SNR for an STBC MIMO system over $\eta - \mu$ fading channels can be given by

$$p_\gamma(\gamma) = \frac{2\sqrt{\pi}\mu^{\mu+\frac{1}{2}}h^\mu}{\Gamma(\mu)H^{\mu-\frac{1}{2}}\overline{\gamma}^{\mu+\frac{1}{2}}}\gamma^{\mu-\frac{1}{2}}e^{\frac{-2\mu h\gamma}{\overline{\gamma}}}I_{\mu-\frac{1}{2}}\left(\frac{2\mu H\gamma}{\overline{\gamma}}\right) \tag{6.169}$$

where H and h are the functions of fading parameter η defined in Chapter 2 and $\mu > 0$ is a fading parameter defined as $\mu = \frac{1}{2V(P_{\eta-\mu})}\left[1 + \left(\frac{H}{h}\right)^2\right] = \mu N_t N_r$. Note that in the above expression, we do not substitute $\mu N_t N_r$ in place of μ to keep the representation simpler. Actually, in all future discussions related to STBC MIMO systems over $\eta - \mu$ fading channels μ shall be interpreted as $\mu N_t N_r$.

Note that $\overline{\gamma}$ is the average SNR and is used as a scaling factor of $\|\mathbf{H}\|_F^2$ in many books to indicate average SNR at the receiver. The PDF of SNR is further used in evaluation of various performance parameters of STBC MIMO systems. Using the fact that received instantaneous SNR follows $\eta - \mu$ distribution, the further analysis follows from the analysis of SISO systems or receiver MRC diversity techniques which are thoroughly investigated by the researchers over various fading channels including $\eta - \mu$ fading channels. In the next part of this section, we will evaluate outage probability and SER of STBC MIMO systems over $\eta - \mu$ fading channels.

6.7.2 Outage probability

Outage probability is already defined in the earlier chapters of this book. Here, we will use the same definition to evaluate the outage probability of STBC MIMO systems. Outage probability gives the proportion of time for which a communication link is out of order. Given the average SNR at the receiver is $\overline{\gamma}$, the outage probability is defined as the probability that the instantaneous receiver SNR is below a predefined threshold, γ_{th}. Thus, outage probability can be given by

$$P_{out}\left(\overline{\gamma},\gamma_{th}\right) = P\left(\overline{\gamma}\left\|\mathbf{H}\right\|_F^2 < \gamma_{th}\right) \tag{6.170}$$

It can be evaluated by substituting equation (6.169) in the above expression as

$$P_{out}\left(\overline{\gamma},\gamma_{th}\right) = \int_0^{\gamma_{th}} \frac{2\sqrt{\pi}\mu^{\mu+\frac{1}{2}}h^{\mu}}{\Gamma\left(\mu\right)H^{\mu-\frac{1}{2}}\overline{\gamma}^{\mu+\frac{1}{2}}}\gamma^{\mu-\frac{1}{2}}e^{\frac{-2\mu h\gamma}{\overline{\gamma}}}I_{\mu-\frac{1}{2}}\left(\frac{2\mu H\gamma}{\overline{\gamma}}\right)d\gamma \tag{6.171}$$

As discussed earlier and also can be observed from above, outage probability is the CDF of the received instantaneous SNR. Thus, from the relation of CDF of $\eta-\mu$ distributed instantaneous SNR in equation (2.27) and Appendix A.8, the outage probability of STBC MIMO systems over $\eta-\mu$ fading channels can be given as

$$P_{out}\left(\overline{\gamma},\gamma_{th}\right) = 1 - Y_{\mu}\left(\frac{H}{h},\sqrt{\frac{2h\mu\gamma_{th}}{\overline{\gamma}}}\right) \tag{6.172}$$

where $Y_{\mu}(x,y)$ is Yacoub's integral [136]. It is defined as

$$Y_{\mu}(x,y) = \frac{\sqrt{\pi\left(1-x^2\right)^{\mu}2^{\frac{3}{2}-\mu}}}{\Gamma(\mu)x^{\mu-\frac{1}{2}}}\int_y^{\infty} e^{-t^2}t^{2\mu}I_{\mu-\frac{1}{2}}\left(tx^2\right)dt \tag{6.173}$$

where $-1 < x < 1$ and $y \geq 0$. The solution to Yacoub's integral can be given by [81]

$$Y_{\mu}(x,y) = 1 - \frac{\left(1-x^2\right)^{\mu}y^{4\mu}}{\Gamma(1+2\mu)}\Phi_2^{(2)}\left(\mu,\mu;1+2\mu;-(1+x)y^2,-(1-x)y^2\right) \tag{6.174}$$

where $\Phi_2^{(2)}$ is the confluent Lauricella function.

6.7.3 Error performance analysis

In this section, we will discuss a method to obtain an expression for error performance analysis of STBC MIMO systems over $\eta-\mu$ fading channels. Error performance analysis is primarily about the probability of error in the detected symbols at the receiver. Outage probability is directly related to error performance of the system but there is no exact relation between these two performance parameters. For orthogonal STBC designs, the instantaneous SNR at the receiver can be given as Frobenius norm of the channel times average SNR as in equation (6.168).

It has already been discussed that γ is an $\eta-\mu$ distributed random variable with fading parameters η and $\mu N_t N_r$ and its PDF can be given as in the expression of equation (6.169). In general, to evaluate error performance of any wireless communication system, the channel is assumed to be constant for a while and the error performance is evaluated for those channel conditions. This probability of error is known as conditional error probability for given channel conditions. Conditional probability of error depends on the instantaneous received SNR which is a random variable on account of the randomly varying channel coefficients. Conditional probability of error for various modulation schemes can be given by

$$P_e(\gamma) = aQ\left(\sqrt{b\gamma}\right) - cQ^2\left(\sqrt{b\gamma}\right) \tag{6.175}$$

where $Q(\cdot)$ is the Gaussian Q function which gives the area under the tail of a Gaussian curve and is defined as $Q(\hbar) = \frac{1}{\sqrt{2\pi}} \int\limits_{\hbar}^{\infty} e^{-\left(\frac{u^2}{2}\right)} du$, γ is the instantaneous received SNR and a, b and c are modulation dependent parameters listed in Table A.1. The average probability of error can be obtained by averaging the conditional error probability over the PDF of received instantaneous SNR. For averaging the conditional error probability, we will use the MGF based approach in this chapter. It can be given as

$$\overline{P}_e = \underbrace{\frac{a}{\pi} \int\limits_0^{\frac{\pi}{2}} M_\gamma\left(\frac{b}{2\sin^2\theta}\right) d\theta}_{I_1} - \underbrace{\frac{c}{\pi} \int\limits_0^{\frac{\pi}{4}} M_\gamma\left(\frac{b}{2\sin^2\theta}\right) d\theta}_{I_2} \tag{6.176}$$

where $M_\gamma(\cdot)$ is the MGF of received instantaneous SNR. MGF of $\eta-\mu$ distributed instantaneous SNR is given as in equation (2.31).

$$M_\gamma(s) = \int_0^\infty p_\gamma(\gamma) e^{-s\gamma} d\gamma \tag{6.177}$$

$$= \frac{\left(4\mu^2 h\right)^\mu}{\left(2(h-H)\mu + s\overline{\gamma}\right)^\mu \left(2(h-H)\mu + s\overline{\gamma}\right)^\mu} \tag{6.178}$$

For the case of $N_t \times N_r$ MIMO STBC systems, the fading parameters are η and $\mu N_t N_r$. However, as discussed earlier, to keep the expressions simpler, the parameter μ needs to be interpreted as $\mu N_t N_r$ in the above expression and all expressions henceforth.

The probability of error can be obtained by evaluating I_1 and I_2. First, we evaluate I_1 which may be obtained by substituting equation (6.178) in equation (6.176) as

$$I_1 = \frac{a}{\pi} \int\limits_0^{\frac{\pi}{2}} \frac{\left(4\mu^2 h\right)^\mu}{\left(2(h-H)\mu + \frac{b\overline{\gamma}}{2\sin^2\theta}\right)^\mu \left(2(h-H)\mu + \frac{b\overline{\gamma}}{2\sin^2\theta}\right)^\mu} d0 \tag{6.179}$$

Similarly, I_2 can be given as

$$I_2 = \frac{c}{\pi} \int_0^{\frac{\pi}{4}} \frac{\left(4\mu^2 h\right)^{\mu}}{\left(2(h-H)\mu + \frac{b\overline{\gamma}}{2\sin^2\theta}\right)^{\mu} \left(2(h-H)\mu + \frac{b\overline{\gamma}}{2\sin^2\theta}\right)^{\mu}} d\theta \tag{6.180}$$

Integrals I_1 and I_2 are definite integrals with finite limits. It is possible to obtain a numerical solution for them. Alternatively, closed form solutions in terms of hypergeometric functions can be obtained by various substitutions as we did in the case of analysis of STBC MIMO systems over Nakagami-m and $\kappa - \mu$ fading channels.

Next, we will obtain solutions for I_1 and I_2.

Solution for I_1

To solve I_1, let us change the variable by substituting $\cos^2\theta = t$. This leads to

$$\begin{aligned} \sin^2\theta = 1 - \cos^2\theta = 1 - t \\ -2\cos\theta\sin\theta d\theta = dt \end{aligned} \tag{6.181}$$

Thus, the limits of integral would change from 0 to 1 and the integrating variable $d\theta$ changes to

$$d\theta = \frac{dt}{-2\sqrt{t}\sqrt{1-t}} \tag{6.182}$$

Therefore, I_1 can now be given as

$$I_1 = \frac{a}{2\pi} \int_0^1 \frac{\left(4\mu^2 h\right)^{\mu} t^{-\frac{1}{2}} (1-t)^{-\frac{1}{2}} dt}{\left(2(h-H)\mu + \frac{b\overline{\gamma}}{2(1-t)}\right)^{\mu} \left(2(h-H)\mu + \frac{b\overline{\gamma}}{2(1-t)}\right)^{\mu}} \tag{6.183}$$

This integral can be simplified in the following form so that it is possible to represent in the standard form of Appell hypergeometric function as

$$\begin{aligned} I_1 =& \frac{2^{2\mu} a \left(4\mu^2 h\right)^{\mu}}{2\pi \left(4\left(h-H\right)\mu + b\overline{\gamma}\right)^{\mu} \left(4\left(h+H\right)\mu + b\overline{\gamma}\right)^{\mu}} \\ &\times \int_0^1 \frac{t^{-\frac{1}{2}} (1-t)^{2\mu-\frac{1}{2}} dt}{\left(1 - \frac{4(h-H)\mu t}{4(h-H)\mu + b\overline{\gamma}}\right)^{\mu} \left(1 - \frac{4(h+H)\mu t}{4(h+H)\mu + b\overline{\gamma}}\right)^{\mu}} \end{aligned} \tag{6.184}$$

$$= \frac{a}{2\pi} M_{\gamma}\left(\frac{b}{2}\right) \int_0^1 \frac{t^{-\frac{1}{2}} (1-t)^{2\mu-\frac{1}{2}} dt}{\left(1 - \frac{4(h-H)\mu t}{4(h-H)\mu + b\overline{\gamma}}\right)^{\mu} \left(1 - \frac{4(h+H)\mu t}{4(h+H)\mu + b\overline{\gamma}}\right)^{\mu}} \tag{6.185}$$

The above integral may be compared with the expression of Appell hypergeometric function of two variables which is defined as [103]

$$F_A^{(1)}(\alpha;\beta,\beta';\gamma;x;y) = \frac{\Gamma(\gamma)}{\Gamma(\alpha)\Gamma(\gamma-\alpha)} \times \int_0^1 t^{\alpha-1}(1-t)^{\gamma-\alpha-1}(1-tx)^{-\beta}(1-ty)^{-\beta'}\,dt \tag{6.186}$$

The above definition of Appell hypergeometric function can be compared with expression (6.185) to obtain the arguments of the function as

$$\alpha = \frac{1}{2} \tag{6.187}$$

$$\beta = 1 \tag{6.188}$$

$$\beta' = 1 \tag{6.189}$$

$$\gamma = 2\mu + 1 \tag{6.190}$$

$$x = \frac{4(h-H)\mu}{4(h-H)\mu + b\overline{\gamma}} \tag{6.191}$$

$$y = \frac{4(h+H)\mu}{4(h+H)\mu + b\overline{\gamma}} \tag{6.192}$$

Therefore, I_1 can be evaluated as

$$I_1 = \frac{a}{2\pi} M_\gamma\left(\frac{b}{2}\right) \frac{\Gamma\left(\frac{1}{2}\right)\Gamma\left(2\mu+\frac{1}{2}\right)}{\Gamma(2\mu+1)} \times F_A^{(1)}\left(\frac{1}{2};1,1;2\mu+1;\frac{4(h-H)\mu}{4(h-H)\mu+b\overline{\gamma}};\frac{4(h+H)\mu}{4(h+H)\mu+b\overline{\gamma}}\right) \tag{6.193}$$

Solution for I_2

One can proceed for solution of I_2 following the steps used for solution of I_1 earlier. But in this case while making the substitution care is taken that the upper and lower limits of integration be converted to 1 and 0 because most special functions such as hypergeometric functions are defined for this limit of finite integrals. So, in this case, let us change the variable by substituting $2\sin^2\theta = t$. This leads to

$$\sin^2\theta = \frac{t}{2}$$

$$\cos^2\theta = 1 - \sin^2\theta = 1 - \frac{t}{2}$$

$$4\sin\theta\cos\theta d\theta = dt \tag{6.194}$$

Thus, the limits of integral would change from 0 to 1 and the integrating variable $d\theta$ changes to

$$d\theta - \frac{dt}{4\sqrt{\frac{t}{2}}\sqrt{1-\frac{t}{2}}} \tag{6.195}$$

Thus, I_2 can now be given as

$$I_2 = \frac{c}{4\pi} \int_0^1 \frac{\left(4\mu^2 h\right)^{\mu} \left(\frac{t}{2}\right)^{-\frac{1}{2}} \left(1 - \frac{t}{2}\right)^{-\frac{1}{2}}}{\left(2(h-H)\mu + \frac{b\overline{\gamma}}{t}\right)^{\mu} \left(2(h-H)\mu + \frac{b\overline{\gamma}}{t}\right)^{\mu}} dt \tag{6.196}$$

Further mathematical simplifications can bring the above expression in the form as follows.

$$I_2 = \frac{c}{2\sqrt{2\pi}} \frac{\left(4\mu^2 h\right)^{\mu}}{\left(b\overline{\gamma}\right)^{2\mu}} \int_0^1 \frac{t^{2\mu-\frac{1}{2}} \left(1 - \frac{t}{2}\right)^{-\frac{1}{2}}}{\left(1 + \frac{2(h-H)\mu t}{b\overline{\gamma}}\right)^{\mu} \left(1 + \frac{2(h+H)\mu t}{b\overline{\gamma}}\right)^{\mu}} dt \tag{6.197}$$

It can be observed that the integral I_2 is now in the form of Lauricella hypergeometric function of three variables which can be given in the integral form as

$$F_D^{(3)}\left(a, b_1, b_2, b_3, c, x, y, z\right) = \frac{\Gamma(c)}{\Gamma(a)\Gamma(c-a)} \times \int_0^1 t^{a-1} (1-t)^{c-a-1} (1-xt)^{b_1} (1-yt)^{b_2} (1-zt)^{b_3} dt \tag{6.198}$$

From the above definition, the arguments of Lauricella hypergeometric function for evaluation of I_2 can be obtained as

$$a = 2\mu + \frac{1}{2} \tag{6.199}$$

$$b_1 = \frac{1}{2} \tag{6.200}$$

$$b_2 = \mu \tag{6.201}$$

$$b_3 = \mu \tag{6.202}$$

$$c = 2\mu + \frac{3}{2} \tag{6.203}$$

$$x = \frac{1}{2} \tag{6.204}$$

$$y = \frac{2(h-H)\mu t}{b\overline{\gamma}} \tag{6.205}$$

$$z = \frac{2(h+H)\mu t}{b\overline{\gamma}} \tag{6.206}$$

Therefore, the integral can be evaluated as

$$I_2 = \frac{c}{2\sqrt{2\pi}} \frac{\left(4\mu^2 h\right)^{\mu}}{\left(b\overline{\gamma}\right)^{2\mu}} F_D^{(3)}\left(2\mu + \frac{1}{2}, \frac{1}{2}, \mu, \mu, 2\mu + \frac{3}{2}, \frac{1}{2}, \frac{2(h-H)\mu t}{b\overline{\gamma}}, \frac{2(h-H)\mu t}{b\overline{\gamma}}\right) \tag{6.207}$$

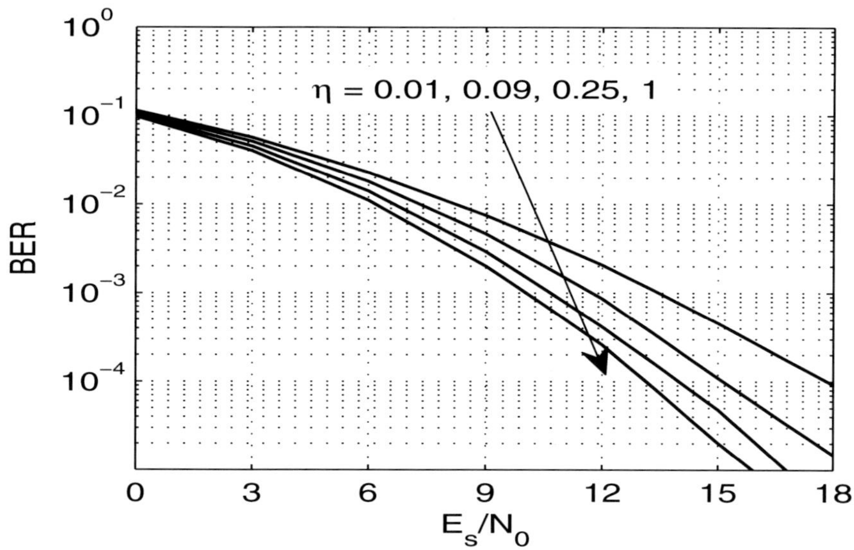

FIGURE 6.3
BER of 2 × 1 Alamouti STBC MIMO systems with BPSK over $\eta - \mu$ fading channels

For convergence of Lauricella hypergeometric function, $|x| < 1$ which is satisfied for all values of average SNR, $|y| < 1$ which is satisfied for all values of average SNR which satisfies $\frac{2(h-H)\mu}{b} < \overline{\gamma}$ and $|z| < 1$ which is satisfied for all values of average SNR which satisfies $\frac{2(h+H)\mu}{b} < \overline{\gamma}$. Thus, in general, the above expression is convergent for $\overline{\gamma} > \frac{2(h+H)\mu}{b}$.

As in the case of error performance analysis of STBC MIMO systems over Nakagami-m and $\kappa - \mu$ fading channels discussed earlier in this chapter, the integrals I_1 and I_2 are in closed form involving hypergeometric functions, Appell hypergeometric function and Lauricella hypergeometric function in this case. Again, the evaluation is to be carried out either by solving the definite integral with finite limits numerically or using the infinite series representation of these functions. Moreover, not all kinds of hypergeometric functions have been implemented in commonly used computational tools and softwares. Thus, in this case too, the computation of error probability can be simplified using Chiani's approximation [16] or Chernoff bound on the Gaussian Q function. This helps because the MGF of $\eta - \mu$ distributed SNR is in elementary functions. It is very easy to evaluate the MGF for any value of SNR rather than computing the complex functions such as hypergeometric functions obtained as a solution of I_1 and I_2.

Bit error rate (BER) results for a 2 × 1 Alamouti STBC MIMO system with binary phase shift keying (BPSK) modulation over $\eta - \mu$ fading channels are shown in Figure 6.3. The BER is shown for various values of η in the range from 0.01 to 1 and $\mu = 1$. BER performance of the STBC system is observed to degrade with the decreasing values of η when $\eta < 1$. Note that η is defined as the ratio of total power of in-phase components to that of quadrature phase components in the multipath clusters. Thus, $\eta = 1$ represents a point of symmetry. The channel model for given value of η will be the same as that with $1/\eta$ and the same value of μ.

In the next section, we will discuss STBC MIMO systems over $\alpha - \mu$ fading channels.

6.8 STBC systems over $\alpha - \mu$ fading channels

The discussion of $\alpha - \mu$ distribution and the fading channel models was already presented in Section 2.4.4 of Chapter 2. There we discussed about the physical model, PDF and CDF of $\alpha - \mu$ distributed envelope and SNR and the techniques to generate samples that follow $\alpha - \mu$ distribution. We will use those discussions and related expressions to evaluate various performance parameters of STBC MIMO systems in this section. In this section, the fading envelope ($|h_{j,i}|$) of the channel coefficients is assumed to be $\alpha - \mu$ distributed and all the links from transmitter to receiver are assumed to be i.i.d. In all the discussions related to performance analysis of STBC MIMO systems over $\alpha - \mu$ fading channels, we will assume $N_t \times N_r$ MIMO systems. It is assumed that an STBC codeword is transmitted over T time slots. So, a codeword for complete STBC transmission can be given by a matrix with dimensions $N_t \times T$. It was already discussed that the STBC codewords can be separately processed at different antennas and then the processed signals can be added together to get the signals equivalent to those received in MRC. From discussions in previous sections of this chapter and Chapter 1 (Section 1.3.2 and expression of equation (1.21)), the instantaneous SNR of the received signal can be given by

$$\gamma = \overline{\gamma}||\mathbf{H}||_F^2 \tag{6.208}$$

where each element of $\mathbf{H}$ is $\alpha - \mu$ distributed.

Note that the effective received signal in equation (6.92) is in the same form of the received signal after MRC in the receiver diversity techniques. In other words, it has the signal part given by $||\mathbf{H}||_F^2 x_i$ which is proportional to the sum of squared magnitudes of the channel coefficients. $||\mathbf{H}||_F^2 x_i$ represents sum of $N_t N_r$ square variants of $\alpha - \mu$ distributed random variables. Note here that the exact distribution of this sum cannot be directly obtained as in the earlier cases of Nakagami-m, $\kappa - \mu$ and $\eta - \mu$ distributed SNR. In the case of $\alpha - \mu$ distributed SNR, one needs to approximate the sum of $\alpha - \mu$ distributed square variates by using various moments as is done in [27]. However, instead of going into those complexities, we can obtain the PDF of received instantaneous SNR as a convolution of PDF of individual link SNRs given that they are mutually independent. This makes the analysis very difficult. So, we represent the MGF of instantaneous SNR of L link MRC system as a function of MGF of instantaneous SNR of single link. The relation can be given as

$$\text{MGF}_L = \text{MGF}_1^L \tag{6.209}$$

where MGF_L is the MGF of instantaneous SNR for a receiver with L branch MRC diversity scheme and MGF_1 is the MGF of instantaneous SNR for a single link system.

To proceed for the analysis of single antenna systems, we refer to the MGF of $\alpha-\mu$ distributed SNR given by the expression of equation (2.52) of Chapter 2.

$$M_\gamma(s) = \int_0^\infty p_\gamma(\gamma)e^{-s\gamma}d\gamma \tag{6.210}$$

$$= \frac{\alpha\mu^\mu}{2\bar{\gamma}^{\frac{\alpha\mu}{2}}} \frac{\sqrt{k}l^{\frac{\alpha\mu-1}{2}}}{(2\pi)^{\frac{l+k-2}{2}} s^{\frac{\alpha\mu}{2}}} G_{l,k}^{k,l}\left(\left(\frac{\mu}{\bar{\gamma}^{\frac{\alpha}{2}}}\right)^k \frac{l^l}{s^l k^k}, \begin{matrix} P\left(l, 1-\frac{\alpha\mu}{2}\right) \\ P(k,0) \end{matrix}\right) \tag{6.211}$$

The average probability of error can be obtained by averaging the conditional error probability over the PDF of received instantaneous SNR. It is observed in all the previous cases that the exact error probability is in the form of complex functions even if the MGF is in the form of simple expressions consisting of elementary mathematical operations. While in this case, the MGF itself is in the form of Meijer's G function. So, in this case, we evaluate approximate error probability.

In Section 6.5, we introduced the following two approximations for Gaussian Q function for positive values of argument.

$$Q(\hbar) \approx \frac{1}{2}e^{-\frac{\hbar^2}{2}} \tag{6.212}$$

$$Q(\hbar) \approx \frac{1}{12}e^{-\frac{\hbar^2}{2}} + \frac{1}{4}e^{-\frac{2\hbar^2}{3}} \tag{6.213}$$

Using the above approximations, the squared Gaussian Q functions may be approximated as

$$Q^2(\hbar) \leq \frac{1}{4}e^{-\hbar^2} \tag{6.214}$$

$$Q^2(\hbar) \leq \frac{1}{144}e^{-\hbar^2} + \frac{1}{16}e^{-\frac{4\hbar^2}{3}} + \frac{1}{24}e^{-\frac{7\hbar^2}{6}} \tag{6.215}$$

These approximations can be used in finding the average error probability of any communication system by averaging the approximated Gaussian Q function over the PDF of received SNR. Now, using equation (6.118) and equation (6.120), the approximated Gaussian Q function is in the form of exponential functions, which when compared with the definition of MGF can be obtained as

$$P_e(\gamma) = aQ\left(\sqrt{b\gamma}\right) - cQ^2\left(\sqrt{b\gamma}\right) \leq \frac{a}{2}e^{-\frac{b\gamma}{2}} - \frac{c}{4}e^{-b\gamma} \tag{6.216}$$

Similarly, using equations (6.119) and (6.121), the approximated error probability can be given as

$$\begin{aligned} P_e(\gamma) &= aQ\left(\sqrt{b\gamma}\right) - cQ^2\left(\sqrt{b\gamma}\right) \\ &\leq \frac{a}{12}e^{-\frac{b\gamma}{2}} + \frac{a}{4}e^{-\frac{2b\gamma}{3}} - \frac{c}{144}e^{-b\gamma} - \frac{c}{16}e^{-\frac{4b\gamma}{3}} - \frac{c}{24}e^{-\frac{7b\gamma}{6}} \end{aligned} \tag{6.217}$$

Taking the average of equation (6.216) and equation (6.217), an approximation for average error probability may be obtained. Here our aim is to make readers familiar with the approach to proceed for the approximations. So, we will proceed further with averaging of equation (6.216) even though equation (6.216) is a looser approximation as compared to equation (6.217). Therefore, average approximate probability of error can be given by

$$\overline{P}_e \leq \int_0^\infty \left(\frac{a}{2}e^{-\frac{b\gamma}{2}} - \frac{c}{4}e^{-b\gamma}\right) p_\gamma(\gamma) d\gamma \tag{6.218}$$

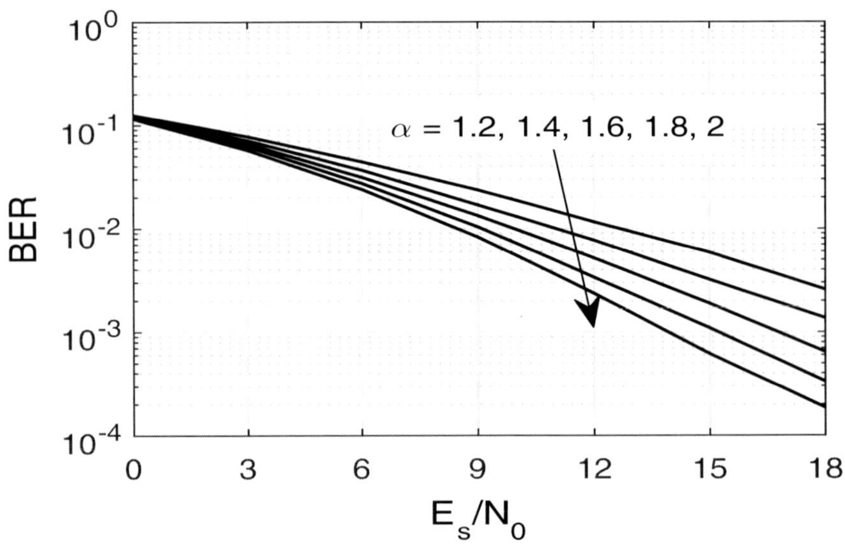

FIGURE 6.4
BER of 2×1 Alamouti STBC MIMO systems with BPSK over $\alpha - \mu$ fading channels

Comparing the above expression with the definition of MGF of a random variable, an upper bound on the average probability of error for single antenna systems over $\alpha-\mu$ fading channels can be given by

$$\begin{aligned}\overline{P}_e \leq & \frac{a}{2}\frac{\alpha\mu^{\mu}}{2\overline{\gamma}^{\frac{\alpha\mu}{2}}}\frac{\sqrt{k}l^{\frac{\alpha\mu-1}{2}}}{(2\pi)^{\frac{l+k-2}{2}}\left(\frac{b}{2}\right)^{\frac{\alpha\mu}{2}}}G_{l,k}^{k,l}\left(\left(\frac{\mu}{\overline{\gamma}^{\frac{\alpha}{2}}}\right)^{k}\frac{l^{l}}{\left(\frac{b}{2}\right)^{l}k^{k}}, \begin{array}{c}P\left(l,1-\frac{\alpha\mu}{2}\right)\\ P(k,0)\end{array}\right)\\ & -\frac{c}{4}\frac{\alpha\mu^{\mu}}{2\overline{\gamma}^{\frac{\alpha\mu}{2}}}\frac{\sqrt{k}l^{\frac{\alpha\mu-1}{2}}}{(2\pi)^{\frac{l+k-2}{2}}b^{\frac{\alpha\mu}{2}}}G_{l,k}^{k,l}\left(\left(\frac{\mu}{\overline{\gamma}^{\frac{\alpha}{2}}}\right)^{k}\frac{l^{l}}{b^{l}k^{k}}, \begin{array}{c}P\left(l,1-\frac{\alpha\mu}{2}\right)\\ P(k,0)\end{array}\right)\end{aligned} \tag{6.219}$$

Note that the above expression is an upper bound on the error probability for a communication system without any diversity technique. With the help of equation (6.209), an upper bound on the probability of error can be evaluated for STBC MIMO systems over $\alpha-\mu$ fading channels.

Among the generalized fading channels discussed in this book, $\alpha-\mu$ channel model is the only model that takes care of the nonlinearities in the fading environment. The parameter, α, shows the non-linearity of the fading environment. The effect of non-linearity on the error performance of STBC MIMO systems is shown in Figure 6.4. It is observed that the BER performance improves with the increasing values of α. This is because a larger value of α models the channels that are more like AWGN or non-fading scenarios ($\alpha \to \infty$ models non-fading environments).

7

MIMO for 5G Mobile Communications

CONTENTS

Multiple-input multiple-output (MIMO) wireless communications we have discussed until now are also known as point-to-point MIMO. There are many disadvantages associated such kinds of MIMO systems. Some of them are listed below.

- Multiplexing gain is not possible for a rank deficit channel matrix like line-of-sight propagation or key hole channel.
- Multiple radio frequency (RF) chains are employed for multiple antennas. Transmit antenna selection (TAS) has simplified such requirements. But TAS needs a feedback path from the receiver to the transmitter.
- Due to size constraints, multiple antennas at the base station are possible but not possible at the mobile station with the present scenario.
- Channel estimation overhead using pilot signals is not manageable for large

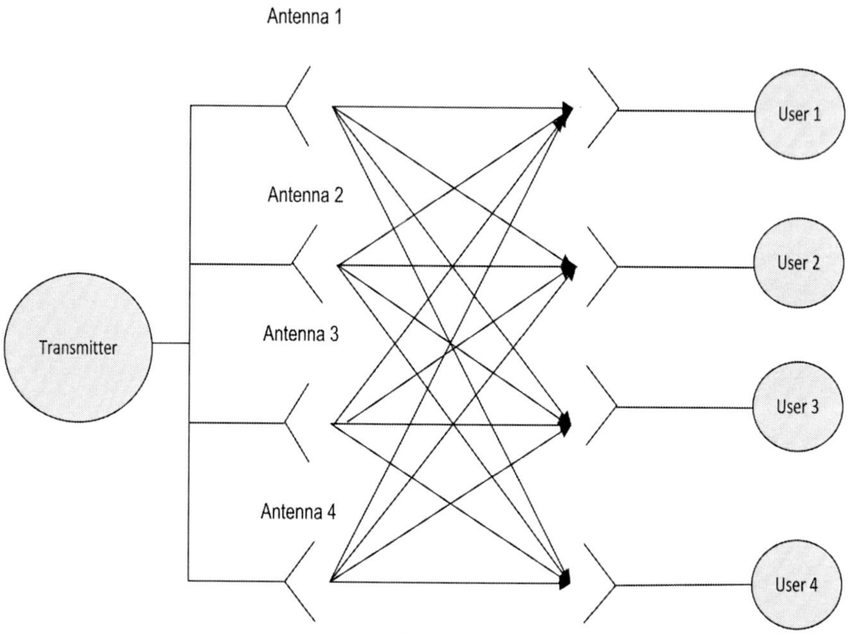

FIGURE 7.1
MU-MIMO system with transmitter employing M=4 antennas serving K=4 users with single antenna (highly simplified model)

scale MIMO systems where the receiver and transmitter supposedly employ hundreds to thousands of antennas.

7.1 Multiuser MIMO

Multiuser MIMO (MU-MIMO) has overcome these shortfalls. Unlike point-to-point MIMO, in MU-MIMO, a base station will serve several users [48,69].

A simplified model for MU-MIMO is shown in Figure 7.1. In this, a transmitter employing four antennas is communicating with four users with a single antenna. It is a highly simplified model. Assume a base station employing M antennas and communicating with K users using a single antenna in a single cell of a cellular network. There are two types of communication here [18,67].

1. Base station (BS) sending data to the mobile users which is popularly known as downlink.
2. Mobile users or Mobile station (MS) sending data to the base station which is also known as uplink.

Let us analyze communication in both cases mentioned above.

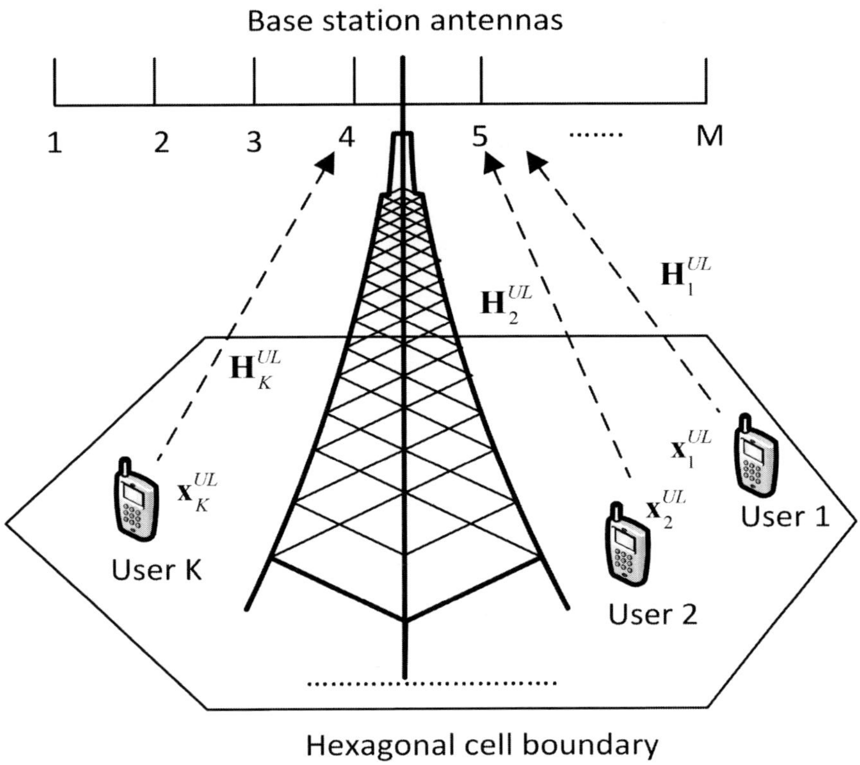

FIGURE 7.2
MU-MIMO system with transmitter employing M antennas serving K users with single antenna (uplink)

7.1.1 Uplink

The channel for uplink transmission is also known as multiple-access channel in the literature. Assume that K users are sending signals to the base station. The signal each user is sending to the base station is denoted by $\mathbf{x}_i^{UL}, i = 1, 2, 3..., K$. Note that for a single antenna user, the signal vector will be a complex scalar quantity. In order to preserve our analysis for any number of antennas at the mobile user, we will continue using the signal from each user at the uplink as a vector quantity. The channel for each user to the base station is expressed as $\mathbf{H}_i^{UL}, i = 1, 2, 3..., K$. The whole process is depicted in Figure 7.2. Then the received signal at the base station can be represented as

$$\mathbf{y}^{UL} = \mathbf{H}^{UL}\mathbf{x}^{UL} + \mathbf{n}^{UL} \tag{7.1}$$

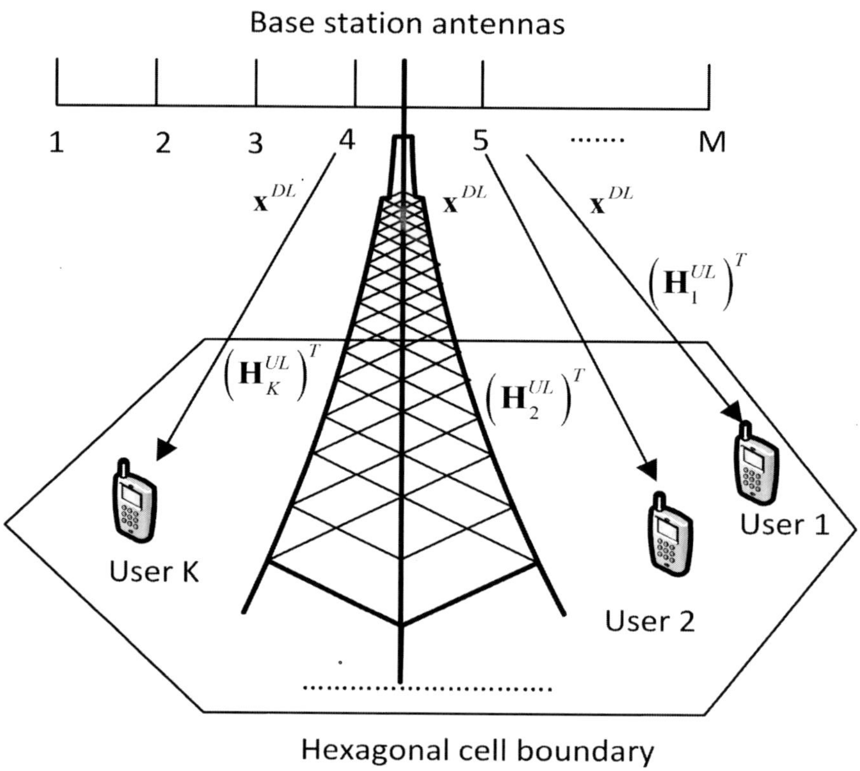

FIGURE 7.3
MU-MIMO system with transmitter employing M antennas serving K users with single antenna (downlink)

where $\mathbf{H}^{UL} = \begin{bmatrix}\mathbf{H}_1^{UL} & \mathbf{H}_2^{UL} & \dots & \mathbf{H}_K^{UL}\end{bmatrix}$

$$\mathbf{x}^{UL} = \begin{bmatrix}\mathbf{x}_1^{UL} \\ \mathbf{x}_2^{UL} \\ \vdots \\ \mathbf{x}_K^{UL}\end{bmatrix}$$

Note the $\mathbf{n}^{UL}$ is the additive white Gaussian noise (AWGN) with zero mean and covariance matrix $\sigma_n^2\mathbf{I}_M$.

7.1.2 Downlink

The channel for downlink transmission is also known as broadcast channel in the literature. Assume that each user as depicted in Figure 7.3 receives independent signal vector $\mathbf{x}^{DL}$ during the broadcast phase. Assume the downlink and uplink signals are transmitted at separate time slots with sufficient guard time intervals between them. For such time division duplex (TDD) mode of operation [11], assuming the reciprocity of the channel, we may write down the received signal at the user terminals as follows.

$$\mathbf{y}^{DL} = (\mathbf{H}^{UL})^T \mathbf{x}^{DL} + \mathbf{n}^{DL} \tag{7.2}$$

where $\mathbf{H}^{UL} = \begin{bmatrix} \mathbf{H}_1^{UL} & \mathbf{H}_2^{UL} & \dots & \mathbf{H}_K^{UL} \end{bmatrix}$

$$\mathbf{y}^{DL} = \begin{bmatrix} \mathbf{y}_1^{DL} \\ \mathbf{y}_2^{DL} \\ \vdots \\ \mathbf{y}_K^{DL} \end{bmatrix}$$

$$\mathbf{n}^{DL} = \begin{bmatrix} \mathbf{n}_1^{DL} \\ \mathbf{n}_2^{DL} \\ \vdots \\ \mathbf{n}_K^{DL} \end{bmatrix}.$$

Note that for a single antenna user, the received signal vector and noise vector for each user will be a complex scalar quantity. In order to preserve our analysis for any number of antennas at the mobile user, we will continue using the signal received as well as noise vector for each user at the downlink as a vector quantity.

7.2 Massive MIMO

Massive MIMO in which base station employs several hundreds to thousands of antennas and mobile users employing fewer antennas like one or two antennas is scalable and has many advantages as mentioned below [70]. We will assume each mobile station (user) employs single antenna for our analysis. Massive MIMO is a special case of MU-MIMO. For massive MIMO as the name suggests, we will assume the number of base station antennas (M) tends to infinity.

- Since mobile user has one or two antennas, much of the signal processing is done at the base station. Mobile station is generally small and receive diversity is not cost-impressive.
- In downlink, using beamforming at the base station, it can direct its beams toward the desired user as depicted in Figure 7.4, thereby minimizing the interference and energy usage. In uplink, base station will separate out the signals received from different users.
- In TDD massive MIMO, for an ideal case, the time required to estimate the channel is not dependent on the number of antennas at the base station and hence it is scalable.

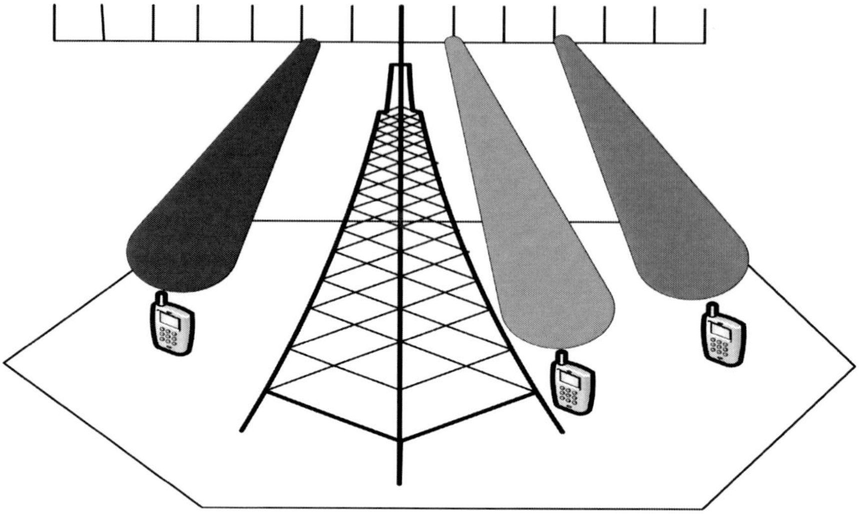

FIGURE 7.4
MU-MIMO system for a single base station employing M=14 antennas serving K=3 users with single antenna (downlink transmission)

Energy and spectral efficiency of moderately large antenna arrays for MU-MIMO has been compared to the single-input single-output (SISO) system. It has been reported that there are significant improvements in both [86]. For an ideal channel state information (CSI) scenario, a mobile user communicating with a massive MIMO base station with much reduced uplink transmit power of just $\frac{1}{M}$ would be to attain the same throughput as that of a single antenna base station. A detailed discussion on the massive MIMO designs has been presented in [85]. A thorough discussion on the massive MIMO has been presented in [9, 17, 71].

7.2.1 Uplink capacity analysis

Let us consider a base station employing M antennas serving K users with a single antenna. Assume the channel coefficient from the ith user to the jth antenna of the base station [63] as

$$h_{ji} = g_{ji}\sqrt{d_i} \tag{7.3}$$

where g_{ji} and d_i are the small scale channel gain coefficients and large scale fading coefficients. The uplink channel for this case may be expressed as

$$\mathbf{H}^{uplink} = \mathbf{G}\mathbf{D}^{1/2} \tag{7.4}$$

where

$$\begin{bmatrix} d_1 & & & \\ & d_2 & & \\ & & \ddots & \\ & & & d_K \end{bmatrix} \text{ and } \begin{bmatrix} g_{11} & g_{12} & g_{13} & \cdots & g_{1K} \\ g_{21} & g_{22} & g_{23} & \cdots & g_{2K} \\ \vdots & \vdots & \vdots & \ddots & \vdots \\ g_{M1} & g_{M2} & g_{M3} & \cdots & g_{MK} \end{bmatrix}$$

If we assume that channels are orthogonal, then

$$(\mathbf{H}^{uplink})^H \mathbf{H}^{uplink} = \mathbf{D}^{1/2}\mathbf{G}^H\mathbf{G}\mathbf{D}^{1/2} = M\mathbf{D} \tag{7.5}$$

The above equation is also known as favorable channel condition. It may be dependent on the type of antenna array and channel environments [87]. But it has been reported in the literature that as the number of antennas at the base station grows large ($M \to \infty >> K$), such favorable conditions are satisfied for different antenna array configurations and channel environments. It may be, due to the vast spatial diversity, the small scale randomness dies out. This needs a further investigation. We assume the favorable channel condition for our analysis. Note that this favorable condition is assumed irrespective of the small scale fading distributions, whether they are classical or generalized fading distributions. Assume equal power is given to each user as $\frac{P}{K}$ during uplink. We may evaluate the capacity for such massive MIMO system of base station with M antennas and single antenna K users for equal power allocation as follows.

$$C_{M>>K}^{uplink} \approx log_2 \det(\mathbf{I}_K + \frac{PM\mathbf{D}}{K\sigma_n^2}) \tag{7.6}$$

The above equation may also be expressed as

$$C_{M>>K}^{uplink} = \sum_{i=1}^{K} log_2(1 + \frac{PMd_i}{K\sigma_n^2}) \tag{7.7}$$

The above capacity may be achieved by a simple linear processing (matched filter) at the base station as follows.

$$(\mathbf{H}^{uplink})^H \mathbf{y}^{uplink} = (\mathbf{H}^{uplink})^H (\mathbf{H}^{uplink}\mathbf{x}^{uplink} + \mathbf{n}^{uplink}) \tag{7.8}$$

If we substitute $(\mathbf{H}^{uplink})^H \mathbf{y}^{uplink} = \tilde{\mathbf{y}}^{uplink}$ and $(\mathbf{H}^{uplink})^H \mathbf{n}^{uplink} = \tilde{\mathbf{n}}^{uplink}$, we have

$$\tilde{\mathbf{y}}^{uplink} = M\mathbf{D}\mathbf{x}^{uplink} + \tilde{\mathbf{n}}^{uplink} \tag{7.9}$$

Note that matched Filter (MF) processing at the base station has decoupled all the users' signals since $\mathbf{D}$ is a diagonal matrix. It means we have K parallel independent Gaussian channels. The signal-to-noise ratio (SNR) for each user's link to the base station can be calculated as $\frac{MPd_i}{K\sigma_n^2}$. Hence MF processing at the base station gives the capacity of equation (7.7).

7.2.2 Downlink capacity analysis

The capacity for the downlink massive MIMO can be calculated similarly. The only difference would be that the base station has the CSI and hence can allocate power adaptively to each user. Assume the power allocation matrix $\mathbf{D}_p$ is as follows.

$$\begin{bmatrix} p_1 & & & \\ & p_2 & & \\ & & \ddots & \\ & & & p_K \end{bmatrix}$$

where $\sum_{i=1}^{K} p_i = P$. We may calculate the capacity for downlink for massive MIMO for favorable channel condition as below.

$$C_{M>>K}^{downlink} \approx \max_{\mathbf{D}_p} log_2 \det(\mathbf{I}_K + \frac{PM\mathbf{D}_P\mathbf{D}}{\sigma_n^2}) \tag{7.10}$$

This capacity for the downlink can be easily achieved for the case when the base station does the simple MF precoding as follows.

$$\mathbf{x}_{prec}^{downlink} = (\mathbf{H}^{uplink})^* \mathbf{D}^{-\frac{1}{2}} \mathbf{D}_p^{\frac{1}{2}} \mathbf{x}^{uplink} \tag{7.11}$$

We can write down the expression for the downlink received signal vector as below.

$$\mathbf{y}^{downlink} = (\mathbf{H}^{uplink})^T \mathbf{x}_{prec}^{downlink} + \mathbf{n}^{downlink} \tag{7.12}$$

Using the expression of (7.11) and channel favorable condition, the above equation may be simplified as follows.

$$\mathbf{y}^{downlink} = M\mathbf{D}^{\frac{1}{2}} \mathbf{D}_p^{\frac{1}{2}} \mathbf{x}_{downlink} + \mathbf{n}^{downlink} \tag{7.13}$$

Note that $\mathbf{D}$ and $\mathbf{D}_p$ are diagonal matrices. Hence the signals for the downlink have been decoupled completely. Also it can be observed that this simple linear precoding at the base station has achieved the capacity of equation (7.10). There have been many assumptions in the above analysis, such as

- The channel is satisfying favorable conditions of orthogonality for a massive MIMO case.
- CSI can be estimated at the base station properly.
- The channel is reciprocal.

But these assumptions may not be true in real life or practical situations. Many researchers are looking into the possible problems in a practical scenario and trying to solve such problems. Note that we have done our analysis for a single cell. But in real life cellular communication is always multicell communication (refer to Figure 7.5). The pilot symbols are transmitted from the users to the base station which will be used for estimating the channels.

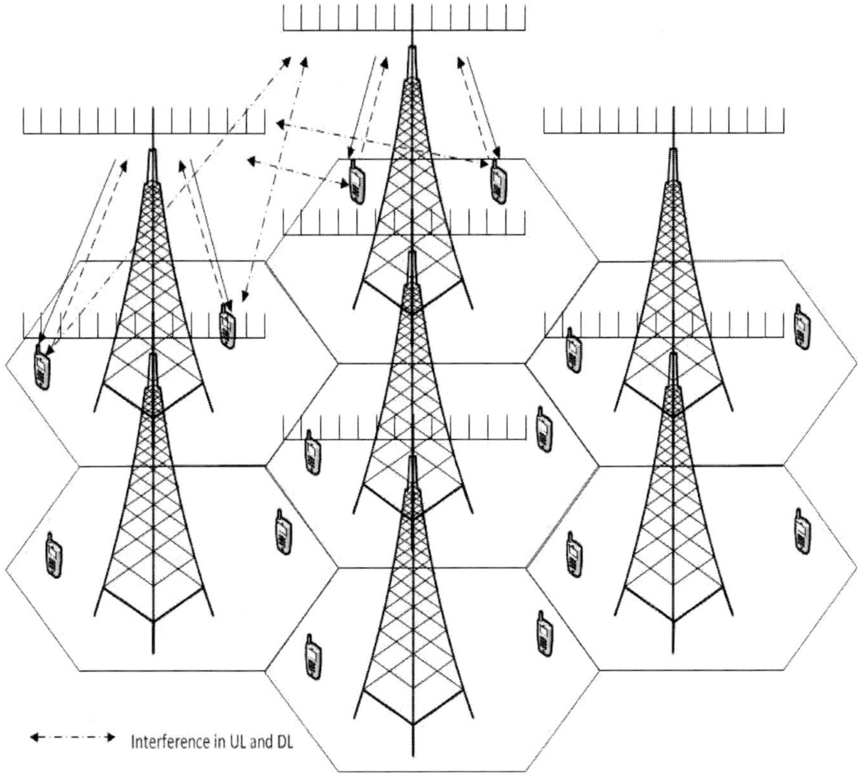

FIGURE 7.5
MU-MIMO-based cellular network (BS equipped with M=14 antennas and single antenna MS or user, each cell has K=2 users for illustration purposes)

This estimated CSI is used by base station in the downlink transmission. But the number of orthogonal pilot symbols is limited for a given period and bandwidth. Hence the pilot symbols are reused in a multicell communication. Hence the signals received by the base station are a linear combination of pilot symbols from the home cell and neighboring cells. There is a high level of interference especially at the cell edge. This is also known as pilot contamination. The MF precoding we have discussed above doesn't work well in multicell MU-MIMO. It has been reported in the literature that zero forcing (ZF), regularized zero forcing (RZF) and minimum mean square error (MMSE) precoder work better in such cases [55]. In RZF, one premultiplys the transmit signal in the downlink by a weight vector defined as [111]

$$\mathbf{W}_{RZF} = \frac{1}{\sqrt{\gamma_l}}(\hat{\mathbf{H}}_l)^H(\hat{\mathbf{H}}_l(\hat{\mathbf{H}}_l)^H + \delta\mathbf{I}_K)^{-1} \tag{7.14}$$

where $\hat{\mathbf{H}}_l = \mathbf{d}_{ll}^{-1}\hat{\mathbf{G}}_l$, $\mathbf{d}_{ll} = \begin{bmatrix} \sqrt{d_{l1l}} & & & \\ & \sqrt{d_{l2l}} & & \\ & & \ddots & \\ & & & \sqrt{d_{lKl}} \end{bmatrix}$ and $\gamma_l = \frac{trace((\hat{\mathbf{H}}_l)^H)\hat{\mathbf{H}}_l}{K}$. In the above equation, $\hat{\mathbf{G}}_l$ is the estimated CSI of users in the lth base station. The particular cases of RZF are:

- $\delta = 0$ (MF precoding).
- $\delta = \infty$ (ZF precoding).
- $\delta = \frac{K\sigma_n^2}{2SNR^{downlink}log_2 M}$ (MMSE precoding [54]).

There are ways of overcoming such things by cooperation among the base stations, which is referred to as coordinated multipoint transmission. There are two types of cooperation (full and partial cooperation). Full cooperation is called network MIMO and partial cooperation is referred to as coordinated beamforming/scheduling [114]. There are twofold advantages of coordinated multipoint transmission (mitigation of intercell interference and augmenting data rate at the edge of cell) [42]. Some of the challenges with massive MIMO are as follows [64].

- It has been assumed that channel for the uplink and downlink are reciprocal. But the hardware at the mobile user and base station may be quite different; in that case, the reciprocal channel assumption may be lost.
- Orthogonal pilots are limited and hence they are reused. But for user at the cell edge, there may be high interference from the adjacent or neighboring cell's user using the same set of pilots. This is called pilot contamination.
- Base station has no knowledge of the channel state information prior to link establishment with a mobile user. Hence downlink beamforming could not be performed. Hence transmit diversity employing space time codes may be used.
- Massive MIMO has highly simplified operation with linear precoding and detection for favorable channel condition. But this favorable channel condition may not always be satisfied depending on the channel propagation conditions, types of antenna arrays employed at the base station, etc. This requires a thorough analysis.

7.2.3 Outage probabilty

It would be good to find an approximate expression for outage probability for each user in the downlink. Such analysis would be of help if we want to design massive MIMO. For instance, we would like to find the number of users a massive base station could serve at a time for a given outage probability. We will use single cell massive MIMO for the analysis. Suppose that the base station is employing MF precoder for the downlink. Assume each user is using a single antenna. For this case, one may express the signal transmitted from the base station as follows [39].

$$\mathbf{x}^{downlink} = \sqrt{\frac{P}{KM}} \sum_{i=1}^{K} (\mathbf{h}_i^{downlink})^H x_i \tag{7.15}$$

The received signal for the ith user may be expressed as

$$\begin{aligned} y_i = & \sqrt{\frac{P}{KM}} \mathbf{h}_i^{downlink} (\mathbf{h}_i^{downlink})^H x_i \\ & + \sqrt{\frac{P}{KM}} \sum_{j=1, j \neq i}^{K} \mathbf{h}_i^{downlink} (\mathbf{h}_j^{downlink})^H x_i + n_i \end{aligned} \tag{7.16}$$

where n_i is AWGN with zero mean and unity variance. The signal-to-interference-plus-noise ratio (SINR) for the ith user is given by

$$SINR_i = \frac{P_u M X_i^2}{1 + P_u Y_i} \tag{7.17}$$

where $P_u = \frac{P}{K}$ is the power per user for equal power allocation, $X_i \triangleq \frac{1}{M} |\mathbf{h}_i^{downlink} (\mathbf{h}_i^{downlink})^H|$ and $Y_i \triangleq \frac{1}{M} \sum_{j=1, j \neq i}^{K} |\mathbf{h}_i^{downlink} (\mathbf{h}_j^{downlink})^H|^2$.

We will assume $\mathbf{h}_1^{downlink}, \mathbf{h}_2^{downlink}, ..., \mathbf{h}_K^{downlink}$ are independent and identically distributed (i.i.d.) and distributed as $CN(0, 1)$. We may assume any generalized distribution as well here and do the analysis. For the assumed distribution here, both X_i and Y_i will be gamma distributed with the different shape parameters and scale factors. The probability density function (PDF) of X_i is

$$f_{X_i}(x) = \frac{x^{M-1} e^{-Mx}}{(\frac{1}{M})^M (M-1)!} \tag{7.18}$$

Hence, $E(X_k^2) = 1 + \frac{1}{M}$. The PDF Y_i is

$$f_{Y_i}(y) = (1 - c_1) \sum_{m=0}^{\infty} c_1^m \frac{y^{c_m - 1} exp(-y/c_3)}{c_3^{c_m} (c_m - 1)!} \tag{7.19}$$

where $c_1 \triangleq \frac{K-1}{\sqrt{M} + K - 2}$, $c_m \triangleq K + m - 1$ and $c_3 \triangleq \frac{\sqrt{M} - 1}{\sqrt{M}}$. The above expression for PDF of Y_i may be simplified as follows.

$$f_{Y_i}(y) = c_4 c_1^{-(K-2)} [e^{-c_4 y} - e^{[-c_5 y]} \sum_{n=0}^{K-3} (c_3 c_1)^2 \frac{y^n}{n!}] \tag{7.20}$$

where $c_4 = \frac{\sqrt{M}}{\sqrt{M+K-2}}$. Therefore the outage probability for a threshold SINR (γ_{th}) can be obtained as

$$P_{outage} = Prob(\frac{P_u M X_i^2}{1 + P_u Y_i} < \gamma_{th}) \tag{7.21}$$

The above expression may be approximated as

$$P_{outage} \approx Prob(\frac{P_u M(1 + 1/M)}{1 + P_u Y_i} < \gamma_{th}) \tag{7.22}$$

With some mathematical manipulations, we may rewrite the above equation as

$$P_{outage} \approx Prob(c_5 < Y_i) \tag{7.23}$$

where $c_5 = \frac{M+1}{\gamma_{th}} - \frac{1}{P_u}$. Hence the approximate outage probability can be simplified as below.

$$P_{outage} \approx c_1^{-(K-2)} e^{-c_2 c_5} - (1 - c_1) \sum_{n=0}^{K-3} \frac{c_1^{n-K+2} \Gamma(n+1, \frac{c_5}{c_3})}{n!} \tag{7.24}$$

where $\Gamma(s,x) = \int_x^\infty t^{(s-1)} e^{-t} dt$ is the upper incomplete gamma function.

7.3 mmWave massive MIMO

mmWave massive MIMO is basically a massive MIMO which has been envisioned to be implemented at mmWave frequency region.

7.3.1 Requirements for 5G mobile

Every generation cellular communication comes in a decade or so [108]: 1G (1981), 2G(1992), 3G (2001) and 4G (2011). Requirements for the next generation 5G mobile communications in comparison with 4G mobile is tabulated in Table 7.1. Even though 5G has not been standardized (it is expected somewhere in 2020), however, International Telecommunication Unit (ITU) has come up with certain requirements for 5G mobile communications [82]. Peak data rate is the best possible data rate a user can get in a network and area traffic capacity is the cumulative data amount the network can serve. Peak data rate is expected to be 20 Gbps and area traffic capacity, 10 Mbps/m^2. Latency of 1 ms is required for future 5G applications like two-way games, tactile internet [40], and virtual and augmented reality (for instance, google glass). 5G is yet to be standardized. It is expected that it will have super high capacity and ultra high speed data rate. Besides 5G devices will consume less power and be relatively cheaper.

TABLE 7.1
Requirements for 5G mobile communications in comparison with 4G mobile.

Parameters	**Units**	**5G**	**4G**
Area traffic capacity	Mbps/m^2	10	0.1
Peak data rate	Gbps	20	1
User experienced data rate	Mbps	100	10
Spectrum efficiency		3×	1×
Energy efficiency		100×	1×
Connection density	devices/km^2	10^6	10^5
Latency	ms	1	10
Mobility	km/h	500	350

Ten pillars for 5G mobile wireless communications [110] are listed below.

- Small cells
- mmWave
- Massive MIMO
- Multi-radio access technology (RAT)
- Self-organizing networks (SON)
- Device-to-device (D2D) communications
- Backhaul
- Energy efficiency (EE)
- New spectrum and sharing
- Radio access network (RAN) virtualization

7.3.2 Three big pillars for 5G mobile

In order to fulfill the requirements of 5G wireless communications, some suggestions are listed below.

- Small cell networks like pico- and femtocells, which enhance the spatial and frequency reuse, will be able to enhance the capacity of the system [137]. Small cells will reduce propagation losses, and increase data rate and energy efficiency. Qualcomm demonstrated that the capacity of the network almost doubles whenever the number of small cells double.
- Spectrum is crowded at microwave frequencies. But mmWave frequencies are relatively free and the available bandwidth is huge and correspondingly at a higher capacity of the system.

- Large antenna arrays at the base station will further enhance the capacity of the system.

The above three technologies are also known as the three big pillars for 5G wireless [4] and they will get a lion's share in augmenting the 5G capacity 1000×. In 5G, networks will be more and more heterogenous. Multi-RAT will enable 5G devices to connect to 5G standard, 3G standard, 4G LTE, WiFi and device-to-device (D2D) communications. There are two techniques which are suggested in the literature for heterogenous networks (HeNets) of macro cells and small cells.

- Biasing means that the preference will be given to the small cells even if it has lower SINR, hence, network traffic of the heavily loaded macrocell could be off-loaded to lightly loaded small cells.

- Muting means basically stopping the transmission from the macrocell when the mobile user is served by the small cells which will enhance SINR of the small cell considerably.

The idea is to have several small cells with base station employing many small antennas connected with low power amplifiers. The aim is to exploit elevation angle along with the azimuth angle so that we can have antenna arrangements in 3-D, which is also known as full dimension MIMO [84]. In fact all these three pillars are interrelated and one of them will influence the other. All three techniques mentioned above will collectively work together to augment capacity by a thousandfold [120]. mmWave band covers 30-300 GHz corresponds to operating wavelength from 1-10 mm [142]. Hence one can incorporate many antennas in a comparatively smaller device since we know that antenna array needs a spacing of at least half the wavelength. Besides, dipole antenna size is usually also half wavelength. Let us make some typical calculations for large scale antenna arrays. Consider a mobile headset of 10 cm length, 1.5 cm thickness and 5 cm width. From our knowledge of antenna arrays, antenna spacing for antenna arrays should be at least $\frac{\lambda}{2} = 15\ cm$ for 1 GHz operating frequency. Assume it is a halfwave dipole antenna we are considering. Hence for one antenna along with the spacing between antenna arrays it occupies 30 cm. So we cannot place more than a single antenna on such a mobile phone at 1 GHz. Consider mmWave communication at 30 GHz. A single antenna needs a space of 1 cm. Now we have the scope for placing at least 10 antennas along length and 5 antennas along breadth. Further, such antenna arrays are allowed to be placed on the top and bottom surface of the mobile headset. Hence a total of 100 antennas can be accommodated on the same mobile headset for 30 GHz wireless communications. Similar calculations will enable us to put thousands of antennas at the base station easily. Therefore large scale antenna arrays are a possibility. The bandwidth of mmWave band is also huge. A 10% bandwidth at 60 GHz is 6 GHz. Hence the channel capacity which is directly proportional to the bandwidth will be quite high. One of the major hurdles in mmWave band is that path loss is very high,

thereby highly dense small cell networks may be deployed. So all the above three suggestions will collectively improve the network capacity in order to meet the demands of 5G wireless.
What is the motivation behind using large scale antenna arrays in mmWave communication? According to the Friis transmission formula, the received power for line of sight (LOS) propagation can be expressed as

$$P_r = P_t G_t G_r (\lambda / 4\pi d)^2 \tag{7.25}$$

where P_r is the received signal power, P_t is the transmitted signal power, G_t is the gain of transmitting antenna, G_r is the gain of receiving antenna, λ is the wavelength of the signal used and d is the distance between the transmitter and receiver.
The path loss at the mmWave is higher than 2. It has been reported in the literature that path loss exponent varies from 1.7-2.5 for LOS and 3.5-6 for non-LOS [108]. Also note that the mmWave is strongly dependent on the blockage since there is less diffraction and specular propagation is more pronounced. It leads to a bimodal channel based on absence or presence of LOS. mmWave beams will be like a flashlight and interference may not play a significant role. In handoffs and initial access in mmWave, it basically needs to align the transmitter (base station) and receiver (mobile user) which is quite challenging as the beams are very narrow. Hence, looking at the above expression for P_r, one can notice that it is directly proportional to the square of the operating wavelength. Hence at higher frequencies like mmWave, the received power will be reduced significantly. For a fixed distance, if we want to increase the received signal power at mmWave, we can increase the gain of the transmitting and receiving antennas. Mobile users are usually small so we will not be able to put many antennas on the mobile headset. But base stations are large we may put hundreds to thousands of antennas. From antenna theory, we know that the gain of the antenna can be increased significantly for large antenna arrays. We could also do transmit beamforming from the base station and direct the beams toward the user, which significantly reduce transmit power and minimize the interference such as shown in Figure 1.4. Major hurdles for mmWave communications are higher path loss and high attenuation due to rainfall, snowfall, fog, foliage and atmospheric absorption [106]. Raindrop size is of the order of millimeters and hence raindrops will scatter mmWave signals. Some typical values for these said attenuations are at a distance of 200 m which is a typical distance for a small cell.

- Loss due to atmospheric absorption due to H_2O and O_2 is 0.02 dB.
- For heavy rainfall at 110 mm/h, a 4-dB loss is observed.
- For heavy snowfall at 10 mm/h and fog with visibility less than 50 m, a 0.1-dB loss is observed.

Propagation tends to be LOS and small scale fading has minimal effect on the received signal and channel impulse response. For LOS propagation, it may be possible to estimate channel by estimating the direction of arrival (DOA), hence pilot contamination issues may be overcome. Since mmWave propagation is predominantly LOS, the channel matrix will be low rank, hence there will be no multiplexing gain for point-to-point MIMO communications. But for multiuser MIMO there can be multiplexing gains from multiple user communication with base station [119]. mmWave signals cannot pass through bricks and concrete walls, hence there should be different cells indoors and outdoors. Backhaul transmission can be made adaptive for mmWave massive MIMO by electronically steering the beams of antenna arrays, hence there won't be the requirement of physically adjusting the antenna alignment of dish antennas in microwave backhaul transmission. While mobile users move around, beamforming weights need to be adjusted adaptively so that antenna beams from the user will be always be directed toward the base station. Complementary metal oxide semiconductor (CMOS) is one of the enabling technologies for mmWave since one can integrate mixers, low-noise amplifier, power amplifier and inter-frequency amplifiers in a single package.

7.3.3 Channel model for 60GHz mmWave WPAN

Let us discuss the mmWave wireless personal area network (WPAN). We will present the IEEE 802.15.3c indoor channel impulse for 60 GHz communications [113].

$$h(t,\theta) = \beta\delta(t)\delta(\theta) + \sum_{c=0}^{C}\sum_{r=0}^{R}\alpha_{r,c}\delta(t - T_c - \tau_{r,c})\delta(\theta - \theta_c - \phi_{r,c}) \tag{7.26}$$

where

- t is the time of arrival (TOA).
- θ is the DOA.
- β is the gain coefficient for the LOS component.
- C is the number of clusters.
- R is the number of rays.
- $\alpha_{r,c}$ is the channel gain coefficient for the rth ray inside the cth cluster which is assumed to be log normal distributed.
- T_c is the TOA of the cth cluster.
- $\tau_{r,c}$ is the TOA of the rth ray inside the cth cluster.
- θ_c is the DOA of the cth cluster.
- $\phi_{r,c}$ is the DOA of the rth ray inside the cth cluster.

7.4 Device-to-device communication for Internet of things

It has been projected by Ericsson that the number of connected internet of things (IoT) will be 50 billion in 2020. Anything that will benefit from connection will be connected. 5G will make the IoT real by allowing devices to connect without human intervention. The list for machine-to-machine (M2M) communication goes on.

- wireless metering
- mobile payments
- smart grid
- critical infrastructure monitoring
- connected home
- smart transportation
- telemedicine

Some possible IoT applications in real life scenarios are mentioned below.

- Smart devices in the home may connect with the smart devices in the neighborhood and cooperate with each other to extinguish fire.
- When people are doing window shopping in shopping malls, information about the location of discounted items may be alerted to customers using mobile clouds.
- Alerts for upcoming meetings will be given to the office goers and files necessary for such meetings may be automatically transferred to the user's device.
- Cars will inform on the user's mobile phone about the status of fuel, engine oil and battery.
- From vehicle-to-infrastructure (V2I) communications, the shortest route to the destination for the cars may be obtained.
- From vehicle-to-vehicle (V2V) communications, situation about the traffic jam or emergency on the road can be obtained in real time.
- It will allow us to experience true real life inversive multimedia experiences using virtual reality and augmented reality anywhere and anytime.
- It will allow remote access to heavy industrial machines to hazardous and inaccessible places.

There will be high demand for remote control of devices like trains, buses and cars. Things we see will be connected and create the world of IoT. Hence the underlying network operations like security, identity, mobility, etc. should be scalable in order to cope with an outburst of IoT. Smart devices will connect to each other and share their information, which is also known as D2D communication. D2D communication basically allows users nearby to communicate directly instead of the long radio hops via base station.

7.4.1 V2V and V2I channel models

From the V2V channel measurement results reported in [78], first delay bins exhibit better than Rayleigh fading, whereas longer delay bins are worse than Rayleigh fading. It has been observed that Weibull distribution fits better than the other fading distribution like Nakagami. The PDF of Weibull distribution is given by

$$f_W(r) = \frac{\beta}{\alpha^\beta} r^{\beta-1} exp(-(\frac{r}{\alpha})^\beta) \tag{7.27}$$

where β is the shape parameter and α is the scale parameter given by $\alpha = \sqrt{\frac{E(r^2)}{\Gamma(1+\frac{2}{\beta})}}$. $\beta = 2$ is a better fit for V2V channels [72]. The Root mean square (RMS) delay spread is fitted to a lognormal distribution. The V2V spectrum is a little bit smoother than the classical Jake spectrum [51]. Table 7.2 shows the channel metrics [73] like path loss exponent (n), mean delay spread and Doppler spread for highway, rural and urban scenarios. The channel metrics for V2I communications for rural macrocell, urban macrocell and microcell are listed in Table 7.3. It has been reported that the Doppler spectrum for rural macrocell follows Jake's spectrum, Urban macrocell Doppler spectrum follows double exponential elevation power spectrum and microcells doppler spectrum follows similar to macrocells.

TABLE 7.2
V2V channel characteristics.

Parameters	Highway	Rural	Urban
n	1.8-1.9	1.8-1.9, 4	1.6-1.7
Mean RMS delay spread (ns)	40-400	20-60	40-300
Mean Doppler spread (Hz)	100	782	30-350

7.4.2 Channel models for mobile D2D cooperative communications

In mobile cooperative D2D communication, multiple scatterers are present at both the transmitter and receiver. Hence a more appropriate channel model is cascaded fading channels [121]. The PDF of n cascaded $\alpha-\mu$ distributions [66]

TABLE 7.3
V2I channel characteristics.

Parameters	**Rural**	**Urban**	**Microcells**
n	2,2.2	3.5	2.3-2.6(LOS), 3.8 Non line-of-sight (NLOS)
Shadowing	6 dB	6-8 dB	Vary widely
Delay spread	100 ns	100-1000 ns	5-100 ns(LOS), 30-500 ns(NLOS)
Angular spread	1^0-5^0	5^0-10^0	20^0

is given by

$$f_{Z_n}(z) = \frac{(2\pi)^{\frac{h-\nu_s}{2}} u\nu_p}{\Gamma_\mu (h_p z)^{\alpha_1} z} G^{\nu_s,0}_{0,\nu_s}\left[\frac{(h_p z)^u}{\nu_e}\,\middle|\, \begin{matrix} - \\ \mathbf{b} \end{matrix}\right] \tag{7.28}$$

where $G^{m,n}_{p,q}\left[z\,\middle|\,\begin{matrix} - \\ - \end{matrix}\right]$ is the Meijer G function. The parameters in the above equation are defined below.

- $Z_n = \prod_{i=1}^n R_i$ is the product of $n \geq 2$ independent $\alpha - \mu$ random variables.
- $h_i = \frac{\mu_i^{\frac{1}{\alpha_i}}}{\hat{r_i}}$ where $\hat{r_i} = (E(R_i^{\alpha_i}))^{\frac{1}{\alpha_i}}$.
- h_p is the product of h_i for i=1,n, i.e., $h_p = \prod_{i=1}^n h_i$.
- ν_1 is the product of q_j for j=2,n, i.e., $\nu_1 = \prod_{j=2}^n q_j$.
- ν_j is the scaled version of ν_1 by a scale factor of $\frac{p_j}{q_j}$ for j=2,n, i.e., $\nu_j = \frac{p_j}{q_j}\nu_1$.
- ν_s is the sum of ν_i for i=1,n, i.e., $\nu_s = \sum_{i=1}^n \nu_i$.
- ν_e is the product of $\nu_i^{\nu_i}$ for i=1,n, i.e., $\nu_e = \prod_{i=1}^n \nu_i^{\nu_i}$.
- ν_p is the product of $\nu_i^{c_i}$ where $c_i = \mu_i + \frac{\alpha_1}{\alpha_i} - \frac{1}{2}$ for i=1,n, i.e., $\nu_p = \prod_{i=1}^n \nu_i^{c_i}$. Note that the ratio $\frac{\alpha_1}{\alpha_j} = \frac{p_j}{q_j}$ for $p_j, q_j \geq 1$ is co-prime integers and j=2,3,...,n.
- Γ_μ is the product of $\Gamma(\mu_i)$ for i=1,n, i.e., $\Gamma_\mu = \prod_{i=1}^n \Gamma(\mu_i)$.
- The product of $\alpha_i\nu_i$ is equal to a constant denoted by u, i.e., $\alpha_1\nu_1 = \alpha_2\nu_2 = \cdots = \alpha_n\nu_n$.
- $\mathbf{b} = \mathbf{b}_1, \mathbf{b}_2, ..., \mathbf{b}_n$ where $\mathbf{b}_i = \frac{\nu_i}{c_i+\frac{1}{2}}, \frac{\nu_i+1}{c_i+\frac{1}{2}}, ..., \frac{\nu_i+c_i-\frac{1}{2}}{c_i+\frac{1}{2}}$.

The cumulative distribution function (CDF) of n cascaded $\alpha - \mu$ distributions is given by

$$F_{Z_n}(z) = \frac{(2\pi)^{\frac{h-\nu_s}{2}} \nu_p}{\Gamma_\mu (h_p z)^{\alpha_1}} G^{\nu_s,1}_{1,\nu_s+1}\left[\frac{(h_p z)^u}{\nu_e}\,\middle|\, \begin{matrix} \frac{1}{\nu_1}+1 \\ \mathbf{b}, \frac{1}{\nu_1} \end{matrix}\right] \tag{7.29}$$

End to end performance analysis of mobile D2D cooperative communication over cascaded $\alpha-\mu$ channel has been reported in the literature [8]. The system model assumes that there are three nodes, viz. source, relay and destination nodes. Relay node assists the source node in transmitting data to the destination node by applying amplify and forward protocol. No direct link path is assumed from the source to the destination. A lower bound on the average symbol error probability for binary phase shift keying (BPSK) has been derived and validated by Monte Carlo simulations.

7.5 Large scale MIMO systems

In large scale MIMO systems, we will consider hundreds of antennas at the transmitter and receiver. Basically, this will be a point-to-point MIMO system like D2D communication mentioned above. For large antenna arrays at the transmitter and receiver of a point-to-point MIMO system, there is a prominent effect also known as channel hardening effect in the literature. It has been observed that channel is no more random for such cases. From the Marcenko-Pastur law from random matrix theory [19], the empirical distribution of the eigenvalues of $\mathbf{H}^H\mathbf{H}$ converges almost surely to the density

$$f(x) = (c)^+\delta(x) + \frac{\sqrt{(x-a)^+(b-x)^+}}{2\pi r x} \tag{7.30}$$

where $c = \frac{r-1}{r}$, $r = \frac{N}{M}$, $a = (1-\sqrt{r})^2$, $a = (1+\sqrt{r})^2$, $(z)^+ = max(z,0)$ and $\mathbf{H}$ is the MIMO channel matrix of dimension $M \times N$ whose elements may have any classical or generalized fading distribution with zero mean and variance $\frac{1}{M}$.

7.5.1 Low complexity MIMO detection

Due to the channel hardening effect, it is possible to employ machine learning-based low complexity algorithms whose performance is comparable to maximum likelihood (ML) detection with reduced complexity. Note that the decoding complexity of an ML decoder is the number of metric computations required to reach the ML decision. The number of metric calculations for ML detection [20] can be computed as $|A_s|^N$ where A_s is the alphabet size for M-ary modulation scheme used and N is the number of transmit antennas. For instance, for a 5×5 MIMO system employing 16-QAM modulation, the number of such metric calculations is $16^5 = 10,48,576$. Hence it is almost impossible to have a detection scheme like ML at the receiver for large scale MIMO systems. Some low complexity algorithms are reported in [141], viz. likelihood ascent search, reactive Tabu search, k-neighborhood search for ZF solution [12], etc. The above-mentioned method actually searches the neighborhood of ZF solution. Similarly, we may initialize the above algorithm with MMSE [47] or lattice reduction aided ZF or MMSE solution and search its neighborhood for a solution with a better cost function. This iterative process will not stop until it reaches the threshold cost value. In [112], it has been mentioned that one may not look for all the neighborhoods, but instead look for some reduced neighborhood which is more likely to have the solution, which is also known as reduced neighborhood search algorithms.

7.5.2 Perfect space time codes

At the transmitter, one can employ perfect space time codes [88]. Such space time codes can achieve full diversity, a non-vanishing determinant for increased spectral efficiency, uniform average transmitted energy per antenna and rate $r \geq 1$. The rate r perfect space time codes for a MIMO system with N transmitting antennas can be constructed as [36]

$$\mathbf{C}_N^{PSTC} = \sum_{k=1}^{r} \frac{1}{\sqrt{\lambda}} \mathbf{D}_k \mathbf{\Gamma}^{k-1} \tag{7.31}$$

where

- $\mathbf{D}_k = diag(z_k, \sigma(z_k), \sigma^2(z_k), \cdots, \sigma^{N-1}(z_k))$.
- λ is a real and scalar value designed to meet the energy constraint.
- γ is a unit magnitude complex number dependent on N.
- $\mathbf{\Gamma} = (\gamma \mathbf{e}_N, \mathbf{e}_1, \mathbf{e}_2, \cdots, \mathbf{e}_{N-2}, \mathbf{e}_{N-1})$ where $\mathbf{e}_i$ represents the ith column of a $N \times N$ identity matrix.

7.5.3 Bounds on capacity

The point-to-point MIMO system for equal power allocation has instantaneous capacity as

$$C = log_2|\mathbf{I}_{R_H} + \frac{P\mathbf{Q}}{\sigma_n^2 N}| bits/s/Hz \tag{7.32}$$

where R_H is the rank of the channel matrix $\mathbf{H}$ and $\mathbf{Q}$ is the complex Wishart channel matrix. For full rank MIMO channel, then $R_H = min(M, N) = m$, we have

$$C = \sum_{i=1}^{m} log_2(1 + \frac{P\lambda_i}{\sigma_n^2 N}) bits/s/Hz \tag{7.33}$$

Since the trace of $\mathbf{Q}$ is equal to the sum of its eigenvalues, we may write the above equation for two cases:

- The worst case analysis when the channel has only one singular value

$$C \geq log_2(1 + \frac{Ptrace(\mathbf{Q})}{\sigma_n^2 N})bits/s/Hz \tag{7.34}$$

- The best case analysis when the singular values of channel are equal

$$C \leq mlog_2(1 + \frac{Ptrace(\mathbf{Q})}{\sigma_n^2 N})bits/s/Hz \tag{7.35}$$

If we normalize the channel gain coefficients to one we have $trace(\mathbf{Q}) = MN$, then the lower and upper bounds on the instantaneous capacity of the channel is given by

$$log_2(1 + \frac{PM}{\sigma_n^2})bits/s/Hz \leq C \leq mlog_2(1 + \frac{nP}{\sigma_n^2 N})bits/s/Hz \tag{7.36}$$

where $n = max(N, M)$.

A

Appendix

CONTENTS

A.1 Parameters for various modulation schemes

TABLE A.1
Modulation parameters for various modulation schemes [13, 45]

Modulation Scheme	a	b	c
BPSK	1	2	0
BFSK	1	1	0
MPSK	2	$2\sin^2\left(\frac{\pi}{M}\right)$	0
MPAM	$2\frac{M-1}{M}$	$\frac{6}{M^2-1}$	0
QPSK or MSK	2	2	1
Coherent DPSK	2	2	2
MQAM	$4\frac{\sqrt{M}-1}{\sqrt{M}}$	$\frac{3}{M-1}$	$4\left(\frac{\sqrt{M}-1}{\sqrt{M}}\right)^2$

Table A.1 enlists the values of parameters for various modulation schemes required for conditional probability of error in (6.132).

A.2 Approximation of Gaussian Q function

In general, Gaussian Q function gives the area under the tail of Gaussian curve. It is frequently used to find the probability of error for wireless communication systems in a noisy environment with Gaussian noise. Gaussian Q function is defined as

$$Q(x) = \frac{1}{\sqrt{2\pi}} \int_{x}^{\infty} e^{-\frac{u^2}{2}} du \tag{A.1}$$

In wireless communication systems, we frequently come across the integrals with Q function as a part of the integrand. This makes integrations difficult many times. So, Q function is approximated using simpler functions. Most commonly, it is approximated in terms of exponential functions. We have used the approximation proposed by Chiani et al. in [16].

$$Q(\hbar) \approx \frac{1}{12} e^{-\frac{\hbar^2}{2}} + \frac{1}{4} e^{-\frac{2\hbar^2}{3}} \tag{A.2}$$

The approximated and exact values of Q function are plotted in Figure A.1 to confirm the validity of the above expression. It is observed that (A.2)

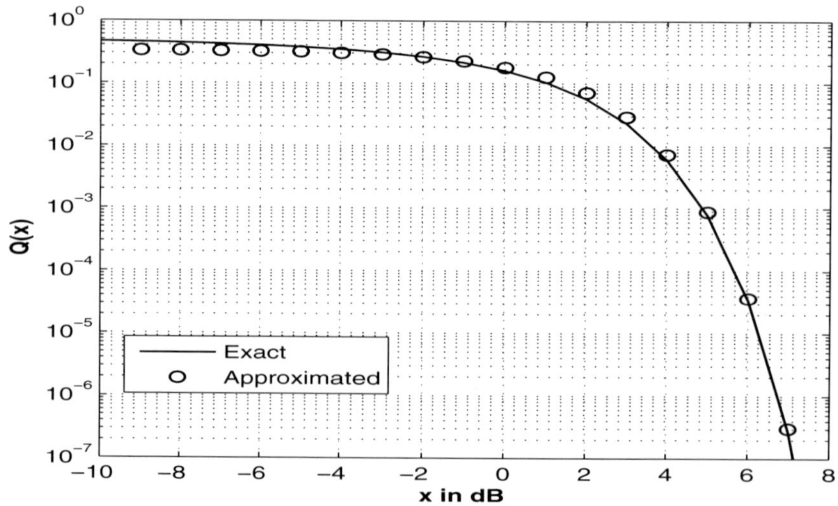

FIGURE A.1
Plot to compare the values of exact and approximated Gaussian Q function

gives a very accurate approximation for all values of x. For $x \geq -2$ dB, the approximated and the exact values of Q function are almost the same.

A.3 Detailed solution for (5.95)

$$I_2 = \frac{2cN}{\pi} \ell \sum_{i,j} \frac{2^{j_\Sigma} H^{2i_\Sigma} q\Gamma(r) \int_0^{\frac{\pi}{4}} \left(\lambda + \frac{b}{2\sin^2\theta} \right)^{-r} d\theta}{h^{N\mu+2i_\Sigma} \prod_{p=1}^{N} (i_p! \Gamma(\mu')) \prod_{p=1}^{N-1} \left((\mu + i_p) (2\mu')_{j_p} \right)} \tag{A.3}$$

The integral in the above expression is of the form

$$\begin{aligned} I &= \int_0^{\frac{\pi}{4}} \left(\lambda + \frac{b}{2\sin^2\theta} \right)^{-r} d\theta \\ &= \lambda^{-r} \int_0^{\frac{\pi}{4}} \left(\frac{\sin^2\theta}{\sin^2\theta + \zeta} \right)^{r} d\theta \end{aligned} \tag{A.4}$$

Substituting $\sin^2\theta = t/2$ in the above integral, the integral limits become 0 to 1 and

$$d\theta = \frac{dt}{4\sqrt{\frac{t}{2}}\sqrt{1 - \frac{t}{2}}}$$

Now, the integral can be represented as

$$\begin{aligned} I &= \lambda^{-r} \int_0^1 \left(\frac{\frac{t}{2}}{\frac{t}{2} + \zeta} \right)^r \frac{dt}{4\sqrt{\frac{t}{2}}\sqrt{1 - \frac{t}{2}}} \\ &= \frac{\lambda^{-r}}{4} \int_0^1 \left(\frac{t}{t + 2\zeta} \right)^r \frac{dt}{\sqrt{\frac{t}{2}}\sqrt{1 - \frac{t}{2}}} \\ &= \frac{\lambda^{-r}}{2\sqrt{2}(2\zeta)^r} \int_0^1 t^{r-\frac{1}{2}} \left(1 + \frac{t}{2\zeta} \right)^{-r} \left(1 - \frac{t}{2} \right)^{-\frac{1}{2}} dt \end{aligned} \tag{A.5}$$

Using [103, (7.2.4.42)], the above integral can be obtained in the form of Appell hypergeometric function as

$$I = \frac{\lambda^{-r}}{2\sqrt{2}(2\zeta)^r} F_A^{(1)} \left(r + \frac{1}{2}, \frac{1}{2}, r, r + \frac{3}{2}, \frac{1}{2}, -\frac{1}{2\zeta} \right) \tag{A.6}$$

Expression (5.98) can be obtained by using the above solution of integral in (5.97).

A.4 CDF of Nakagami-m distributed SNR

CDF of Nakagami-m distributed SNR can be obtained by integrating the PDF of Nakagami-m distributed SNR. The PDF of Nakagami-m distributed SNR can be given by

$$p_\gamma(\gamma) = \frac{m^m \gamma^{m-1}}{\bar{\gamma}^m \Gamma(m)} e^{\frac{-m\gamma}{\bar{\gamma}}} \tag{A.7}$$

Thus, CDF can be obtained as

$$\begin{aligned} P_\gamma(y) &= \int_0^y \frac{m^m \gamma^{m-1}}{\bar{\gamma}^m \Gamma(m)} e^{\frac{-m\gamma}{\bar{\gamma}}} d\gamma \\ &= \frac{m^m}{\bar{\gamma}^m \Gamma(m)} \int_0^y \gamma^{m-1} e^{\frac{-m\gamma}{\bar{\gamma}}} d\gamma \end{aligned} \tag{A.8}$$

The above integral can be solved by using the following definition of lower incomplete gamma function

$$\int_0^x t^{n-1} e^{-\mu t} dt = \mu^{-n} \gamma_{inc}(n, \mu x) \tag{A.9}$$

So, the CDF of Nakagami-m distributed SNR can be given as

$$P_\gamma(y) = \frac{\gamma_{inc}\left(m, \frac{my}{\bar{\gamma}}\right)}{\Gamma(m)} \tag{A.10}$$

Many times, average SNR is assumed to be unity for the ease of computations. In this case, the average SNR is adjusted by the signal power and the factor $\frac{E_b}{N_0}$ is used to indicate average SNR. In such cases, the above expression needs to be modified accordingly.

A.5 MGF of Nakagami-m distributed SNR

MGF of a random variable is defined as

$$M_\gamma(s) = E\left(e^{-s\gamma}\right) = \int_0^\infty e^{-s\gamma} p_\gamma(\gamma)\, d\gamma \tag{A.11}$$

So, the MGF of Nakagami-m distributed SNR can be given as

$$\begin{aligned} M_\gamma(s) &= \int_0^\infty e^{-s\gamma} \frac{m^m \gamma^{m-1}}{\bar{\gamma}^m \Gamma(m)} e^{\frac{-m\gamma}{\bar{\gamma}}}\, d\gamma \\ &= \frac{m^m}{\bar{\gamma}^m \Gamma(m)} \int_0^\infty \gamma^{m-1} e^{-\left(\frac{m}{\bar{\gamma}}+s\right)\gamma} d\gamma \end{aligned} \tag{A.12}$$

The integral in the above expression can be evaluated by using the Gamma function which is defined as

$$\int_0^\infty t^{n-1} e^{-\mu t} dt = \mu^{-n} \Gamma(n) \tag{A.13}$$

Thus, the MGF of Nakagami-m distributed SNR can be given as

$$M_\gamma(s) = \frac{m^m}{\bar{\gamma}^m} \left(\frac{m}{\bar{\gamma}} + s\right)^{-m} \tag{A.14}$$

It can be further simplified as

$$M_\gamma(s) = \left(\frac{m}{m + s\bar{\gamma}}\right)^{-m} \tag{A.15}$$

A.6 CDF of $\kappa - \mu$ distributed SNR

CDF of $\kappa - \mu$ distributed SNR can be obtained by integrating the PDF of $\kappa - \mu$ distributed SNR. The PDF of $\kappa - \mu$ distributed SNR can be given by

$$p_\gamma(\gamma) = \frac{\mu(1+\kappa)^{\frac{\mu+1}{2}}}{\kappa^{\frac{\mu-1}{2}} e^{\mu\kappa} \bar{\gamma}^{\frac{\mu+1}{2}}} \gamma^{\frac{\mu-1}{2}} e^{-\frac{\mu(1+\kappa)}{\bar{\gamma}}\gamma} I_{\mu-1}\left(2\mu\sqrt{\frac{\kappa(1+\kappa)\gamma}{\bar{\gamma}}}\right) \tag{A.16}$$

Thus, CDF can be obtained as

$$P_\gamma(y) = \int_0^y \frac{\mu(1+\kappa)^{\frac{\mu+1}{2}}}{\kappa^{\frac{\mu-1}{2}} e^{\mu\kappa} \bar{\gamma}^{\frac{\mu+1}{2}}} \gamma^{\frac{\mu-1}{2}} e^{-\frac{\mu(1+\kappa)\gamma}{\bar{\gamma}}} I_{\mu-1}\left(2\mu\sqrt{\frac{\kappa(1+\kappa)\gamma}{\bar{\gamma}}}\right) d\gamma \quad \text{(A.17)}$$

The above integral can be solved by taking complementary CDF as follows.

$$P_\gamma(y) = 1 - \int_y^\infty \frac{\mu(1+\kappa)^{\frac{\mu+1}{2}}}{\kappa^{\frac{\mu-1}{2}} e^{\mu\kappa} \bar{\gamma}^{\frac{\mu+1}{2}}} \gamma^{\frac{\mu-1}{2}} e^{-\frac{\mu(1+\kappa)\gamma}{\bar{\gamma}}} I_{\mu-1}\left(2\mu\sqrt{\frac{\kappa(1+\kappa)\gamma}{\bar{\gamma}}}\right) d\gamma \quad \text{(A.18)}$$

To solve the above integral, take the following substitution.

$$\frac{\mu(1+\kappa)\gamma}{\bar{\gamma}} = \frac{t^2}{2} \quad \text{(A.19)}$$

which further leads to

$$d\gamma = \frac{\bar{\gamma}}{\mu(1+\kappa)} t dt \quad \text{(A.20)}$$

and the lower limit of integral gets converted from y to $\sqrt{\frac{2\mu(1+\kappa)y}{\bar{\gamma}}}$. So, the integral can be given as

$$\begin{aligned} P_\gamma(y) &= 1 - \int_{\sqrt{\frac{2\mu(1+\kappa)y}{\bar{\gamma}}}}^\infty \frac{\mu(1+\kappa)^{\frac{\mu+1}{2}}}{\kappa^{\frac{\mu-1}{2}} e^{\mu\kappa} \bar{\gamma}^{\frac{\mu+1}{2}}} \left(\frac{\bar{\gamma}}{2\mu(1+\kappa)}\right)^{\frac{\mu-1}{2}} \\ &\quad \times t^{\mu-1} e^{-\frac{t^2}{2}} I_{\mu-1}\left(2\mu\sqrt{\frac{\kappa(1+\kappa)t^2}{2\mu(1+\kappa)}}\right) \frac{\bar{\gamma}}{\mu(1+\kappa)} t dt \\ &= 1 - \int_{\sqrt{\frac{2\mu(1+\kappa)y}{\bar{\gamma}}}}^\infty t\left(\frac{t}{\sqrt{2\kappa\mu}}\right)^{\mu-1} e^{-\frac{t^2}{2} - \mu\kappa} I_{\mu-1}\left(t\sqrt{2\mu\kappa}\right) dt \end{aligned} \quad \text{(A.21)}$$

The integral in the above expression is in the form of Marcum Q function. So, the CDF of $\kappa - \mu$ distributed SNR can be given as

$$P_\gamma(y) = 1 - Q_\mu\left(\sqrt{2\mu\kappa}, \sqrt{\frac{2\mu(1+\kappa)y}{\bar{\gamma}}}\right) \quad \text{(A.22)}$$

A.7 MGF of $\kappa - \mu$ distributed SNR

MGF of a random variable is defined as

$$M_\gamma(s) = E\left(e^{-s\gamma}\right) = \int_0^\infty e^{-s\gamma} p_\gamma(\gamma)\, d\gamma \tag{A.23}$$

So, the MGF of $\kappa - \mu$ distributed SNR can be given as

$$M_\gamma(s) = \int_0^\infty e^{-s\gamma} \frac{\mu(1+\kappa)^{\frac{\mu+1}{2}}}{\kappa^{\frac{\mu-1}{2}} e^{\mu\kappa} \bar{\gamma}^{\frac{\mu+1}{2}}} \gamma^{\frac{\mu-1}{2}} e^{-\frac{\mu(1+\kappa)\gamma}{\bar{\gamma}}} I_{\mu-1}\left(2\mu\sqrt{\frac{\kappa(1+\kappa)\gamma}{\bar{\gamma}}}\right) d\gamma \tag{A.24}$$

The above integral can be evaluated by representing Bessel function using infinite series as below.

$$I_\nu(\omega) = \sum_{n=0}^\infty \frac{1}{n!\Gamma(\nu+n+1)} \left(\frac{\omega}{2}\right)^{\nu+2n} \tag{A.25}$$

Using the above representation of Bessel function, the integral can be solved as

$$M_\gamma(s) = \frac{\mu(1+\kappa)^{\frac{\mu+1}{2}}}{\kappa^{\frac{\mu-1}{2}} e^{\mu\kappa} \bar{\gamma}^{\frac{\mu+1}{2}}} \int_0^\infty \gamma^{\frac{\mu-1}{2}} e^{-\gamma\left(\frac{\mu(1+\kappa)}{\bar{\gamma}}+s\right)} \sum_{n=0}^\infty \frac{\left(\mu\sqrt{\frac{\kappa(1+\kappa)\gamma}{\bar{\gamma}}}\right)^{\mu-1+2n}}{n!\Gamma(\mu+n)} d\gamma \tag{A.26}$$

$$= \frac{\mu(1+\kappa)^{\frac{\mu+1}{2}}}{\kappa^{\frac{\mu-1}{2}} e^{\mu\kappa} \bar{\gamma}^{\frac{\mu+1}{2}}} \sum_{n=0}^\infty \frac{\left(\mu\sqrt{\frac{\kappa(1+\kappa)}{\bar{\gamma}}}\right)^{\mu-1+2n}}{n!\Gamma(\mu+n)} \int_0^\infty \gamma^{\mu+n-1} e^{-\gamma\left(\frac{\mu(1+\kappa)}{\bar{\gamma}}+s\right)} d\gamma \tag{A.27}$$

The integral in the above expression can be evaluated using the definition of Gamma function which is given as

$$\int_0^\infty t^{n-1} e^{-\mu t} dt = \mu^{-n}\Gamma(n) \tag{A.28}$$

Now, the MGF of $\kappa - \mu$ distributed SNR can be given as

$$M_\gamma(s) = \frac{\mu(1+\kappa)^{\frac{\mu+1}{2}}}{\kappa^{\frac{\mu-1}{2}} e^{\mu\kappa} \bar{\gamma}^{\frac{\mu+1}{2}}} \sum_{n=0}^\infty \frac{\left(\mu\sqrt{\frac{\kappa(1+\kappa)}{\bar{\gamma}}}\right)^{\mu-1+2n}}{n!\Gamma(\mu+n)} \frac{\Gamma(\mu+n)}{\left(\frac{\mu(1+\kappa)}{\bar{\gamma}}+s\right)^{\mu+n}} \tag{A.29}$$

It can be further simplified to

$$M_\gamma(s) = \frac{\mu(1+\kappa)^{\frac{\mu+1}{2}}}{\kappa^{\frac{\mu-1}{2}} e^{\mu\kappa} \bar{\gamma}^{\frac{\mu+1}{2}}} \frac{\left(\frac{\mu^2\kappa(1+\kappa)}{\bar{\gamma}}\right)^{\frac{\mu-1}{2}}}{\left(\frac{\mu(1+\kappa)}{\bar{\gamma}}+s\right)^{\mu}} \sum_{n=0}^{\infty} \frac{\left(\frac{\mu^2\kappa(1+\kappa)}{\mu(1+\kappa)+s\bar{\gamma}}\right)^n}{n!} \tag{A.30}$$

$$= \frac{(\mu(1+\kappa))^{\mu}}{(\mu(1+\kappa)+s\bar{\gamma})^{\mu}} e^{\left(\frac{\mu^2\kappa(1+\kappa)}{\mu(1+\kappa)+s\bar{\gamma}}\right)-\mu\kappa} \tag{A.31}$$

This expression is the same as reported in [38].

A.8 CDF of $\eta-\mu$ distributed SNR

CDF of $\eta-\mu$ distributed SNR can be obtained by integrating the PDF of $\eta-\mu$ distributed SNR. The PDF of $\eta-\mu$ distributed SNR can be given by

$$p_\gamma(\gamma) = \frac{2\sqrt{\pi}\mu^{\mu+\frac{1}{2}}h^{\mu}}{\Gamma(\mu) H^{\mu-\frac{1}{2}}\bar{\gamma}^{\mu+\frac{1}{2}}} \gamma^{\mu-\frac{1}{2}} e^{\frac{-2\mu h\gamma}{\bar{\gamma}}} I_{\mu-\frac{1}{2}}\left(\frac{2\mu H\gamma}{\bar{\gamma}}\right) \tag{A.32}$$

Thus, CDF can be obtained as

$$P_\gamma(y) = \int_0^y \frac{2\sqrt{\pi}\mu^{\mu+\frac{1}{2}}h^{\mu}}{\Gamma(\mu) H^{\mu-\frac{1}{2}}\bar{\gamma}^{\mu+\frac{1}{2}}} \gamma^{\mu-\frac{1}{2}} e^{-\frac{2\mu h\gamma}{\bar{\gamma}}} I_{\mu-\frac{1}{2}}\left(\frac{2\mu H\gamma}{\bar{\gamma}}\right) d\gamma \tag{A.33}$$

$$= 1 - \frac{2\sqrt{\pi}\sqrt{h}\left(\frac{\mu}{\bar{\gamma}}\right)^{\mu+\frac{1}{2}}}{\Gamma(\mu)\left(\frac{H}{h}\right)^{\mu-\frac{1}{2}}} \int_y^{\infty} \gamma^{\mu-\frac{1}{2}} e^{-\frac{2\mu h\gamma}{\bar{\gamma}}} I_{\mu-\frac{1}{2}}\left(\frac{2\mu H\gamma}{\bar{\gamma}}\right) d\gamma \tag{A.34}$$

This can be further simplified by taking the following substitution for required change of variables.

$$\frac{2\mu h\gamma}{\bar{\gamma}} = t^2$$

This will change the lower limit of the integral from y to $\sqrt{\frac{2\mu h y}{\bar{\gamma}}}$ and the other variables as follows.

$$\gamma = \frac{\bar{\gamma}}{2\mu h}t^2 \tag{A.35}$$

$$d\gamma = \frac{\bar{\gamma}}{\mu h}t\,dt \tag{A.36}$$

Now, the CDF of $\eta - \mu$ distributed SNR can be given as

$$P_\gamma(y) = 1 - \frac{2\sqrt{\pi}\sqrt{h}\left(\frac{\mu}{\bar{\gamma}}\right)^{\mu+\frac{1}{2}}}{\Gamma(\mu)\left(\frac{H}{h}\right)^{\mu-\frac{1}{2}}} \int\limits_{\sqrt{\frac{2\mu h y}{\bar{\gamma}}}}^{\infty} \left(\frac{\bar{\gamma}}{2\mu h}\right)^{\mu-\frac{1}{2}} t^{2\mu-1} e^{-t^2} I_{\mu-\frac{1}{2}}\left(\frac{H}{h}t^2\right)\frac{\bar{\gamma}}{\mu h} t dt \tag{A.37}$$

$$= 1 - \frac{2^{\frac{3}{2}-\mu}\sqrt{\pi}h^{-\mu}}{\Gamma(\mu)\left(\frac{H}{h}\right)^{\mu-\frac{1}{2}}} \int\limits_{\sqrt{\frac{2\mu h y}{\bar{\gamma}}}}^{\infty} t^{2\mu} e^{-t^2} I_{\mu-\frac{1}{2}}\left(\frac{H}{h}t^2\right) dt \tag{A.38}$$

The integral in the above expression is given by Yacoub's integral which is defined as

$$Y_\mu(x,y) = \frac{\sqrt{\pi}\left(1-x^2\right)^{\mu} 2^{\frac{3}{2}-\mu}}{\Gamma(\mu)x^{\mu-\frac{1}{2}}} \int_y^{\infty} e^{-t^2} t^{2\mu} I_{\mu-\frac{1}{2}}\left(tx^2\right) dt \tag{A.39}$$

Thus, the CDF of $\eta - \mu$ distributed SNR can be given as

$$P_\gamma(y) = 1 - Y_\mu\left(\frac{H}{h}, \sqrt{\frac{2\mu h y}{\bar{\gamma}}}\right) \tag{A.40}$$

A solution to Yacoub's integral has been reported in [81]. It is given in the form of confluent Lauricella hypergeometric function as

$$Y_\mu(a,b) = 1 - \frac{\left(1-a^2\right)^{\mu} b^{4\mu}}{\Gamma(1+2\mu)} \Phi_2\left(\mu, \mu; 1+2\mu; -(1+a)b^2, -(1-a)b^2\right) \tag{A.41}$$

A.9 MGF of $\eta - \mu$ distributed SNR

MGF of a random variable is defined as

$$M_\gamma(s) = E\left(e^{-s\gamma}\right) = \int\limits_0^{\infty} e^{-s\gamma} p_\gamma(\gamma) d\gamma \tag{A.42}$$

So, the MGF of $\kappa - \mu$ distributed SNR can be given as

$$M_\gamma(s) = \int\limits_0^{\infty} e^{-s\gamma} \frac{2\sqrt{\pi}\mu^{\mu+\frac{1}{2}} h^{\mu}}{\Gamma(\mu) H^{\mu-\frac{1}{2}} \bar{\gamma}^{\mu+\frac{1}{2}}} \gamma^{\mu-\frac{1}{2}} e^{\frac{-2\mu h \gamma}{\bar{\gamma}}} I_{\mu-\frac{1}{2}}\left(\frac{2\mu H \gamma}{\bar{\gamma}}\right) d\gamma \tag{A.43}$$

$$= \frac{2\sqrt{\pi}\mu^{\mu+\frac{1}{2}} h^{\mu}}{\Gamma(\mu) H^{\mu-\frac{1}{2}} \bar{\gamma}^{\mu+\frac{1}{2}}} \int\limits_0^{\infty} \gamma^{\mu-\frac{1}{2}} e^{-\gamma\left(\frac{2\mu h}{\bar{\gamma}}+s\right)} I_{\mu-\frac{1}{2}}\left(\frac{2\mu H \gamma}{\bar{\gamma}}\right) d\gamma \tag{A.44}$$

The integral in the above expression is in the form $\int x^p e^{-qx} I_\nu (rx)\, dx$. It can be solved using [102, 2.15.3.2]

$$\int_0^\infty x^{q-1} e^{-px} I_\nu(cx) dx = p^{-q-\nu} \left(\frac{c}{2}\right)^\nu \frac{\Gamma(\nu+q)}{\Gamma(\nu+1)} {}_2F_1\left(\frac{\nu+q}{2}, \frac{\nu+q+1}{2}, \nu+1, \frac{c^2}{p^2}\right) \tag{A.45}$$

Comparing the above integral with equation (A.44), we get the constants as $q = \mu + \frac{1}{2}$, $p = \left(\frac{2\mu h}{\overline{\gamma}} + s\right)$, $\nu = \mu - \frac{1}{2}$ and $c = \frac{2\mu H}{\overline{\gamma}}$. Equation (A.45) can be further simplified for $q = \nu + 1$ [102, 2.15.3.2] which is valid in the present case. Thus, it can be given by

$$\int_0^\infty x^{q-1} e^{-px} I_\nu (cx)\, dx = \frac{(2c)^\nu \Gamma\left(\nu + \frac{1}{2}\right)}{\sqrt{\pi} \left(p^2 - c^2\right)^{\nu + \frac{1}{2}}} \tag{A.46}$$

MGF of $\eta - \mu$ distributed SNR can be obtained by substituting the values of q, p, ν and c in the above expression and equation (A.44). After further simplifications, the expression of MGF can be given as

$$M_\gamma(s) = \frac{\left(4\mu^2 h\right)^\mu}{\left(2\mu(h-H) + s\overline{\gamma}\right)^\mu \left(2\mu(h+H) + s\overline{\gamma}\right)^\mu} \tag{A.47}$$

This expression is the same as reported in [38].

A.10 CDF of $\alpha - \mu$ distributed SNR

CDF of $\alpha - \mu$ distributed SNR can be obtained by integrating the PDF of $\alpha - \mu$ distributed SNR. The PDF of $\alpha - \mu$ distributed SNR can be given by

$$p_\gamma(\gamma) = \frac{\alpha \mu^\mu \gamma^{\frac{\alpha\mu}{2}-1}}{2\Gamma(\mu)\overline{\gamma}^{\frac{\alpha\mu}{2}}} e^{-\mu\left(\frac{x}{\overline{\gamma}}\right)^{\frac{\alpha}{2}}} \qquad \mu > 0,\ \alpha > 0 \tag{A.48}$$

Thus, CDF can be obtained as

$$P_\gamma(y) = \int_0^y \frac{\alpha \mu^\mu \gamma^{\frac{\alpha\mu}{2}-1}}{2\Gamma(\mu)\overline{\gamma}^{\frac{\alpha\mu}{2}}} e^{-\mu\left(\frac{x}{\overline{\gamma}}\right)^{\frac{\alpha}{2}}} d\gamma \tag{A.49}$$

The above integral can be brought into the form of incomplete Gamma function by taking the substitution $\left(\frac{\gamma}{\overline{\gamma}}\right)^{\frac{\alpha}{2}} = t$. This substitution converts the upper limit of the integral from y to $\left(\frac{y}{\overline{\gamma}}\right)^{\frac{\alpha}{2}}$ and

$$d\gamma = \frac{2\overline{\gamma}}{\alpha} t^{\left(\frac{2}{\alpha}-1\right)} dt \tag{A.50}$$

Thus, the integral for CDF of $\alpha - \mu$ distributed SNR gets converted into the following form.

$$P_\gamma(y) = \frac{2}{\alpha}\int\limits_0^{\left(\frac{y}{\overline{\gamma}}\right)^{\frac{\alpha}{2}}} \frac{\alpha\mu^\mu}{2\Gamma(\mu)\overline{\gamma}^{\frac{\alpha\mu}{2}}} e^{-\mu t}\left(t^{\frac{2}{\alpha}}\overline{\gamma}\right)^{\left(\frac{\alpha\mu}{2}-1\right)}\overline{\gamma}t^{\left(\frac{2}{\alpha}-1\right)}dt \tag{A.51}$$

$$= \int\limits_0^{\left(\frac{y}{\overline{\gamma}}\right)^{\frac{\alpha}{2}}} \frac{\mu^\mu}{\Gamma(\mu)} e^{-\mu t} t^{\mu-1} dt \tag{A.52}$$

Now, the following relation can be used to evaluate CDF of $\alpha - \mu$ distributed SNR.

$$\int\limits_0^x t^{n-1}e^{-\mu t}dt = \mu^{-n}\gamma_{inc}(n, \mu x) \tag{A.53}$$

So, for $\alpha - \mu$ distributed instantaneous SNR, CDF can be given in the form of lower incomplete Gamma function as

$$P_\gamma(y) = \frac{\gamma_{inc}\left(\mu, \mu\left(\frac{y}{\overline{\gamma}}\right)^{\frac{\alpha}{2}}\right)}{\Gamma(m)} \tag{A.54}$$

B

MATLAB® Codes for Generating Results

CONTENTS

B.1 MATLAB scripts for plotting PDF of envelope and SNR of generalized fading channels and verification by generating channel coefficients

B.1.1 Nakagami-m fading distribution

```
%MATLAB script to plot PDF of Nakagami-m distributed envelope and
%power/SNR coefficinets. It also generates Nakagami-m distributed
%coefficinets and the PDF of generated coefficients is compared
%with the PDF obtained using the mathematical expression.
clc;
clear all;
m =3.15;mu=m/2; % value of fading parameter
N=10^6;      % Number of samples generated for verification of PDF
Nr=1;
Nt=N;
sn=[-1 1]; %variable to define sign of the generated coefficients
%% Nak-m function starts
n=sn((rand(Nr,Nt)>0.5)+1).*sqrt(gamrnd(mu,1/mu,Nr,Nt)/2) ...
     +j*sn((rand(Nr,Nt)>0.5)+1).*sqrt(gamrnd(mu,1/mu,Nr,Nt)/2);
%%Instructions for obtaining uniformly distributed phase
phi=2*pi*rand(Nr,Nt);
n=abs(n);
H=(n).*cos(phi)+j*(n).*sin(phi);
omega = mean(abs(H).^2)
ksdensity(abs(H)); hold on; % to obtain PDF of generated envelope
ksdensity(abs(H).^2); % to obtain PDF of power of the generated
                      % envelope
%% Nakagami-m envelop PDF
x = 0.01:0.01:3;
m=2*mu;
a=m^m;
b=x.^(2*m-1);
c=m*(x.^2)/omega;
d=exp(-c);
f=omega^(m-1);
pdf_e=(2.*a.*b.*d)./(f.*gamma(m));
hold on;
plot(x, pdf_e);
%% Nakagami-m power or SNR PDF
b=x.^(m-1);
c=m*x/omega;
d=exp(-c);
f=omega^mu;
pdf_p=(a.*b.*d)/(f*gamma(m));
hold on;
plot(x, pdf_p);
```

B.1.2 $\kappa - \mu$ fading distribution

```
% Matlab script to verify the PDF of kappa-mu distributed
% envelope and the power/SNR.
% This program is also for generation of kappa-mu distributed
%samples with any real positive value of kappa and half integer
%values of mu.
% It verifies the PDF of generated coefficients with those
% obtained by analytical expressions.
clc;
clear all;
kappa=3.0000001; % values of fading parameters
mu =3.50;
N=10^6;
m = sqrt( kappa/((kappa+1))) ; % Mean of in phase and quadrature
                                %component
s = sqrt( 1/(2*(kappa+1)) );%variance of in phase and quadrature
                                %component
ni=zeros(1,N); nq=zeros(1,N);
for j=1:2*mu
    if mod(j,2)==1
        norm1=m/sqrt(2)+s*randn(1,N);
        ni=ni+norm1.^2;
    else
        norm2=m/sqrt(2)+s*randn(1,N);
        nq=nq+norm2.^2;
    end
end
h=sqrt(ni+nq)/sqrt(mu);
theta = 2*pi*rand(1,N);
h=h.*cos(theta)+sqrt(-1)*h.*sin(theta);
%% To obtain non uniform phase of the generated coefficients
% sn=[-1 1];
% h=(sn((rand(1,N)>0.5)+1).*sqrt(ni) ...
    %+sn((rand(1,N)>0.5)+1).*sqrt(-nq))/sqrt(mu);
h_abs=abs(h);
ksdensity(h_abs.^2); hold on; %Plots the PDF of envelope
ksdensity(h_abs);  %Plots the PDF of power
%% Kappa-Mu envelop PDF
bins=200;
[lp, X] = hist(h_abs, bins);
alpha = X(2)-X(1):X(2)-X(1):max(X);
omega = mean(h_abs.^2);
a=(1+kappa)^((mu+1)/2);
b=alpha.^mu;
c=mu*(1+kappa)*(alpha.^2)/omega;
d=exp(-c);
e=kappa^((mu-1)/2);
f=omega^((mu+1)/2);
g=2*mu*sqrt(kappa*(1+kappa)/omega);
p=besseli(mu-1,g*alpha);
pdf_e=(2.*mu.*a.*b.*d.*p)./(e.*f.*exp(mu*kappa));
hold on;
plot(alpha, pdf_e);
```

```
%% Kappa-Mu power PDF
a=(1+kappa)^((mu+1)/2);
b=alpha.^((mu-1)/2);
c=mu*(1+kappa)*(alpha)/omega;
d=exp(-c);
e=kappa^((mu-1)/2);
f=omega^((mu+1)/2);
g=2*mu*sqrt(kappa*(1+kappa)*alpha/omega);
p=besseli(mu-1,g);
pdf_p=(mu.*a.*b.*d.*p)/(e*f*exp(mu*kappa));
hold on;
plot(alpha, pdf_p);
```

B.1.3 $\eta - \mu$ fading distribution

```
% Matlab script to verify the PDF of eta-mu distributed envelope
%and the power/SNR.
% This program is also for generation of eta-mu distributed
% samples with any real positive value of eta and half integer
%values of mu.
% It verifies the PDF of generated coefficients with those
% obtained by analytical expressions.
clc;
clear all;
N=10^6; % Number of coefficinets generated for PDF verification
mu=2.5; % Fading parameter values
eta=0.2500000001;
coeff=sqrt(eta); % standard deviation of quadrature component
ni=zeros(1,N);
nq=zeros(1,N);
for j=1:2*mu
        norm1=randn(1,N);
        ni=ni+norm1.^2;
        norm2=coeff*randn(1,N);
        nq=nq+norm2.^2;
end
h=(sqrt(ni)+sqrt(-nq))/sqrt(2*(mu+eta*mu));
h_abs=abs(h);
theta = 2*pi*rand(1,N);
h=h_abs.*cos(theta)+sqrt(-1)*h_abs.*sin(theta);
h_abs=abs(h);
ksdensity(h_abs); hold on; % To plot the PDF of envelope
ksdensity(h_abs.^2); % To plot the PDF of power
%% Eta-Mu envelop PDF
bins=200;
omega = mean(h_abs.^2);
[N,X] = hist(h_abs/sqrt(omega),bins);
alpha = 0:X(2)-X(1):max(X);
a=(2+eta+1/eta)/4;
A=(1/eta-eta)/4;
b=alpha.^(2*mu);
c=((mu/omega)^(mu+0.5))*(a^(mu));
d=(2*mu*a/omega).*(alpha.^2);
```

```
e=exp(-d);
f=A^(mu-1/2);
g=(2*mu*A/omega).*(alpha.^2);
p=besseli(mu-(1/2),g);
pdf_e=(4.*c.*e.*p.*b.*sqrt(pi))./(f*gamma(mu));
hold on;
plot(alpha,(pdf_e));

%% Eta-Mu power PDF
a=(2+eta+1/eta)/4;
A=(1/eta-eta)/4;
b=(alpha/A).^(mu-0.5);
c=((mu/omega)^(mu+0.5))*(a^(mu));
d=(2*mu*a/omega).*(alpha);
e=exp(-d);
g=(2*mu*A/omega).*(alpha);
p=besseli(mu-1/2,g);
pdf_p=(2.*c.*e.*p.*b.*sqrt(pi))./(gamma(mu));
hold on;
plot(alpha, abs(pdf_p));
```

B.1.4 $\alpha - \mu$ fading distribution

```
% Matlab script to verify the PDF of alpha-mu distributed
% envelope and the power/SNR.
% This program is also for generation of alpha-mu distributed
% samples with any real positive value of alpha and real positive
%values of mu.
% It verifies the PDF of generated coefficients with those
% obtained by analytical expressions.
clc;
clear all;
alpha=3.5; % fading parameter values
mu =2.15;m=mu/2;
N=10^5; % Number of samples generated for PDF verification
Nr=1;
Nt=N;
sn=[-1 1];
%% alpha-mu function starts
n=zeros(Nr,Nt);
n=sn((rand(Nr,Nt)>0.5)+1).*sqrt(gamrnd(m,1/m,Nr,Nt)/2) ...
    +j*sn((rand(Nr,Nt)>0.5)+1).*sqrt(gamrnd(m,1/m,Nr,Nt)/2);
n=n.^(2/alpha);
omega = (mean(abs(n).^alpha))^(1/alpha);
ksdensity(abs(n)); hold on; % To plot PDF of envelope
ksdensity(abs(n).^2); %To plot PDF of power
%% Use the following to obtain alpha-mu distributed coefficients
%% with uniformy distributed phase
% phi=2*pi*rand(Nr,Nt);
% nn=abs(n);
% H=(nn).*cos(phi)+j*(nn).*sin(phi);
% omega = (mean(abs(H).^alpha))^(1/alpha)

```

```
%% alpha-Mu envelop PDF
x = 0.01:0.01:3;
am=alpha*mu;
a=(mu)^(mu);
b=x.^(am-1);
c=(x/omega).^alpha;
d=exp(-mu*c);
f=omega^(am);
pdf_e=(alpha*a*b.*d)./(f.*gamma(mu));
hold on;
plot(x, pdf_e);

%% alpha-Mu power PDF
c=(x/omega).^(alpha/2);
d=exp(-mu*c);
e=omega^(am/2);
p=x.^(am/2-1);
pdf_p=(alpha*a.*p.*d)/(2*gamma(mu)*e);
hold on;
plot(x, pdf_p);
```

B.2 MATLAB functions for generating channel coefficients of generalized fading channels

B.2.1 Nakagami-m fading distribution

```
function chan = nak_m_channel(m,M,N);
%function to generate Nakagami-m distributed channel coefficients
%for all real positive values of fading parameter, m
%This function generates coefficients with uniformly distributed
%phase and non-uniformly distributed phase as well.
mu=m/2;
sn=[-1 1];
%% Nak-m function starts
chan=sn((rand(M,N)>0.5)+1).*sqrt(gamrnd(mu,1/mu,M,N)/2) ...
     +j*sn((rand(M,N)>0.5)+1).*sqrt(gamrnd(mu,1/mu,M,N)/2);
%% comment the following instructions for non-uniformly
%%distributed phase of the channel coefficients
n = abs(chan);
phi=2*pi*rand(M,N);
H=n.*cos(phi)+j*n.*sin(phi);
chan = H;
```

B.2.2 $\kappa - \mu$ fading distribution

```
function chan = kappa_mu_channel(kappa,mu,M,N)
% Function to generate kappa-mu distributed channel coefficients
% with integer values of 2*mu and all real positive values of
% kappa
% To generate coefficients with uniformly distributed phase
m = sqrt( kappa/((kappa+1))) ;
s = sqrt( 1/(2*(kappa+1)) );
ni=zeros(M,N); nq=zeros(M,N);
for j=1:2*mu
    if mod(j,2)==1
        norm1=m/sqrt(2)+s*randn(M,N);
        ni=ni+norm1.^2;
    else
        norm2=m/sqrt(2)+s*randn(M,N);
        nq=nq+norm2.^2;
    end
end
h=sqrt(ni+nq)/sqrt(mu);
theta = 2*pi*rand(M,N);
chan=h.*cos(theta)+sqrt(-1)*h.*sin(theta);

%% Use following instructions for coefficients with non-uniformly
%% distributed phases
% sn=[-1 1];
% chan=(sn((rand(M,N)>0.5)+1).*sqrt(ni)+sn((rand(M,N)>0.5)+1)...
%       .*sqrt(-nq))/sqrt(mu);
```

B.2.3 $\eta - \mu$ fading distribution

```
function chan = eta_mu_channel(eta,mu,M,N)
% Function to generate eta-mu distributed channel coefficients
% with integer values of 2*mu and all real positive values of eta
% To generate coefficients with uniformly distributed phase
coeff=sqrt(eta);
ni=zeros(M,N);
nq=zeros(M,N);
for j=1:2*mu
        norm1=randn(M,N);
        ni=ni+norm1.^2;
        norm2=coeff*randn(M,N);
        nq=nq+norm2.^2;
end
chan=(sqrt(ni)+sqrt(-nq))/sqrt(2*(mu+eta*mu));
%% For uniformly distributed phase of generated coefficients use
%%following instructions
h_abs=abs(chan);
theta = 2*pi*rand(M,N);
chan=h_abs.*cos(theta)+sqrt(-1)*h_abs.*sin(theta);
```

B.2.4 $\alpha - \mu$ fading distribution

```
function chan = a_mu_channel(alpha,mu,M,N)
%function to generate alpha-mu distributed channel coefficients
%for all real positive values of fading parameters, alpha and mu
%This function generates coefficients with uniformly distributed
%phase and non-uniformly distributed phase as well.
m=mu/2;
sn=[-1 1];
%% alpha-mu function starts
n=sn((rand(M,N)>0.5)+1).*sqrt(gamrnd(m,1/m,M,N)/2) ...
      +j*sn((rand(M,N)>0.5)+1).*sqrt(gamrnd(m,1/m,M,N)/2);
chan=n.^(2/alpha);
%% comment the following instructions for non-uniformly
%% distributed phase of the channel coefficients
phi=2*pi*rand(M,N);
nn=abs(chan);
H=(nn).*cos(phi)+j*(nn).*sin(phi);
chan = H;
```

B.3 MATLAB functions for evaluating MGF of received SNR after MRC

B.3.1 Nakagami-m fading distribution

```
function pe=MGF_nak_m(m,NR,SNR,b)
%   %SNR is mean(received_signal_to_noise_ratio);
%   m is Nakagami-m fading parameter
%   NR is number of branches used for MRC
%   b is the factor depending on the modulation scheme used
  SNR1=10.^(SNR/10);
  m1=NR*m;
  s=b*SNR1;
  a=m1+s;
  mgf_nakm=(m1./a).^m1;
pe=mgf_nakm;
```

B.3.2 $\kappa - \mu$ fading distribution

```
function pe=MGF_kappa_mu(kappa,mu,NR,SNR,b)
%   SNR is mean(received_signal_to_noise_ratio);
%   kappa and mu are fading parameters
%   NR is number of branches used for MRC
```

```
%    b is the factor depending on the modulation scheme used
   SNR1=10.^(SNR/10);
   mu1=NR*mu;
   s=b*SNR1;
   for i=1:length(SNR)
       a=((mu1)*(1+kappa));
       b=((mu1)*(1+kappa)+s(i));
       c=(a/b)^(mu1);
   d=(mu1)^2*kappa*(1+kappa);
   e=(mu1)*(1+kappa)+s(i);
   f=exp((d/e)-(mu1)*kappa);
   mgf_kappa_mu1(i)=c*f;
   end
pe=mgf_kappa_mu1;
```

B.3.3 $\eta - \mu$ fading distribution

```
function pe=MGF_eta_mu(eta,mu,NR,SNR,b)
%    SNR is mean(received_signal_to_noise_ratio);
%    eta and mu are fading parameters
%    NR is number of branches used for MRC
%    b is the factor depending on the modulation scheme used
   SNR1=10.^(SNR/10);
   mu1=NR*mu;
   s=b*SNR1;
    h = 0.25 * (2 + eta + 1/eta);
    H = 0.25 * (1/eta - eta);
   for i=1:length(SNR)
   a=4*h*((mu1)^2);
   b=2*mu1*(h-H)+s(i);
   c=2*mu1*(h+H)+s(i);
   d=a/(b*c);
   mgf_e_mu(i)=d^mu1;
   end
pe=mgf_e_mu;
```

B.3.4 $\alpha - \mu$ fading distribution

Note that the MGF of $\alpha - \mu$ distributed SNR is already discussed in Chapter 2 while discussing the physical model. It can be given by (2.52) as follows.

$$M_\gamma(s) = \frac{\alpha\mu^\mu}{2\bar{\gamma}^{\frac{\alpha\mu}{2}}} \frac{\sqrt{k}l^{\frac{\alpha\mu-1}{2}}}{(2\pi)^{\frac{l+k-2}{2}} s^{\frac{\alpha\mu}{2}}} G_{l,k}^{k,l}\left(\left(\frac{\mu}{\bar{\gamma}^{\frac{\alpha}{2}}}\right)^k \frac{l^l}{s^l k^k}, \begin{matrix} P\left(l, 1-\frac{\alpha\mu}{2}\right) \\ P(k,0) \end{matrix}\right) \tag{B.1}$$

It shall be noted here that the MGF of $\alpha - \mu$ distributed SNR is in the form of Meijer's G function. To the best of our knowledge, Meijer's G function is not implemented in the current versions of MATLAB toolboxes. So, it is not possible to directly evaluate Meijer's G function in MATLAB. However, current versions of MATLAB are equipped with an interface that integrates *Mathematica*® notebook files with itself. And Meijer's G function is implemented in *Mathematica.* A notebook file can be opened using command "mupad" in the MATLAB command window and a notebook file will open which can execute *Mathematica* instructions and this can solve our purpose of evaluating MGF of $\alpha - \mu$ distributed SNR.

B.4 MATLAB script for SER of spatially multiplexed MIMO system

```
%% MATLAB code for simulation of spatial multiplexing systems
clc;
clear all;
%% specifications and parameters
N=3*10^4; %Number of symbols for which simulation has to continue
Nt = 4; % Number of transmitter antennas
Nr = 3; % Number of receiver antennas
M  = 2; % Modulation order
lut = pskmod([0:M-1],M); % Look up table
Eavg = sum(abs(lut).^2)/M; % Average energy of constellation
N_lut = lut/sqrt(Eavg); % Normalized look up table
EbNo = 0:2:20; % Range of SNR for simulation

TA2=zeros(Nt,M^Nt);
for m=1:M^Nt
    for n=1:Nt
        for p=1:Nt
        TA2(p,m)= N_lut(mod(floor((m-1)/(M^(Nt-p))),M)+1)/sqrt(Nt);
        end
    end
end

for c = 1:1:length(EbNo)
    error(c)= 0;
for u = 1:N
    H = 1/sqrt(2)*(randn(Nr,Nt)+1i*(randn(Nr,Nt)));
                                                %Rayleigh fading
    n = 1/sqrt(2)*(randn(Nr,1)+1i*randn(Nr,1));
%% Spatial Multiplexing
    x=round((M-1)*rand(Nt,1));
    x_q = pskmod(x,M)/sqrt(Nt*Eavg);      % digital modulation
    y =  H*x_q;  % faded signal
    r = y+10^(-EbNo(c)/20)*n; % Received noisy signal
%% Detection  and error calculation
    d = repmat(r,1,M^Nt)-H*TA2;
    nor = sum(abs(d).^2);
    [mi,ind] = min(nor);
    hat_x_q = TA2(:,ind);
```

```
    err = hat_x_q ~= x_q;
    error(c)=error(c)+(sum(err)~=0);
end
end
errorrate = error/(N); % symbol error rate
semilogy(EbNo,errorrate);
grid on
hold on
```

B.5 MATLAB script for SER of SM MIMO system

```
%% MATLAB code for simulation of SM MIMO systems
clc;
clear all;
sn=[-1 1];
%% specifications and parameters
N=10^5; % Number of symbols for which simulation has to continue
Nt = 4; % Number of transmitter antennas
Nr = 4; % Number of receiver antennas
M  = 4; % Modulation order for signal constellation
Na=log2(Nt); % Number of bits transmitted as antenna index
lut = pskmod([0:M-1],M); % Look up table
Eavg = sum(abs(lut).^2)/M; % Average energy of constellation
N_lut = lut/sqrt(Eavg); % Normalized look up table
EbNo = 0:2:16; % Range of SNR for simulation
eta=1;
mu = 1;
a_SNR=1;
for c = 1:1:length(EbNo)
    error(c)= 0;
     errora(c)= 0;
      errors(c)= 0;
for u = 1:N
%% channel coefficient matrix generation
    w = 1/sqrt(2)*(randn(Nr,Nt)+1i*(randn(Nr,Nt)));
                                            %Rayleigh fading
%    K=0; %Rician K factor
%    w = sqrt(K/(K+1))*ones(Nr,Nt)+sqrt(1/(K+1))*w; %Rician Fading
%   w in the RHS of Rician fading channel generation comes from
% Rayleigh fading channel generation
%    w =nak_m_channel(m,Nr,Nt);           %Nakagami-m fading
%    w = eta_mu_channel(1,1.5,Nr,Nt);      % eta-mu fading
%    w = kappa_mu_channel(0,2,Nr,Nt);      % kappa-mu fading
%    w = a_mu_channel(2,2,Nr,Nt);       % alpha-mu fading
    H = w;
    n = 1/sqrt(2)*(randn(Nr,1)+1i*randn(Nr,1));
%% Spatial Modulation
    x=zeros(Nt,1);
    x_i = randi([0,1],log2(Nt)+log2(M),1); % SM symbol
```

```
        x_a = x_i(1:Na);            % bits from spatial constellation
        a_i = s_mapping(x_a);       % antenna index
        x_s1 = x_i(Na+1:end);        % bits from signal constellation
        x_s = s_mapping(x_s1)-1;     % bits to decimal
        x_q = pskmod(x_s,M)/sqrt(Eavg);      % digital modulation
        x(a_i)=x_q;

        y =  H*x;  % faded signal
        r = y+10^(-EbNo(c)/20)*n; % Received noisy signal
%% Detection
        for m=1:M
            z(:,:,m)=H*N_lut(m);
            d(:,:,m)=repmat(r,1,Nt)-z(:,:,m);
            nor(m,:)=sum(abs(d(:,:,m)).^2);
        end
        [mi,ins]=min(nor);
        [mis,ina]=min(mi);
        d_a=ina; %antenna index
        d_s=ins(ina)-1; %transmitted symbol
%% error calculation
        e_a = (d_a~=a_i); %error in antenna index
        errora(c)=errora(c)+((e_a~=0));
        e_s = (d_s~=x_s); %error in symbol detection
        errors(c)=errors(c)+((e_s)~=0);
        error(c)=error(c)+((e_a+e_s)~=0); % error in SM symbol
end
end
errorrate = error/(N); % symbol error rate
semilogy(EbNo,errorrate);
grid on
hold on
```

B.5.1 Function for mapping the incoming bits to an SM symbol

The sequence of incoming bits is required to be mapped to a transmit antenna and a signal constellation point of the digital modulation scheme being used. The following MATLAB function is prepared for generalized SM mapping, i.e. it can be used to map the incoming bits to an antenna for any number of transmit antennas and a signal constellation point for any order of modulation for the digital modulation scheme under consideration.

```
function x=s_mapping(y);
l=length(y);
de=0;
for k=1:l
    de=de+y(end-k+1)*2^(k-1);
end
x=de+1;
```

B.6 MATLAB script for SER of 2×2 Alamouti STBC MIMO system

```
% MATLAB script for simulation of 2x2 Alamouti STBC MIMO system
% Some system parameters
% Es : Energy per symbol per transmitter antenna per time slot
% No : One sided noise spectral density
% EsNo : ratio Es/No
clc;
clear all;
N = 10^3; %Number of STBC codeword transmission
Nt = 2; %Number of transmitter antennas
Nr = 2; %Number of receiver antennas
T = 2 ; %Number of time slots per STBC codeword
K = 2; %Number of symbols per STBC codeword
M = 4; %Modualtion order for M-ary modualtion scheme
sym_lut = pskmod([0:M-1],M); %Look up table of modulated symbols
E_avg = sum(abs(sym_lut).^2)/M;%Average energy per constellation
N_lut = sym_lut/E_avg; %Normalized look up table
EsNo = 0:5:30; %EsNo range in dB
for c=1:length(EsNo)
   error(c)=0;
   sigma=sqrt(1/(10^(EsNo(c)/10)));
   for j=1:N
       smbols = randi([0 M-1], 1, K); % K symbols at the source
       mod_sym = pskmod(smbols,M)/sqrt(Nt*E_avg);
                                  % Digitally modulated K symbols
       stbc_cw = [mod_sym(1) -conj(mod_sym(2)); ...
                              mod_sym(2) conj(mod_sym(1))];
                              % STBC codeword for two transmit antennas

       H=1/sqrt(2)*(randn(Nr,Nt)+1i*randn(Nr,Nt));
       H_eq1 = [H(1,1),H(1,2);conj(H(1,2)),-conj(H(1,1)}];
       H_eq2 = [H(2,1),H(2,2);conj(H(2,2)),-conj(H(2,1))];
       R=H*stbc_cw + 1/sqrt(2)*sigma*(randn(Nr,T)+1i*randn(Nr,T));
       R=transpose(R);
       [row,col]=size(R);
       %R(:,col/2+1:end)=conj(R(:,col/2+1:end));
       R(row/2+1:end,:)=conj(R(row/2+1:end,:));
       R_hat=transpose(conj(H_eq1))*R(:,1) ...
                               +transpose(conj(H_eq2))*R(:,2);
       det_sym=pskdemod(R_hat/sum(sum(abs(H).^2)),M);
       for k=1:length(smbols)
           if smbols(k) ~= det_sym(k)
               error(c) = error(c)+1;
           end
       end
   end
end
errorrate = error/(K*N);
semilogy(EsNo,errorrate);
hold on; grid on;
```

Note that the operation of conjugate transpose can also be performed on matrices with complex entries by A' in MATLAB instead of using the functions $transpose(conj(A))$. Transpose of a matrix with complex elements can be obtained using the function $transpose(R)$ as in line '28' of the above script or simply using the operator $'$ with dot as $R.'$.

B.7 MATLAB script for SER of TAS MIMO system with MRC at the receiver

```
% MATLAB script file to simulate TAS/MRC MIMO Systems.
% Evaluates SER performance for M-PSK modulation schemes over
% generalized fading channels.
% The same script is useful even for other modulation schemes,
% only the instruction involving modulation schemes need to be
% changed.
% The functions to generate fading coefficients have to be used
% as per the requirements.

clc;
clear all;
N=10^3;
M=2; % For modulation order
EbNo=[0:1:15]; % Range of SNR for simulation
eta=1;  %Fading parameters
mu=0.5;
N_tot=2; %Number of antennas at the transmitter
Nt=1;   % Antennas to be selected at thr transmitter. In this case,
        % it is always 1
Nr=2; % Number of antennas at the receiver
lut = pskmod([0:M-1],M); % Look up table
Eavg = sum(abs(lut).^2)/M; % Average energy of constellation
N_lut = lut/sqrt(Eavg); % Normalized look up table

  for c=1:length(EbNo)
      SNRdB=EbNo(c);
      error(c)=0;
      sigma=sqrt(0.5/(10^(SNRdB/10)));
      for i=1:N
          data_bits=randi([0 1],log2(M),Nt); %Bits for transmission
          bits2sym = s_mapping(data_bits)-1; %Bits to M-ary symbols
          x_t = pskmod(bits2sym,M)/sqrt(Eavg); %PSK modulation.
           % This modulation scheme has to be the same as the one
           % used for look up table at the beginning.
     %% channel coefficient matrix generation
     %  Hr = 1/sqrt(2)*(randn(Nr,N_tot)+1i*(randn(Nr,N_tot)));
                                                  %Rayleigh fading
     %  K=0; w = sqrt(K/(K+1))*ones(Nr,N_tot)+sqrt(1/(K+1))*Hr;
                                                  % Rician Fading
```

```
    %  Hr =nak_m_channel(mu,Nr,N_tot);         % Nakagami-m fading
    Hr = eta_mu_channel(eta,mu,Nr,N_tot); % eta mu fading channel
    %   Hr = kappa_mu_channel(kappa,mu,Nr,N_tot);
                                                % kappa-mu fading
    %   Hr = a_mu_channel(alpha,mu,Nr,N_tot);   % alpha-mu fading
        for sel=1:N_tot
            mesure(sel)=sum(abs(Hr(:,sel)).^2);
        end
        [sel_ant,ind]=max(mesure);
        H=Hr(:,ind);
        Habs = sum(abs(H).^2);
        y = H*x_t +  sigma*(randn(Nr,Nt)+j*randn(Nr,Nt));
        y_det = transpose(conj(H)*y;
        det_sym = pskdemod(y_det/Habs,M); %M-PSK demodulation
        if det_sym~=bits2sym
            error(c)=error(c)+1;
        end
    end
   end% end of FOR loop for SNR
   SER = error/N;
   semilogy(EbNo,SER), hold on
```

References

[1] V. Aggarwal. Jagadish Chandra Bose: The real inventor of Marconis wireless receiver. *The Ancient Wireless Association Journal*, 47(3):50–54, July 2006.

[2] S. M. Alamouti. A simple transmit diversity technique for wireless communications. *IEEE Journal on Selected Areas in Communications*, 16(8):1451–1458, Oct 1998.

[3] M. S. Alouini and A. J. Goldsmith. Capacity of Rayleigh fading channels under different adaptive transmission and diversity-combining techniques. *IEEE Transactions Vehicular Technology*, 48(4):1165–1181, July 1999.

[4] J. G. Andrews, S. Buzzi, W. Choi, S. Hanly, A. Lozano, A. C. K. Soong, and J. Zhang. What will 5G be? *IEEE Journal on Selected Areas in Communications*, 32(6):1065–1082, 2014.

[5] O. S. Badarneh and R. Mesleh. Spatial modulation performance analysis over generalized $\eta - \mu$ fading channels. In 24th *IEEE International Symposium on Personal Indoor and Mobile Radio Communications (PIMRC)*, pages 886–890, Sept 2013.

[6] L. Bai and J. Choi. *Low Complexity MIMO Detection.* Springer Science & Business Media, 2012.

[7] E. Basar, U. Aygolu, E. Panayirci, and H. V. Poor. Space-time block coded spatial modulation. *IEEE Transactions on Communications*, 59(3):823–832, March 2011.

[8] A. Bhowal and R. S. Kshetrimayum. End to end performance analysis of m2m cooperative system over cascaded $\alpha - \mu$ channels. In *Proceedings of the International Conference on COMmunication Systems & NETworkS*, pages 1–7. IEEE, 2017.

[9] E. Bjornson, E. G. Larsson, and T. L. Marzetta. Massive MIMO: Ten myths and one critical question. *IEEE Communications Magazine*, 54(2):114–123, 2016.

[10] P. K. Bondyopadhyay. Under the glare of a thousand suns — the pioneering works of Sir J. C. Bose. *Proceedings of the IEEE*, 86(1):218–224, Jan 1998.

[11] R. Brandt. *Coordinated Precoding for Multicell MIMO Networks.* KTH, Sweden, 2014.

[12] M. Chaudhury, N. K. Meena, and R. S. Kshetrimayum. Local search based near optimal low complexity detection for large MIMO system. In *Proceedings of the IEEE International Conference on Advanced Networks and Telecommunications Systems*, pages 1–5. IEEE, 2016.

[13] Y. Chen and C. Tellambura. Distribution functions of selection combiner output in equally correlated Rayleigh, Rician, and Nakagami-m fading channels. *IEEE Transactions on Communications*, 52(11):1948–1956, 2004.

[14] Z. Chen, Z. Chi, Y. Li, and B. Vucetic. Error performance of maximal-ratio combining with transmit antenna selection in flat Nakagami-m fading channels. *IEEE Transactions on Wireless Communications*, 8(1):424–431, Jan 2009.

[15] Z. Chen, J. Yuan, and B. Vucetic. Analysis of transmit antenna selection/maximal-ratio combining in Rayleigh fading channels. *IEEE Transactions on Vehicular Technology*, 54(4):1312–1321, July 2005.

[16] M. Chiani, D. Dardari, and M. K. Simon. New exponential bounds and approximations for the computation of error probability in fading channels. *IEEE Transactions on Wireless Communications*, 2(4):840–845, 2003.

[17] T. V. Chien and E. Bjornson. Massive MIMO communications. In W. Xiang, K. Zheng, and X. Xuemin, editors, *5G Mobile Communications*, chapter 4, pages 77–116. Springer, 2017.

[18] Y. S. Cho, J. Kim, W. Y. Yang, and C.-G. Kang. *MIMO-OFDM Wireless Communications with MATLAB.* John Wiley & Sons, New York, 2010.

[19] A. Chockalingam and B. S. Rajan. *Large MIMO Systems.* Cambridge University Press, Cambridge, 2014.

[20] J. Choi. *Optimal Combining and Detection.* Cambridge University Press, Cambridge, 2010.

[21] S. T. Chung and A. J. Goldsmith. Degrees of freedom in adaptive modulation: A unified view. *IEEE Transactions on Communications*, 49(9):1561–1571, Sept 2001.

[22] W. H. Chung and C. Y. Hung. Multi-antenna selection using space shift keying in MIMO systems. In *Vehicular Technology Conference (VTC Spring), 2012 IEEE 75th*, pages 1–5, May 2012.

[23] R. Cogliatti, R. A. Amaral de Souza, and M. D. Yacoub. Practical, highly efficient algorithm for generating κ-μ and η-μ variates and a near-100% efficient algorithm for generating α-μ variates. *IEEE Communications Letters*, 16(11):1768–1771, 2012.

[24] A. F. Coskun and O. Kucur. Performance analysis of joint single transmit and receive antenna selection in Nakagami-m fading channels. *IEEE Communications Letters*, 15(2):211–213, Feb 2011.

[25] A. F. Coskun and O. Kucur. Performance analysis of joint single transmit and receive antenna selection in non-identical nakagami-m fading channels. *IET Communications*, 5(14):1947–1953, Sept 2011.

[26] D. B. Da Costa and M. D. Yacoub. Moment generating functions of generalized fading distributions and applications. *IEEE Communications Letters*, 12(2):112–114, Feb 2008.

[27] D. B. Da Costa, M. D. Yacoub, and J. Filho. Highly accurate closed-form approximations to the sum of α-μ variates and applications. *IEEE Transactions on Wireless Communications*, 7(9):3301–3306, Sept 2008.

[28] J. W. Craig. A new, simple and exact result for calculating the probability of error for two-dimensional signal constellations. In *Conference Record, Military Communications in a Changing World, IEEE Military Communications Conference, MILCOM '91*, pages 571–575 vol.2, Nov 1991.

[29] M. D. Tolga and A. Ghrayeb. *Coding for MIMO Communication Systems*, John Wiley & Sons, New York, 2007.

[30] D. Benevides da Costa and M. D. Yacoub. The η - μ joint phase-envelope distribution. *IEEE Antennas and Wireless Propagation Letters*, 6:195–198, 2007.

[31] H. A. David and H. N. Nagaraja. *Order Statistics*, 3rd ed. Wiley Interscience, New York, 2003.

[32] M. Di Renzo and H. Haas. Bit error probability of space modulation over nakagami-m fading: Asymptotic analysis. *IEEE Communications Letters*, 15(10):1026–1028, Oct 2011.

[33] M. Di Renzo and H. Haas. Bit error probability of SM-MIMO over generalized fading channels. *IEEE Transactions on Vehicular Technology*, 61(3):1124–1144, March 2012.

[34] D. Dixit and P. R. Sahu. Performance of L-branch MRC receiver in $\eta - \mu$ and $\kappa - \mu$ fading channels for qam signals. *IEEE Wireless Communications Letters*, 1(4):316–319, 2012.

[35] P. Elia, B. A. Sethuraman, and P. V. Kumar. Perfect space-time codes with minimum and non-minimum delay for any number of antennas. In *International Conference on Wireless Networks, Communications and Mobile Computing*, volume 1, pages 722–727. June 2005.

[36] P. Elia, B. A. Sethuraman, and P. V. Kumar. Perfect space-time codes for any number of antennas. *IEEE Transactions on Information Theory*, 53(11):3853–3868, Nov 2007.

[37] A. Erdelyi. *Higher Transcendental Functions*, volume 1, McGraw-Hill Book Company, New York, 1953.

[38] N. Y. Ermolova. Moment generating functions of the generalized $\eta-\mu$ and $\kappa-\mu$ distributions and their applications to performance evaluations of communication systems. *IEEE Communication Letters*, 12(7), 2008.

[39] C. Feng, Y. Jing, and S. Jin. Interference and outage probability analysis for massive MIMO downlink with MF precoding. *IEEE Signal Processing Letters*, 23(3):366–370, 2016.

[40] G. P. Fettweis. The tactile internet: Applications and challenges. *IEEE Vehicular Technology Magazine*, 9(1):64–70, 2014.

[41] G. J. Foschini. Layered space-time architecture for wireless communication in a fading environment when using multi-element antennas. *Bell Labs Technical Journal*, 1(2):41–59, Autumn 1996.

[42] G. J. Foschini, K. Karakayali, and R. A. Valenzuela. Coordinating multiple antenna cellular networks to achieve enormous spectral efficiency. *IEE Proceedings Communications*, 153(4):548–555, 2006.

[43] G. J. Foschini and M. J. Gans. On limits of wireless communications in a fading environment when using multiple antennas. *Wireless Personal Communications*, 6(3):311–335, 1998.

[44] Lord Rayleigh F.R.S. XII. On the resultant of a large number of vibrations of the same pitch and of arbitrary phase. *Philosophical Magazine and Journal of Science [Series 5]*, 10(60):73–78, 1880.

[45] A. Goldsmith. *Wireless Communications*, Cambridge University Press, New Delhi, 2005.

[46] I. S. Gradshteyn and I. M. Ryzhik. *Table of Integrals, Series, and Products*, 7th ed. Academic Press, New York, 2007.

[47] S. Gupta, J. Yadav, and R. S. Kshetrimayum. Low complexity detection algorithm for Alamouti like STBC for large MIMO systems. In *Proceedings of the IEEE International Conference on Electronics, Computing and Communication Technologies*, pages 1–6. IEEE, 2015.

[48] H. Huang, C. B. Papadias, and S. Venkatesan. *MIMO Communication for Cellular Networks.* Springer, New York, 2012.

[49] C.-C. Hung, C.-T. Chiang, N.-Y. Yen, and R.-C. Wu. Outage probability of multiuser transmit antenna selection/maximal-ratio combining systems over arbitrary nakagami-m fading channels. *IET Communications*, 4(1):63–68, Jan 2010.

[50] H. Jafarkhani. *Space-Time Coding: Theory and Practice.* Cambridge University Press, Cambridge, 2005.

[51] W. C. Jakes. *Microwave Mobile Communications.* Wiley-IEEE Press, New York, 1994.

[52] M. Jankiraman. *Space-Time Codes and MIMO Systems.* Artech House, Boston, 2004.

[53] J. Jeganathan, A. Ghrayeb, and L. Szczecinski. Spatial modulation: Optimal detection and performance analysis. *IEEE Communications Letters*, 12(8):545–547, 2008.

[54] M. Joham, W. Utschick, and J. Noseek. Linear transmit processing in MIMO communication systems. *IEEE Transactions on Signal Processing*, 53(8):2700–2712, 2013.

[55] J. Jose, A. Ashikhmin, T. L. Marzetta, and S. Vishwanath. Pilot contamination and precoding in multi-cell TDD systems. *IEEE Transactions on Wireless Communications*, 10(8):2640–2651, 2011.

[56] Hyungsoo Kim, Hyounkuk Kim, N. Kim, H. Park, S. Seo, and J. Choi. Efficient transmit antenna selection for correlated MIMO channels. In *Proceedings IEEE Wireless Communications and Networking Conference, WCNC*, pages 1–5, April 2009.

[57] M. Koca and H. Sari. Performance analysis of spatial modulation over correlated fading channels. In *IEEE Vehicular Technology Conference (VTC Fall)*, pages 1–5, Sept 2012.

[58] M. Kulkarni, L. Choudhary, B. Kumbhani, and R. S. Kshetrimayum. Performance analysis comparison of transmit antenna selection with maximal ratio combining and orthogonal space time block codes in equicorrelated Rayleigh fading multiple input multiple output channels. *IET Communications*, 8(10):1850–1858, July 2014.

[59] S. Kumar, G. Chandrasekaran, and S. Kalyani. Analysis of outage probability and capacity for $\kappa-\mu/\eta-\mu$ faded channel. *IEEE Communications Letters*, 19(2):211–214, Feb 2015.

[60] B. Kumbhani and R. S. Kshetrimayum. Outage probability analysis of spatial modulation systems with antenna selection. *Electronics Letters*, 50(2):125–126, Jan 2014.

[61] B. Kumbhani and R. S. Kshetrimayum. Analysis of TAS/MRC based MIMO systems over $\eta - \mu$ fading channels. *IETE Technical Review*, 32(4):252–259, 2015.

[62] B. Kumbhani and R. S. Kshetrimayum. Performance analysis of MIMO systems with antenna selection over generalized $\kappa - \mu$ fading channels. *IETE Journal of Research*, 62(1):45–54, 2016.

[63] L. Lu, G. Y. Li, A. L. Swindlehurst, A. Ashikhmin, and R. Zhang. An overview of massive MIMO: Benefits and challanges. *IEEE Journal on Selected Topics in Signal Processing*, 8(5):742–758, 2014.

[64] E. Larsson, O. Edfors, F. Tufvesson, and T. Marzetta. Massive MIMO for next generation wireless systems. *IEEE Communications Magazine*, 52(2):186–195, Feb 2014.

[65] K. Lee. Doubly ordered sphere decoding for spatial modulation. *IEEE Communications Letters*, 19(5):795–798, May 2015.

[66] E. J. Leonardo and M. D. Yacoub. The product of $\alpha - \mu$ variates. *IEEE Wireless Communications Letters*, 4(6):637–640, 2016.

[67] X. Ma and Q. Zhou. Massive MIMO and Its Detection. In M. M. da Silva and F. A. Monteiro, editors, *MIMO Processing for 4G and beyond*, chapter 10, pages 449–471. CRC Press, Boca Raton, FL, 2014.

[68] A. M. Magableh and M. M. Matalgah. Moment generating function of the generalized α-μ distribution with applications. *IEEE Communications Letters*, 13(6):411–413, June 2009.

[69] T. L. Marzetta. Noncooperative cellular wireless with unlimited number of base station antennas. *IEEE Transactions Wireless Communications*, 9(11):3590–3600, 2010.

[70] T. L. Marzetta. Massive MIMO: An introduction. *Bell Labs Technical Journal*, 20(1):11–22, 2015.

[71] T. L. Marzetta, E. G. Larsson, H. Yang, and H. Q. Ngo. *Fundamentals of Massive MIMO*. Cambridge University Press, Cambridge, 2016.

[72] W. Matolak and J. Frolik. Worse-than-Rayleigh fading: Experimental results and theoretical models. *IEEE Communications Magazine*, 49(4):140–146, 2011.

[73] C. F. Mecklenbrauker, A. F. Molisch, J. Karedal, F. Tufvessan, A. Pair, L. Bernado, T. Zemen, O. Klemp, and N. Czink. Vehicular channel characterization and its implications for wireless system design and performance. *Proceedings of the IEEE*, 99(7):1189–1212, 2011.

[74] S. R. Meraji. Performance analysis of transmit antenna selection in Nakagami-m fading channels. *Wireless Personal Communications*, 43(2):327–333, Oct 2007.

[75] R. Mesleh, O. S. Badarneh, A. Younis, and H. Haas. Performance analysis of spatial modulation and space-shift keying with imperfect channel estimation over generalized $\eta - \mu$ fading channels. *IEEE Transactions on Vehicular Technology*, 64(1):88–96, Jan 2015.

[76] R. Mesleh, H. Haas, C. W. Ahn, and S. Yun. Spatial modulation — a new low complexity spectral efficiency enhancing technique. In *Proceedings First International Conference on Communications and Networking in China*, pages 1–5, 2006.

[77] R. Y. Mesleh, H. Haas, S. Sinanovic, C. W. Ahn, and S. Yun. Spatial modulation. *IEEE Transactions on Vehicular Technology*, 57(4):2228–2241, July 2008.

[78] A. F. Molisch, F. Tufvesson, J. Karedal, and C. F. Mecklenbrauker. A survey on vehicle-to-vehicle propagation channels. *IEEE Communications Magazine*, 16(6):12–22, 2009.

[79] A. F. Molisch and M. Z. Win. MIMO systems with antenna selection. *IEEE Microwave*, 5(1):46–56, March 2004.

[80] A. F. Molisch, M. Z. Win, Y.-S. Choi, and J. H. Winters. Capacity of MIMO systems with antenna selection. *IEEE Transactions on Wireless Communications*, 4(4):1759–1772, 2005.

[81] D. Morales-Jimenez and J. F. Paris. Outage probability analysis for η-μ fading channels. *IEEE Communications Letters*, 14(6):521–523, June 2010.

[82] S. Mumtaz, J. Rodriguez, and D. Linglong. *mmWave Massive MIMO*. Elsevier, London, 2017.

[83] M. Nakagami. The m-distribution — a general formula of intensity distribution of rapid fading. In W. C. Hoffman, editor, *Statistical Methods in Radio Wave Propagation*, pages 3–36. Pergamon, Oxford, 1960.

[84] Y.-H. Nam, B. L. Ng, K. Sayana, Y. Li, J. Zhang, Y. Kim, and J. Lee. Full-dimension MIMO (FD-MIMO) for next generation cellular technology. *IEEE Communications Magazine*, 51(6):172–179, 2013.

[85] H. Q. Ngo. *Massive MIMO: Fundamentals and System Designs.* Linköping University, Linköping, 2015.

[86] H. Q. Ngo, E. G. Larsson, and T. L. Marzetta. Energy and spectral efficiency of very large multiuser MIMO systems. *IEEE Transactions on Communications*, 61(4):1436–1449, 2013.

[87] H. Q. Ngo, E. G. Larsson, and T. L. Marzetta. Aspects of favorable propagation in massive MIMO. In *Proceedings of the European Signal Processing Conference*, pages 1–5. IEEE, 2014.

[88] F. Oggier, G. Rekaya, J.-C. Belfiore, and E. Viterbo. Perfect space time block codes. *IEEE Trans. Information Theory*, 52(9):3885–3902, 2006.

[89] S. Panic, M. Stefanovic, J. Anastasov, and P. Spalevic. *Fading and Interference Mitigation in Wireless Communications.* CRC Press, Boca Raton, FL, 2013.

[90] A. K. Papazafeiropoulos and S. A. Kotsopoulos. Generalized phase-crossing rate and random FM noise for $\alpha-\mu$ fading channels. *IEEE Transactions on Vehicular Technology*, 59(1):494–499, Jan 2010.

[91] A. Papoulis and S. U. Pillai. *Probability, Random Variables, and Stochastic Processes.* Tata McGraw-Hill Education, 2002.

[92] J. F. Paris. Nakagami-q (Hoyt) distribution function with applications. *Electronics Letters*, 45(4):210–211, Feb 2009.

[93] A. Paulraj and T. Kailath. Increasing capacity in wireless broadcast systems using distributed transmission/directional reception (DTDR). *U.S. Patent*, 5 345 599, 1993.

[94] A. Paulraj, R. Nabar, and D. Gore. *Introduction to Space-Time Wireless Communications.* Cambridge University Press, Cambridge, 2003.

[95] J. Pena-Martin, J. Romero-Jerez, and C. Tellez-Labao. Performance of selection combining diversity in $\eta-\mu$ fading channels with integer values of μ. *IEEE Transactions on Vehicular Technology*, PP(99):1–1, 2014.

[96] J. P. Pena-Martin, J. M. Romero-Jerez, and C. Tellez-Labao. Performance of TAS/MRC wireless systems under Hoyt fading channels. *IEEE Transactions on Wireless Communications*, 12(7):3350–3359, July 2013.

[97] K. Peppas, F. Lazarakis, A. Alexandridis, and K. Dangakis. Error performance of digital modulation schemes with MRC diversity reception over $\eta-\mu$ fading channels. *IEEE Transactions on Wireless Communications*, 8(10):4974–4980, Oct 2009.

[98] V. J. Phillips. The 'Italian Navy coherer' affair: A turn-of-the-century scandal. *IEE Proceedings A — Science, Measurement and Technology*, 140(3):175–185, May 1993.

[99] N. Pillay and H. Xu. Comments on "antenna selection in spatial modulation systems". *IEEE Communications Letters*, 17(9):1681–1683, Sept 2013.

[100] N. Pillay and H. J. Xu. Comments on "signal vector based detection scheme for spatial modulation". *IEEE Communications Letters*, 17(1):2–3, Jan 2013.

[101] J. G. Proakis. *Digital Communications*, 4th ed. McGraw-Hill, New York, 1995.

[102] A. P. Prudnikov, I. A. Brychkov, and O. I. Marichev. *Integrals and Series. Volume 2, Special Functions.* Gordon and Breach Science Publ., Amsterdam, Paris, New York, 1986.

[103] A. P. Prudnikov, I. A. Brychkov, O. I. Marichev, and G. G. Gould. *Integrals and Series. Volume 3, More Special Functions.* Gordon and Breach Science Publ., Amsterdam, Paris, New York, 1986.

[104] X. Qiu and K. Chawla. On the performance of adaptive modulation in cellular systems. *IEEE Transactions on Communications*, 47(6):884–895, June 1999.

[105] R. Rajashekar, K. V. S. Hari, and L. Hanzo. Antenna selection in spatial modulation systems. *IEEE Communications Letters*, 17(3):521–524, 2013.

[106] T. Rappaport, R. W. Heath Jr., R. C. Daniels, and J. N. Murdock. *Millimeter Wave Wireless Communications.* Prentice Hall, New Jersey, 2014.

[107] T. S. Rappaport. *Wireless Communications: Principles and Practice*, volume 2. Prentice Hall, New Jersey, 1996.

[108] T. S. Rappaport, S. Sun, R. Mayzus, H. Zhao, Y. Azar, K. Wang, G. N. Wong, J. K. Schulz, M. Samimi, and F. Gutierrez. Millimeter wave mobile communications for 5G cellular: It will work! *IEEE Access*, 1(5):335–349, 2013.

[109] S. O. Rice. Statistical properties of a sine wave plus random noise. *The Bell System Technical Journal*, 27(1):109–157, Jan 1948.

[110] J. Rodriguez. *Fundamentals of 5G Mobile Networks.* John Wiley & Sons, New York, 2015.

[111] F. Rusek, D. Persson, B. K. Lau, E. G. Larsson, T. M. Marzetta, O. Edfors, and F. Tufvesson. Scaling up MIMO: Opportunities and challenges with very large arrays. *IEEE Signal Processing Magazine*, 30(1):40–60, 2013.

[112] A. K. Sah and A. K. Chaturvedi. Reduced neighborhood search algorithms for low complexity detection in MIMO systems. In *Proceedings of the IEEE Global Communications Conference*, pages 1–6. IEEE, 2015.

[113] H. Sawada, Y. Shoji, and C.-S. Choi. Proposal of novel statistic channel model for milli meterwave WPAN. In *Proceedings of the Asia Pacific Microwave Conference*, pages 1–4. IEEE, 2006.

[114] N. Seifi, J. Zhang, R. W. Heath Jr., T. Svensson, and M. Coldrey. Coordinated 3D beamforming for interference managament in cellular networks. *IEEE Transactions on Wireless Communications*, 13(10):5396–5410, 2016.

[115] H. Shin and J. H. Lee. On the error probability of binary and m-ary signals in Nakagami-m fading channels. *IEEE Transactions on Communications*, 52(4):536–539, April 2004.

[116] M. K. Simon and M. S. Alouini. *Digital Communication over Fading Channels*, 2nd ed. Wiley, New York, 2005.

[117] Eo W. Stacy. A generalization of the gamma distribution. *The Annals of Mathematical Statistics*, pages 1187–1192, 1962.

[118] G. L. Stuber. *Principles of Mobile Communication*, 2nd ed. Kluwer Academic Publishers, Dordrecht, 2002.

[119] S. Sun, T. S. Rappaport, R. W. Heath Jr., A. Nix, and S. Rangan. MIMO for millimeter-wave wireless communications: Beamforming, spatial multiplexing, or both? *IEEE Communications Magazine*, 52(12):110–121, 2014.

[120] A. L. Swindlehurst, E. Ayanoglu, P. Heydari, and F. Capolino. Millimeter-wave massive MIMO: The next wireless revolution? *IEEE Communications Magazine*, 52(9):56–62, 2014.

[121] B. Talha and M. Patzold. Channel models for mobile-to-mobile cooperative communication systems. *IEEE Vehicular Technology Magazine*, 6(2):33–43, 2011.

[122] V. Tarokh, H. Jafarkhani, and A. R. Calderbank. Space-time block codes from orthogonal designs. *IEEE Transactions on Information Theory*, 45(5):1456–1467, 1999.

[123] V. Tarokh, H. Jafarkhani, and A. R. Calderbank. Space-time block coding for wireless communications: Performance results. *IEEE Journal on Selected Areas in Communications*, 17(3):451–460, 1999.

[124] V. Tarokh, A. Naguib, N. Seshadri, and A. R. Calderbank. Space-time codes for high data rate wireless communication: Performance criteria in the presence of channel estimation errors, mobility, and multiple paths. *IEEE Transactions on Communications*, 47(2):199–207, 1999.

[125] V. Tarokh, N. Seshadri, and A. R. Calderbank. Space-time codes for high data rate wireless communication: Performance criterion and code construction. *IEEE Transactions on Information Theory*, 44(2):744–765, 1998.

[126] S. Thoen, L. Van der Perre, B. Gyselinckx, and M. Engels. Performance analysis of combined transmit-SC/receive-MRC. *IEEE Transactions on Communications*, 49(1):5–8, Jan 2001.

[127] W. H. Tranter, T. S. Rappaport, K. L. Kosbar, and K. S. Shanmugan. *Principles of Communication Systems Simulation with Wireless Applications*, volume 1. Prentice Hall, New Jersey, 2004.

[128] D. Tse and P. Viswanath. *Fundamentals of Wireless Communication.* Cambridge University Press, New York, 2005.

[129] B.-Y. Wang and W.-X. Zheng. BER performance of transmitter antenna selection/receiver-MRC over arbitrarily correlated fading channels. *IEEE Transactions on Vehicular Technology*, 58(6):3088–3092, July 2009.

[130] J. Wang, S. Jia, and J. Song. Signal vector based detection scheme for spatial modulation. *IEEE Communications Letters*, 16(1):19–21, Jan 2012.

[131] P. W. Wolniansky, G. J. Foschini, G. D. Golden, and R. Valenzuela. V-BLAST: An architecture for realizing very high data rates over the rich-scattering wireless channel. In *Proceedings URSI International Symposium on Signals, Systems, and Electronics*, pages 295–300, 1998.

[132] M. D. Yacoub. The $\eta - \mu$ distribution: A general fading distribution. In *Proceedings IEEE* 52nd *Vehicular Technology Conference, 2000*, volume 2, pages 872–877, 2000.

[133] M. D. Yacoub. The $\kappa - \mu$ distribution: A general fading distribution. In *Proceedings IEEE* 54th *Vehicular Technology Conference, 2001*, volume 3, pages 1427–1431, 2001.

[134] M. D. Yacoub. The α-μ distribution: A general fading distribution. In *The* 13th *IEEE International Symposium on Personal, Indoor and Mobile Radio Communications, 2002*, volume 2, pages 629–633, Sept 2002.

[135] M. D. Yacoub. The α-μ distribution: A physical fading model for the Stacy distribution. *IEEE Transactions on Vehicular Technology*, 56(1):27–34, Jan 2007.

[136] M. D. Yacoub. The $\kappa - \mu$ distribution and the $\eta - \mu$ distribution. *IEEE Antennas and Propagation Magazine*, 49(1):68–81, 2007.

[137] H. Yang and Q. S. Quek. *Massive MIMO Meets Small Cell Backhaul and Cooperation.* Springer, New York, 2016.

[138] L. Yang. Performance analysis of transmit antenna selection with MRC over correlated fast-fading channels. In *Proc. 4th International Conference on Wireless Communications, Networking and Mobile Computing, WiCOM '08*, pages 1–4, Oct 2008.

[139] P. Yang, Y. Xiao, Y. L. Guan, S. Li, and L. Hanzo. Transmit antenna selection for multiple-input multiple-output spatial modulation systems. *IEEE Transactions on Communications*, 64(5):2035–2048, May 2016.

[140] P. Yang, Y. Xiao, Y. Yu, and S. Li. Adaptive spatial modulation for wireless MIMO transmission systems. *IEEE Communications Letters*, 15(6):602–604, 2011.

[141] S. Yang and L. Hanzo. Fifty years of MIMO detection: The road to large-scale MIMOs. *IEEE Communications Surveys and Tutorials*, 17(4):1941–1988, 2015.

[142] T. Yilmaz, G. Gokkoca, and O. B. Akan. Millimeter wave communication for 5g iot applications. In C. X. Mavromoustakis, G. Mastorakis, and J. M. Batalla, editors. *Internet of Things (IoT) in 5G Mobile Technologies*, chapter 4, pages 37–53. Springer, 2016.

[143] A. Younis, N. Serafimovski, R. Mesleh, and H. Haas. Generalised spatial modulation. In *Conference Record of the Forty Fourth Asilomar Conference on Signals, Systems and Computers (ASILOMAR)*, pages 1498–1502, 2010.

[144] A. Younis, S. Sinanovic, M. Di Renzo, R. Mesleh, and H. Haas. Generalised sphere decoding for spatial modulation. *IEEE Transactions on Communications*, 61(7):2805–2815, July 2013.

[145] J. Zheng. Signal vector based list detection for spatial modulation. *IEEE Wireless Communications Letters*, 1(4):265–267, Aug 2012.

[146] L. Zheng and D. N. C. Tse. Diversity and multiplexing: A fundamental tradeoff in multiple-antenna channels. *IEEE Transactions on Information Theory*, 49(5):1073–1096, 2003.

Index